Dr. Sam K. Baldoo

BIOTECHNOLOGY IN THE FEED INDUSTRY

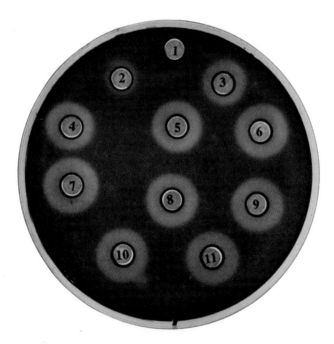

Determination of in-feed enzymes using the zone diameter of substrate hydrolysed. Well No.1: Buffer, No.2: Feed extract devoid of enzyme, Wells No.3–7: Spiked with various levels of enzyme (0.25, 0.5, 1.0, 1.5, 2.0 kg/tonne). Wells No.8–11 were loaded with an extract obtained from a commercial feed sample to which enzyme had been added by the manufacturer at a rate of 1 kg/tonne

Biotechnology in the Feed Industry

Proceedings of
Alltech's Tenth Annual Symposium

Edited by TP Lyons and KA Jacques

NOTTINGHAM
University Press

Nottingham University Press
Sutton Bonington Campus
Loughborough, Leicestershire LE12 5RD, United Kingdom

NOTTINGHAM

© Copyright 1994 by
Alltech Inc

All rights reserved. No part of this publication
may be reproduced in any material form
(including photocopying or storing in any
medium by electronic means and whether or not
transiently or incidentally to some other use of
this publication) without the written permission
of the copyright holder except in accordance with
the provisions of the Copyright, Designs and
Patents Act 1988. Applications for the copyright
holder's written permission to reproduce any part
of this publication should be addressed to the
publishers

ISBN 1897676514

Typeset by The Midlands Book Typesetting Company, Loughborough, Leicestershire, England
Printed and bound in Great Britain

TABLE OF CONTENTS

**BIOTECHNOLOGY IN THE FEED INDUSTRY:
1994 AND BEYOND**   1
A panorama of techniques, processes and products to address
animal production problems today and tomorrow
T.P. Lyons
Alltech Inc., Nicholasville, Kentucky USA

SECTION 1:
ENZYMES: PRACTICAL APPLICATIONS FOR BIOLOGICAL CATALYSTS IN ANIMAL NUTRITION AND WASTE MANAGEMENT

**PLANT POLYSACCHARIDES – THEIR PHYSIOCHEMICAL
PROPERTIES AND NUTRITIONAL ROLES IN
MONOGASTRIC ANIMALS**   51
G. Annison and M. Choct
*CSIRO Division of Human Nutrition, Glenthorne Laboratory
Majors Road, O'Halloran Hill, SA, Australia*

**THE USE OF ENZYMES IN DESIGNING A PERFECT
PROTEIN SOURCE FOR ALL ANIMALS**   67
S.L. Woodgate
Beacon Research Ltd., Market Harborough, Leicestershire, UK

**MANIPULATION OF FIBRE DEGRADATION:
AN OLD THEME REVISITED**   83
A. Chesson
Rowett Research Institute, Bucksburn, Aberdeen, UK

**THE UNITED STATES MARKET FOR FEED ENZYMES:
WHAT OPPORTUNITIES EXIST?**   99
C.W. Newman
*Montana Agricultural Experiment Station, Montana State
University, Bozeman, Montana, USA*

Contents

REGISTRATION OF ENZYMES AND BIOLOGICAL
PRODUCTS AROUND THE WORLD:
CURRENT ANALYTICAL METHODS FOR FEED ENZYMES
AND FUTURE DEVELOPMENTS   117
R. Power and G. Walsh
European Biosciences Research Centre
National University of Ireland, Galway, Ireland

SECTION 2:
USE OF NOVEL BIOMOLECULES TO ACTIVATE AND MAXIMIZE HEALTH AND PRODUCTION

DE NOVO DESIGNED SYNTHETIC PLANT STORAGE
PROTEINS: ENHANCING PROTEIN QUALITY OF PLANTS
FOR IMPROVED HUMAN AND ANIMAL NUTRITION   129
J.M. Jaynes
Demeter Biotechnologies, Ltd., Research Triangle Park
North Carolina, USA

POTENTIAL FOR MANIPULATING THE
GASTROINTESTINAL MICROFLORA:
A REVIEW OF RECENT PROGRESS   155
S.A. Martin
Departments of Animal and Dairy Science and Microbiology
University of Georgia, Athens, Georgia, USA

MANNAN-OLIGOSACCHARIDES: NATURAL POLYMERS
WITH SIGNIFICANT IMPACT ON THE GASTROINTESTINAL
MICROFLORA AND THE IMMUNE SYSTEM   167
K. Newman
North American Biosciences Center, Nicholasville, Kentucky, USA

STUDY OF THE USE OF YUCCA EXTRACT (DE-ODORASE)
IN THE FATTENING OF BOARS   175
K. Ender, G. Kuhn, K. Nürnberg
Research Institute for the Biology of Agricultural Animals,
Dummerdorf D-2551, Germany

SECTION 3:
ENVIRONMENTAL ASPECTS OF ANIMAL PRODUCTION: THE DEPENDENCE OF LIVESTOCK PRODUCTION ON, AND IMPACT ON, ENVIRONMENTAL QUALITY

WATER QUALITY AND NUTRITION FOR DAIRY CATTLE   183
D.K. Beede
Dairy Science Department, University of Florida
Gainesville, Florida, USA

Contents

ASCITES – A METABOLIC CONDITION? 199
J.D. Summers
*Department of Animal and Poultry Science, University of Guelph
Guelph, Ontario, Canada*

WATER – FORGOTTEN NUTRIENT AND NOVEL DELIVERY
SYSTEM 211
P.H. Brooks
*Seale-Hayne Faculty of Agriculture, Food and Land Use
University of Plymouth, Devon, UK*

A BIOLOGICAL APPROACH TO COUNTERACT
AFLATOXICOSIS IN BROILER CHICKENS AND DUCKLINGS
BY THE USE OF *SACCHAROMYCES CEREVISIAE*
CULTURES ADDED TO FEED 235
G. Devegowda, B.I.R. Aravind, K. Rajendra, M.G. Morton, A.
Baburathna and C. Sudarshan
*Department of Poultry Science, University of Agricultural Sciences
Bangalore, India*

SHRIMP FARMING: A BREAKTHROUGH IN CONTROLLING
NITROGEN METABOLISM AND MINIMIZING WATER
POLLUTION 247
C. Wacharonke
*Animal Health Department Diethelm Trading Co., Ltd.
Bangkok, Thailand*

SECTION 4:
TRACE MINERALS: THE ROLE OF MINERAL PROTEINATES IN IMMUNITY, REPRODUCTION AND PERFORMANCE

UNDERSTANDING STRESS IN CATTLE 255
C. Nockels
*Colorado State University,
Fort Collins, Colorado 80523, USA*

STRESS EFFECTS ON CHROMIUM NUTRITION OF HUMANS
AND FARM ANIMALS 267
R.A. Anderson
*Vitamin and Mineral Nutrition Laboratory, Beltsville Human
Nutrition Research Center, U.S. Department of Agriculture, ARS
Beltsville, Maryland, USA*

Contents

ORGANIC CHROMIUM: A NEW NUTRIENT FOR STRESSED ANIMALS 275
D.N. Mowat
Department of Animal and Poultry Science
University of Guelph
Guelph, Ontario, Canada

TRANSITION METALS, OXIDATIVE STATUS, AND ANIMAL HEALTH: DO ALTERATIONS IN PLASMA FAST-ACTING ANTIOXIDANTS LEAD TO DISEASE IN LIVESTOCK? 283
J.K. Miller
Animal Science Department, University of Tennessee,
Knoxville, Tennessee, USA
F.C. Madsen
Suidae Technology, Greensburg, Indiana, USA

CHOOSE A NEW PATH: BORON BIOLOGICALS 303
B.F. Spielvogel
Boron Biologicals, Inc., Raleigh, North Carolina, USA

ORGANIC SELENIUM SOURCES FOR SWINE – HOW DO THEY COMPARE TO INORGANIC SELENIUM SOURCES? 323
D.C. Mahan
Animal Science Department, Ohio State University,
Columbus, Ohio, USA

INDEX 335

DISTRIBUTORS AROUND THE WORLD 341

BIOTECHNOLOGY IN THE FEED INDUSTRY: 1994 AND BEYOND

A panorama of techniques, processes and products to address animal production problems today and tomorrow

T.P. LYONS
Alltech Inc., Nicholasville, Kentucky USA

Introduction

In a recent interview on the subject of the growth of biotech companies, Drew commented that seventeen years after the usually acknowledged beginning of the biotech era there were now 1200 companies employing an estimated 80,000 people. While the industry remains dominated by companies that make therapeutic products (66%), still a solid 9% is represented by the agricultural sector with 1992 sales estimated at just over $1 billion. It is in this area of agriculture that some of the most significant, albeit less glamorous, advances have been made.

Alltech, founded in 1980, is typical of the newly-started companies in many ways. Now in its fourteenth year, this comparative teenager now employs 250 people with products shipped to 60 countries around the world. Research and developement are concentrated at two bioscience centers, with a third center soon to be opened in China. Market driven, the company, through its own international marketing center, is in touch with grassroots developments in its field, and is probably qualified to speculate on where this aspect of agribusiness is going.

The impact of biotechnology upon animal nutrition has been significant, as any review of topics at recent conferences on animal production will indicate. Topics include such areas as immune modulation, enzyme biocatalysts, growth hormones, diet specific yeast culture supplements, aflatoxin binders, ammonia scavengers, glucose tolerance factor, chromium yeast, selenomethionine, pollution control, and mineral proteinates. To the uninformed reader it would appear that the feeding of vitamins, minerals, protein and energy are no longer of interest to nutritionists. While clearly this is not the case, the focus has definitely shifted and will continue to do so as the new science of biotechnology allows us to solve health and management problems while at the same time boosting productivity and reducing pollution.

The role of regulatory agencies, whether it be in allowing a new product to be used, as in the case of the recent US approval of BST, or in curtailing for environmental reasons use of an essential nutrient such as selenium, must also be considered. A review therefore of the impact

Table 1.   Research and development interests in biotechnology applications for agriculture.

Boron-based amino acids and peptides	Organic selenium
Plant derived biomolecules	Substrate specific enzymes
Toxin adsorbents	Ideal proteins
GI microflora modifiers	Organic chromium
Mineral bioplexes	Diet specific yeast cultures
Noxious gas binders	Plant and microbial-derived immune modulators

of biotechnology on animal production is probably appropriate. So vast, however, is the area that it would be impossible to discuss each and every new development. Accordingly, this brief review will only address some of the major topics with which our group is involved (Table 1).

Boron biochemistry: possibilities for new organic compounds for animal feed applications

Carbon, when chemically bound with nitrogen, hydrogen, oxygen and other elements, makes up the fundamental molecules of life such as amino acids, protein and nucleic acids. Indeed, carbon chemistry is often referred to as the chemistry of life. In recent years, scientists have begun to reflect on the properties of an element that may have virtually untapped commercial potential. This element is boron. Boron lies next to carbon in the periodic table; and structurally boron and carbon are very similar. Boron, with the atomic number five, has three electrons in its valence shell, while carbon, with the atomic number six, has four valence electrons. Both form simple compounds with hydrogen, and both obey the Lewis Octet rule. A fundamental difference, however, is that the boron species will have an overall negative charge while the carbon is electrically neutral (Figure 1).

Chemists quickly lost interest in boron compounds due to their negative charge and the sensitivity of three-coordinate boron compounds to water. In recent years, however, one company has been exploiting the biological activity of molecules based on four-coordinate boron compounds due to their similarity with four coordinate carbon compounds (Spielvogel, 1993).

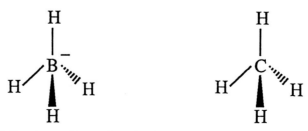

Figure 1   Both carbon and boron form simple compounds with hydrogen

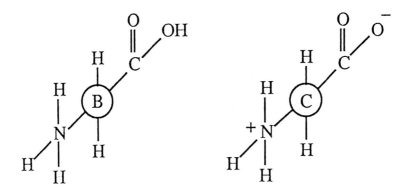

Figure 2 Boroglycine™

Biochemicals based on boron enable production of powerful new compounds for therapeutic, agricultural and animal health applications. Initial efforts were directed toward synthesis of a boron analogue of glycine. The carbon atom in glycine, the simpliest amino acid, can easily be replaced with boron (Figure 2). Both glycine and boronated glycine are white crystalline solids at room temperature, soluble in water and nearly indentical in molecular weight (75.07 compared with 74.88). Boronated glycine, however, has several unique properties which merit exploitation. It can more easily cross membrane barriers because it is uncharged. It is also remarkably stable in air and water.

Among the commercial applications of Boroglycine which exploit the ability of this compound to cross membrane barriers is manipulation of critical cellular events such as bone formation or in therapies for osteoporosis. It is also now known that boron analogs can have anticancer and antibacterial properties. The ideal therapy for cancer or for the manipulation of microbial growth might consist of a compound that would selectively kill either the cancer cell or one particular microorganism. Since many microorganisms have unique requirements for either one amino acid or a peptide based on that amino acid; the analog, which might be identical in all chemical characteristics except for containing boron, might, in effect, be a 'poison pill' for that particular organism (Spielvogel, 1993).

Boron is just one element in a possible new path for choosing biologicals; a path that is bringing us back to nature in our search.

Plants as sources of biomolecules

The plant kingdom has a long history of producing drugs. Opium from poppies provided the analgesics morphine and codeine. A humble periwinkle from Madagascar is the source of the most potent antileukemia drugs known today, the Catharanthus alkaloids (Hylands, 1993).

Plants are also used as sources of starter chemicals which can be chemically modified to form valuable therapeutic compounds. The Pacific yew tree, *Taxus brevifolia*, provides a diterpine called taxol, which is effective in ovarian cancer treatment. While initially taxol was derived from the Pacific yew tree, now the yew tree cells can be cultured in fermentation tanks where the cells excrete taxol. In the future, these bark-derived cells will result in a lower cost alternative to the use of the bark itself.

Taxol is a good example of how first generation products, derived from plants, can often, through the work of a chemist and biochemist, be brought to more active second and third generation derivatives. The taxol itself is poorly soluble in water and had to be adminstered as an oily injection formulation, which itself unfortunately produced secondary toxic effects.

By understanding the structure of taxol, synthesis of a more soluble derivative was possible, which nevertheless retained the original plant extract activity. With the advent of plant cell culture, a new reproducible supply of diverse plant materials for evaluation becomes available. Plant cell culture has extended the range of compounds, and, at the same time, expedited the process of discovering new useful compounds. A case in point is cytokines, which have been known to stimulate plant performance, and are now being evaluated for their animal effects.

Today several hundred thousand species of plants are known, yet only a small percentage of these have ever been examined for their potential to provide natural compounds with human or animal health and(or) performance benefits. As animal scientists better understand the factors that affect growth and performance, new and simple tests can be established. With these tests, screening techniques can be set up in the laboratory, and new biologically active molecules detected, examined, and identified.

YUCCA SCHIDIGERA EXTRACT AND AMMONIA CONTROL

Scientists have been looking for many years at compounds produced by a desert plant called *Yucca schidigera*. Gaining an understanding of this plant's ability to survive under arid desert conditions has led to a better understanding of its nitrogen metabolism. This plant contains a number of glycomponents known to be involved in nitrogen metabolism. In particular, one component has been identified which has the ability to bind or buffer ammonia.

This plant, used for many years as a forage in Mexico and the desert southwestern US, has also been reported to assist in reducing odors in livestock and industrial waste management systems. By investigating the biochemistry and, in particular, by refuting the previously held theory that the plant inhibited urease, biochemists were able to identify the ammonia binding capacity of one particular glycomponent. This component has now been standardized, and indeed is capable of being synthesized in the laboratory. This discovery also led to a quality control method which has done much to improve the reliability of field response.

Availability of an ammonia control agent that performs consistently has had a significant impact on the feed industry. More and more companies are utilizing the material at 120 grams per tonne to reduce ammonia in both confinement units and lagoons with an indirect effect on both health and performance of animals and humans. The relationship between addition of the yucca extract and pig performance and with ascites in broilers serves to illustrate the impact of ammonia control and(or) nitrogen binding.

Poultry, ammonia and ascites

A major concern in the poultry industry is the increase over the past several years in losses due to ascites. In many parts of the world, mortality due to ascites can run as high as 8–12%. While ascites was at one time associated primarily with high altitudes, it is now recognized as a significant problem at any altitude and increases during the winter months.

Many factors contribute to ascites morbidity, including an association with high levels of ammonia. In a study where mortality due to ascites was artificially elevated due to curtains separating treatment reps, it was demonstrated that 120 grams of the standardized yucca extract De-Odorase per tonne significantly reduced both total and ascites mortality (Arce et al., 1994; Table 2). Further discussion on this topic may be found in the chapter by Dr Summers.

Table 2. Effect of De-Odorase on performance of heavy broilers through 56 days[1].

	Control	De-Odorase
Weight, g	2427	2447
Feed intake, g	4969	4980
FCR	2.08	2.07
Mortality, %		
General	27.05[b]	20.91[a]
Ascites	18.86[b]	13.78[a]

[a,b]Means differ, P<0.01
[1]Adapted from Arce et al., 1994.

Yucca extract has also reduced ammonia in layer units with subsequent increases in productivity and improvements health status related to air quality. The resulting litter, possibly due to lower free ammonia levels, is reported to have fewer flies. Higher beetle populations were noted in association with the latter observation (Crober, personal communication). Lower ammonia evolution is also an advantage when spreading the litter on fields. Fertilizer value is actually enhanced since less ammonia is lost to the atmosphere.

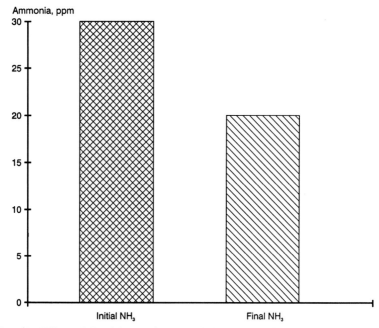

Figure 3 Effect of De-Odorase (*Yucca schidigera* extract) on ammonia levels in a confinement facility for fattening pigs

Pigs, ammonia and skatole

Reductions in atmospheric ammonia due to yucca extract have also been reported in association with improvements in pig performance. In a study with fattening pigs in Ireland Cole and Tuck (1994) reported that ammonia declined from 30 to 20 ppm by week five (Figure 3). This was associated with an improvement in daily gain (Table 3).

A fascinating, if less well understood aspect of yucca extract has been reported by Ender's group in Germany. This group reported reductions in male taint associated with boar meat when De-Odorase was used at recommended levels. Boars convert feed approximately 0.25 units more efficiently between 20 and 90 kgs liveweight than do castrated pigs. Assuming comparative efficiencies of 2.8 and 2.55 for castrates and boars, respectively; a savings of 17 kgs feed per pig is realized (Table 4). On a farm with 100 sows producing 23 pigs per sow per annum, the savings in feed by finishing boars instead of castrates is over 40 tonnes!

Table 3. Effect of De-Odorase addition to the diet on liveweight gain of fattening pigs.

	Control	De-Odorase
Starting weight, kg	32.00	30.23
End weight, kg	91.67	91.88
Days on test	73	73
Daily liveweight gain, kg	0.820	0.850

Table 4. Feed savings realized when feeding boars instead of castrates.

	Boars	Castrates	Difference
Feed efficiency	2.55	2.8	
Feed used per pig, kg	178.5	196	17
Feed used 100 sows @23 pigs per sow per annum, kg	450,800	410,550	40,250

Clearly use of the yucca extract product is providing benefits for the environment, animal health and performance. Reductions in atmospheric ammonia are desirable from an environmental perspective due to the perception of noxious odors in the surrounding community. Lower ammonia in the confinement building itself improves health of livestock which in turn boosts performance and efficiency. Additionally, since an increasingly dim view is being taken of castration by animal rights opinion-makers, the value of avoiding it while saving on feed should be considered.

Yucca extract and prawn culture
On shrimp farms a serious ecosystem imbalance problem due to ammonia typically occurs at least once every prawn culture cycle. During this water quality crisis plankton populations collapse, ammonia rises to toxic levels and dissolved oxygen falls. Additionally, feed intake and growth are depressed and mortality rate sharply increases. Immediate attention must focus on correcting environmental conditions, especially decreasing ammonia and increasing dissolved oxygen as these are the most critical factors threatening survival.

Addition of ammonia-binding *Yucca schidigera* extract (De-Odorase) has proven to be of considerable benefit both in preventing water quality problems when added in pre-treatment plans and in treating crisis conditions in prawn ponds. Field trials with De-Odorase were conducted in both "Crisis Prevention" and "Crisis Treatment" situations in ten commercial ponds (each 6,400 m^2). The prevention program consisted of spraying De-Odorase onto the water surface at a rate to supply 0.3 ppm on Day 1 and every 15 days until harvest. Water quality crises were successfully prevented with 0.3 ppm De-Odorase. The control groups had to be harvested prior to reaching desired market weight due to continuous problems with disease and persistantly high mortality rates (Table 5). Control groups experienced water quality crises (plankton collapse and water color change) two to three times

Table 5. Effect of De-Odorase treatment on production parameters in commercial prawn production.

	Control groups	Treated groups
Average prawn size, #/kg	92	38
Average survival rate, %	28	65
Days to harvest	72	120

Table 6. Effect of De-Odorase on recovery time from water quality crisis and mortality.

	NH_3-N, ppm		Mortality rate during crisis	Hours to recovery[2]
	Initial	After[1]		
De-Odorase	8.0	5.0	Low	24
Control	8.0	10.0	High	No recovery
De-Odorase	3.0	3.0	Low	12
Control pond	3.0	5.0	Moderate	No recovery
De-Odorase	3.0	1.0	Low	12
Control pond	5.0	5.0	High	No recovery
De-Odorase	2.0	0.5	Low	8
Control pond	3.0	5.0	Moderate	No recovery
De-Odorase	5.0	3.0	Moderate	12
Control pond	5.0	5.0	High	No recovery

[1]After treatment with De-Odorase.
[2]Recovery after treatment was determined by whether shrimp returned to the bottom of the pond.

during the production cycle while no imbalances were experienced in the ponds treated with De-Odorase at 0.3 ppm. Treated ponds were harvested following a full production cycle.

Treatment of a water quality crisis with De-Odorase was investigated during plankton collapse. Shrimp showed obvious signs of crisis by swimming near the water surface. There were five different cases in different farms and areas. In each case one pond served as a control while a second was treated with De-Odorase. In general, treating ponds undergoing water quality crises reduced ammonia levels and mortality rates (Table 6). Recovery, defined by the shrimp returning to the bottom of the pond, required 8 to 24 hours after treatment. The researchers concluded that De-Odorase, even under high NH_3 conditions, increased survival rate of the shrimp. The most effective treatment was 0.5 ppm applied three times at six hour intervals.

More recently, prevention programs have focused on additon of De-Odorase to the feed instead of directly to the water.

Roles of plants and microorganisms: modifying the gut microflora and immune modulation

MANNAN-OLIGOSACCHARIDES

Control of intestinal pathogens has long been a goal of both livestock producer and microbiologist. Methods used to achieve this goal include addition of antibiotics to feed, acidification of drinking water, competitive exclusion of pathogens using probiotics or chemical probiosis with selected sugars, and antimicrobial levels of minerals such as zinc oxide and copper sulfate.

More recently complex carbohydrates based on fructose, mannose and galactose have been investigated for their impact on gut microbiology with the goal of lessening the impact of pathogen challenge to the young animal or bird. While the response to these complex sugars by the animal is not entirely understood, three possible modes of action have been proposed. These include:

- Stimulation of immune response
- Blocking colonization of pathogens
- Provision of nutrients that cannot be used by pathogens

Carbohydrates occur in nature in a remarkable number of forms. Carbohydrates are integral parts of all nucleic acids, conjugates with proteins (glycoproteins) and lipids (glycolipids) and in free form (ranging in size from monosaccharides to enormous polysaccharides). While it is well documented how carbohydrates act as metabolic sources of energy or as structural elements involved in maintaining the morphological features of cells and organs, only in recent years has it clearly emerged that carbohydrates also possess additional biological functions (Parekh, 1993). These recently described functions include:

1. Hormonal activity: a carbohydrate may enhance or retard activity of particular hormone
2. Intercellular adhesion: carbohydrates of particular shapes are recognized by equally specific carbohydrate receptors
3. The adhesion of viruses to microbes or target cells
4. In modulating the shape of proteins specific to their functional and immunological activity

It is clear that carbohydrates physically dominate the surfaces of eukaryotic cells and the whole extracellular matrix. Indeed, virtually all cell surfaces are glycosylated with attached carbohydrates often being structurally the dominate components of the resulting glycoproteins.

Since a particular oligosaccharide structure attached to a protein can influence its therapeutic profile, biotechnologists have begun now to engineer carbohydrates and test them for these activities to discover novel pharmaceuticals. An idea of the complexity of carbohydrates compared with amino acids can be seen if we consider the possibilities of three amino acids versus three monosaccharides. Three amino acids can be assembled into six different tripeptides; and these would have a relatively ill-defined conformation. In contrast, three monosaccharides with a similar weight of 400 could be assembled into 64 different trisaccharides with a well defined conformation.

Using the analogy of building a defence/offence structural scaffolding, we should use carbohydrates rather than amino acids. On this scaffolding of carbohydrates could be mounted a particular enzyme, protein or fat to form a myriad of different immune modulators or to attach enzyme activators or deactivators.

Stimulation of immune response
Among the most exciting sugars recently studied are those extracted from the cell wall of *Saccharomyces cerevisiae*. Use of yeast cell wall

oligosaccharides to stimulate immunity and modify gut microflora via the diet adds a new chapter to a long history of use of yeast by mankind. Like cattle in the animal kingdom yeast have been cultivated by man for food and drink (albeit unwittingly until the last century). From leavening of bread to production of beverages containing ethyl alcohol, probably the earliest anesthetic, yeasts have truly been mankind's best microbial friend. Surprisingly, it has been just over 100 years since Pasteur demonstrated in 1866 substantial participation of live yeast in the fermentation process. Since that time, a small number of species of the genus *Saccharomyces* have been produced in amounts measuring millions of tons. It is the genus *Saccharomyces*, and more specifically, unique strains of *Saccharomyces cerevisiae*, that are responsible for many of the advances in agricultural applications of biotechnology.

Interest in this application of yeast centers on the cell wall. Glucan, mannan and chitin are the main components of yeast cell wall (Pfaff, 1984). The basic composition of the wall consists of mannan (30%), glucan (30%) and protein (12.5%). While the ratio of one component to another remains relatively constant from strain to strain, the degree of mannan phosphorylation and the interaction among the mannan, glucan and protein components varies. The glucan is a polysaccharide with glucose bonds connected in β1,6 and β1,3 linkages (Peat et al., 1958). Glucan is thought to make up the matrix of the cell wall which is covered by another layer of mannose sugars (Figure 4). These mannose sugars are arranged in a highly branched chain of manno-pyranoside residues. The linkages in the backbone of this chain are α–1,6 with side

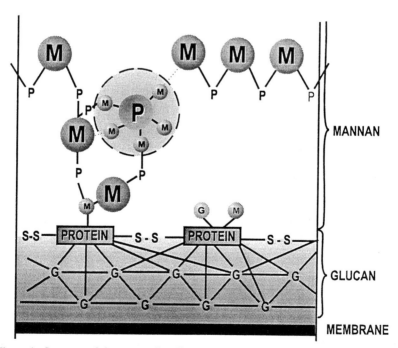

Figure 4 Structure of the yeast cell wall

chains bound by α–1,2 and α–1,3. The cell wall has powerful antigenic stimulating properties, and it is well-established that this property is a characteristic of the mannan chain (Ballou, 1970). Differences in mannan structure exist among strains, and yeast strains can be differentiated based on the antigenicity of extracted mannan. This antigenicity can be increased by the addition of acetyl groups or by increasing the degree of phosphorylation.

Yeast cell wall material is remarkably stable to acid digestion. Fractions are known to survive passage through the stomach or abomasum. It is this ability to pass through acid digestion undisturbed that may account for the product's biological activity in such a wide range of species. In one of the first experiments, mannan was extracted from the cell wall by washing with dilute alkali, and a preparation of mannan-free cell wall was fed to salmon smolts. This preparation, a glucan–protein complex, was incorporated into salmon feed given for six, eight or 12 weeks. Another group was given the standard diet. Five weeks after ending the glucan feeding period fish were challenged in the standard furunculosis co-habitant test. Mortalities were reduced by 28% (Onarheim, 1992). Both response to the challenge model and reduced mortality under natural outbreaks of furunculosis suggested enhanced immunity. Studies with trout fry in commercial testing ponds also pointed toward enhanced immunity in response to gluco-mannan in the diet. Trout fry exposed to cold water pathogens had mortalities up to 25% between weights of 1 g and 7 g. The inclusion of 7 kg of mannan-oligosaccharide reduced this to 1% over the same period (Figure 5).

The uniqueness of mannan as a possible immunomodulator also resides in the fact that each strain of yeast has its own mannan. It is known that the immunodominant side chains of mannan consisted of four types:

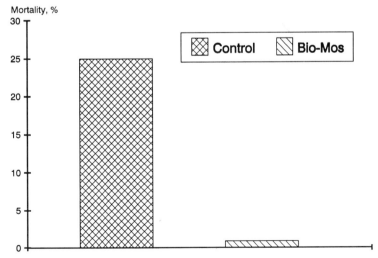

Figure 5 Effect of Bio-Mos on mortality of trout fry challenged with cold water pathogens

mannotetraose, mannotriose, mannobiose and mannose. Techniques exist today using these specific differences to identify individual yeasts. In the future, we can use these differences to produce a range of mannan-oligosaccharides to suit different feeds and to bring about different results.

The ability of mannan-oligosaccharides from a *Saccharomycese cerevisiae* strain to stimulate immune response was addressed in a recent series of experiments which examined both effects of the mannan on phagocytic activy in vitro and response of broiler chicks challenged with salmonella. Phagocytosis is the process by which bacteria are ingested and killed by various classes of cells called leukocytes or phagocytes. Phagocytes are a diverse family of white blood cells which have many immunological functions. The phagocytes, or macrophages, engulf and kill microorganisms. Phagocytes kill ingested microbes by producing biological oxidants such as superoxide ion, hydrogen peroxide, singlet oxygen, hydroxyl radical and enzymes. Because these oxidant species are produced from oxygen, their production is associated with a surge in metabolic activity within the phagocytes. Such increases in phagocytic activity can be quantitatively measured by adding a compound such as luminol (5–amino-2, 3–dihydro-1,4– phthalazinedione) to phagocytosing cells in biological fluids. Luminol reacts with oxidizing species to produce measurable amounts of light at a peak wavelength of 425 nm (Figure 6). Stimulation of phagocytes can, therefore, be measured by monitoring light emissions using a luminometer. Using this technique significant increases in phagocytic activity were observed when mannan-oligosaccharide was incubated in peripheral blood from three-month-old male Wistar rats (Figure 7). The mannan clearly exerted an immune response.

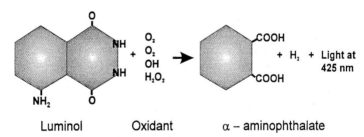

Figure 6 The interaction of luminol with biological oxidants to produce light

To further test whether this immune response could help chicks survive a *Salmonella* challenge, one hundred chicks (1 day old) from a local hatchery were screened for *Salmonella* infection by the hatchery and declared *Salmonella* free. Subsequently chicks were randomly assigned to two groups of fifty birds. One group received the standard hatchery feed while the other received an identical feed supplemented with a commercial mannan-oligosaccharide product at a rate of 1 g/kg. Cloacal swabs taken on this day revealed the presence of wild-type *Salmonella* infection. It was observed that the birds fed the mannan-oligosaccharide were better able to withstand the *Salmonella* challenge (Table 7).

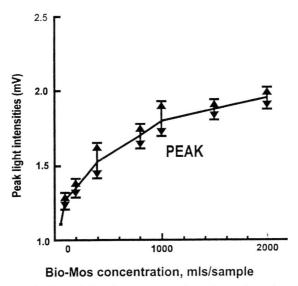

Figure 7 Luminal-enhanced chemiluminescence of rat phagocytes activated with increasing concentrations of Bio-Mos.

Stimulation of disease resistance using mannan-oligosaccharide is not limited to animals. Recent studies from the British Cereals Authority have demonstrated that insoluble fractions of yeast act to stimulate phytoalexins (disease resistance mechanisms) against mildew in agricultural crops (Newton et al., 1993). With the increasing pressure to reduce fungicide use and(or) application rates, the advent of a natural biological control system could be very welcome. In the study, mannan fractions were sprayed on to barley leaves 24 hours prior to inoculation with mildew. The control leaves molded in the normal way, but the treated leaves did not. Additionally, the rate of papilla formation and phenylalanine ammonia lyase (PAL) activity were increased on treated leaves. In field trials the mannan-oligosaccharide reduced mildew infection and increased yield in both spring and winter barley (Triumph cultivar, Figure 8). Though application of the extract did not provide the complete control elicited by the full rate of fungicide, the appropriate use of a resistance stimulant is preventative. Integrating use of yeast-derived

Table 7. The protective effect of mannan-oligosaccharide (Bio-Mos) against a natural and experimental *Salmonella* infection[1]. *Salmonella* detection at sacrifice (birds/group)

GROUP	Wild-type *Salmonella* detected by cloacal swab on day 10 (Birds/Groups)	Wild-type strain (Novobiocin^R)	Experimental strain (Nalidixic acid^R)
Control	11/50	Caecum 38/50 Organs 42/50	Caecum 6/50 Organs 6/50
Bio-Mos	10/50	Caecum 9/50 Organs 14/50	Caecum 0/50 Organs 0/50

[1]Experimental infection was per os at day 10 (6 x 10^7 C.F.U. (*S. enteritidis*)/bird).

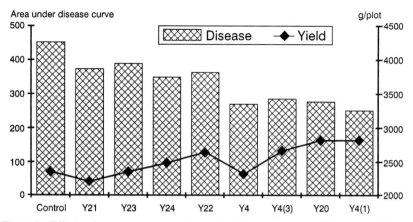

Figure 8 Effects of various mannan-oligosaccharide extracts on mildew and yield of Triumph barley. Adapted from Newton et al., 1993

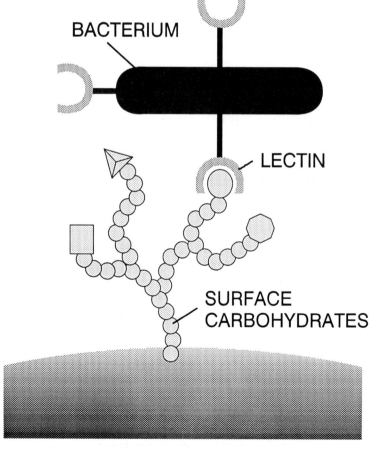

Figure 9 Attachment of bacteria to the intestinal epithelium via lectins

resistance elicitors in a spray regime or mixture with reduced rates of fungicide use may provide a viable means of reducing necessary inputs in agricultural systems (Newton et al., 1993).

Blocking colonization of pathogens
Another possible mode of action for mannan-based oligosaccharides involves interference with colonization of intestinal pathogens. Cell surface carbohydrates are primarily responsible for cell recognition (Sharon and Lis, 1993). At the simplest level is the role of carbohydrates in blood types which are differentiated by cell coat sugars. Bacteria have lectins (proteins or glycoproteins) on the cell surface that recognize specific sugars and allow the cell to attach to that sugar. These sugars can be found on the epithelial cell surface. Binding of *Salmonella, Escherichia coli* and *Vibrio cholera* has been shown to be mediated by a mannose-specific lectin-like substance on the bacterial cell surface (Figure 9). If dietary mannan-oligosaccharides can occupy potential baterial binding sites on intestinal epithelium, then colonization may be precluded. Alternatively, mannans may bind bacterial cell surface sites and pass through the digestive tract without harming the host.

Binding or adsorption of pathogens can be demonstrated quite easily. When *E. coli* or *Salmonella* are incubated with mannan-oligosaccharide, a distinct clumping occurs and pathogens are removed from the solution (Figure 10). Mannose in its simplest form is known to reduce colonization

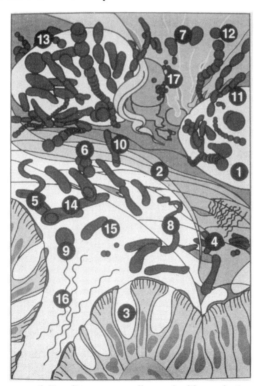

Figure 10 Adherence of *E. coli* to mannan-oligosaccharide product

of various bacteria including coliforms and *Salmonella*. These microorganisms have Type I fimbrae, which adhere to mannose sugars or mannan chains. Since mannose, while effective, is expensive and difficult to deliver to the GI tract, a protected mannose chain has been investigated as a practical alternative.

Mannan-oligosaccharides starve pathogens of essential nutrients
The ability of mannan-oligosaccharide to resist bacterial digestion may also explain some of the observed effects. It has been demonstrated that when selected pathogens are incubated with mannan, this complex carbohydrate does not support growth (Table 8). Ironically, benevolent species such as lactobacilli and bifidobacteria appear to have the enzyme complexes necessary to use this sugar. Indirectly therefore the gut population might shift toward more desirable species. This technique is widely used in control of human diarrhea (Spiegel et al., 1994).

Table 8. Effect of mannan-oligosaccharide on growth of various pathogenic and beneficial intestinal bacteria.

Pathogens		Beneficial species	
Escherichia coli	Depressed	*Bifidobacterium longum*	Stimulated
Salmonella typhimurium	Depressed	*Lactobacillus casei*	Stimulated
Clostridium botulinum	Depressed	*Lactobacillus acidophilus*	Stimulated
Clostridium sporogenes	Depressed	*Lactobacillus delbrekii*	Stimulated

MANNAN-OLIGOSACCHARIDES (MOS) EFFECTS ON ANIMAL PERFORMANCE AND HEALTH ON COMMERCIAL FARMS

The use of oligosaccahrides has not only stimulated considerable interest among researchers; it has been viewed with considerable enthusiam by commercial livestock producers. Since the product is completely natural, a surprisingly responsive market has existed. Trials with virtually all species have been undertaken both on farms and at research institutes.

Poultry
Practical evaluation of MOS in poultry fed diets has involved comparison with antibiotic programs, particularly virginiamycin in the U.S. and Canada. A recent trial conducted in Canada examined the effects of MOS under commercial production conditions. In this trial, 22,000 broilers were used to test the effect of MOS on performance of commercial broilers when added to starter, grower and finisher diets at 2 kg per ton. Reductions in condemnations, improvements in liveweight for age and livability combined to increase profit per bird significantly (Table 9).

Calves
Preruminant calves also appear to respond to the oligosaccharide. Performance and general health of pre-ruminant calves given milk

Table 9. Effect of MOS (Bio-Mos) on performance of broilers at a commercial farm.

	Control	Bio-Mos
Birds started, n	11,160	11,270
Birds marketed, n	11,160	11,204
Condemned, n	53	48
Livability, %	99	99.4
% condemned (including parts)	1.499	1.345
Market age, d	38	42
Feed used, kg	37,970	44,560
Average wt., kg	1.853	2.103
FCR	1.837	1.891
Adjusted to equal weight[1]		
Age	38	37
Adjusted wt.[1]	1.853	1.853
FCR	1.837	1.811
Profit($CA) per kg	0.408	0.428
per bird	0.748	0.892
per ft^2	0.831	0.995

[1]Adjusted assuming 50 g/d gain and a reduction of 0.016 per day of age in FCR

Table 10. Effect of MOS (Bio-Moss) on performance of Holstein bull calves.

	Control	Bio-Moss	Std. error
Number of calves	15	14	
Initial weight, kgs	45.90	45.49	
Weekly gain, lbs			
Day 0–7	0.741	0.309	
Day 7–14	2.122	2.582	
Day 14–21	3.700	3.977	
Day 21–28	3.650	4.482	
Day 28–35	4.073	6.182	
Total gain, kgs	12.682[a]	16.909[b]	1.95
Total starter intake, kgs	16.33	19.30	1.41

[ab] $P<.07$

replacer with or without MOS were monitored through weaning. Calves given MOS had significantly faster rates of gain at 35 days (Newman et al., 1993, Table 10). The improvement in performance was associated with a reduced incidence of bacterial pneumonia (common to the facility and responsive to gentamycin) during weeks four and five (Figure 11).

Pigs

In an effort to evaluate oligosaccharides in pigs under commercial conditions a number of field studies were conducted. Coliform scours and related performance losses are a recurring problem on many pig farms. Antibiotic programs reduce, but do not eliminate, the problem. Mannan-oligosaccharides have been tested in commercial settings in

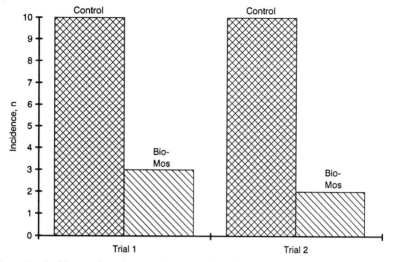

Figure 11 Incidence of respiratory disease in calves fed diets with or without Bio-Moss

Table 11. Effect of adding Bio-Mos to a diet with Banminth and Mecadox on performance of starter boars and gilts.[1]

	Boars		Gilts	
	BanMec	BanMec + MOS	BanMec + MOS	BanMec
Number of pigs	18	18	18	18
Days on feed	7	7	7	7
Start wt., lbs	22.72	23.11	22.83	21.67
End wt., lbs	28.28	29.06	28.16	27.22
Gain, lbs	5.56	5.95	5.33	5.55
Daily gain[2], lbs	0.794	0.85	0.761	0.794

[1]Started July 26 ended August 2, 1993
[2]Average feed cost per lb gain, boars and gilts: Ban/Mec, 0.362; BanMec + BMos, 0.337.

Table 12. Effect of adding Bio-Mos to a diet containing Banminth and Mecadox on performance of mixed boars and gilts[1]

	Banminth/ Mecadox	Banminth/ Bio-Mos
Number of animals at start	36	36
Days on feed	14	14
Start weight, lbs	17.72	17.36
End weight, lbs	28.22	28.14
Gain, lbs	10.50	10.78
Daily gain, lbs	0.75	0.77
Feed consumed, lbs	16.36	15.83
Feed cost per pig, $	3.25	3.29
lbs feed/lbs gain	1.56	1.47
Feed cost/lb gain, $	0.319	0.305

[1]King, personal communication

starter diets both with and without antibiotics. In a series of field trials where MOS was added to diets either with Banminth/Meccadox or with no added antibiotic the optimum approach was a combination of oligosaccharide and antibiotic. The effects of combining the two were improved performance, fewer scouring problems and better appetites (Tables 11 and 12).

Rabbits
Post-weaning enteritis is a major source of economic loss to the rabbit industry with mortality sometimes as high as 40%. Enteritis is caused by pathogen proliferation when abrupt weaning removes the antibacterial and acidifying influence of the dam's milk. As the etiology of this enteritis syndrome is similar to diarrheas in other species, mannan-oligosaccharides have been investigated in rabbit starter diets. In field studies in Spain addition of Bio-Mos at 2 kgs/t improved feed efficiency and reduced post-weaning mortality (Rosell, personal communication; Table 13).

Oligosaccharides are clearly an exciting area of investigation and bring many questions about mode of action. The next generation of probiotics/probiosis will focus on the ability of different microorganisms to produce and(or)metabolize various sugars. This approach looks at feed from more than its nutritional value by recognizing that the chemical compounds we use as energy sources have different functions. This approach, classifying foods based on more than nutritional attributes, has been done by the Japanese for human foods (Figure 12). At present,

Table 13. Effect of Bio-Mos on performance and mortality of weanling rabbits[1]

	Control	Bio-Mos
Liveweight, kg		
Weaning	0.590	0.610
35 days post-weaning	1.710	1.845
Average daily gain, g	38	41
Feed intake, g/d	116	120
Feed conversion	3.20	3.05
Total mortality, %	8.1	4.5

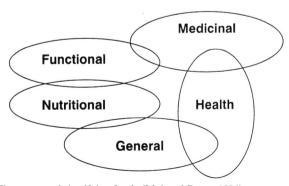

Figure 12 Five means of classifying foods (Mul and Perry, 1994)

Table 14. Sources of oligosaccharides

Fructo-oligosaccharides derived from wheat
Jerusalem artichoke (contains inulin)
Bananas
Onions/garlic
Legume seeds
Mannan-oligosaccharides derived from yeast cell wall

[1]Adapted from Mul and Perry, 1994

examination of carbohydrates and(or) oligosaccharides for functional values has resulted in several oligosaccharide-based products (Table 14).

Selenium: a mineral under fire in nutrition, immunity and the environment

Few minerals have received as much press, both good and bad, as selenium (Se) has over the past few years. Though toxic and carcinogenic at comparatively low levels, Se has long been known to be an essential micronutrient with vital roles in the body's antioxidant system and immune defense (Table 15). Researchers now recognize that its role is wider than originally assumed; and recent work has demonstrated the function of Se in preventing tumor growth. Concern, however, about selenium has arisen from several different perspectives in recent years. In parts of Europe where soils are deficient the concentration of Se in human foods has declined due to less use of cereal grains imported from North America. In the US a problem with Se in the environment in one locality in one state has resulted in a nation-wide restriction on Se inclusion in feeds to a level below that known to be optimum for health. Japanese authorities have banned use of sodium selenite largely due to concerns about toxicity to humans handling the mineral. For these different reasons a need exists for a new means of providing biologically active Se.

Table 15. Results of selenium-deficiency in pigs and ruminants

Pigs	Ruminants
Hepatic necrosis	White muscle disease in newborns
Pale skeletal muscle color ("white muscle disease")	Muscle weakness, buckling in older cattle, sheep
MMA in sows	Myocardial damage
Myocardial damage ("mulberry heart")	Unthriftiness
Low tolerance to stress, Fe injections	Reproductive problems including retained placenta, poor conception rate,
Reproductive problems	Weak and stillborn calves

SELENIUM IN HUMAN HEALTH AND NUTRITION

A short introduction to use of Se in humans may help us understand its role in animals. Research in the 1980s demonstrated the role of Se in preventing certain debilitating diseases including cancer. This grew out of a broader concept that Se was involved as part of an overall resistance mechanism against disease. The knowledge of the anti-cancer properties of Se is not new. Over forty years ago, Wisconsin investigators showed a pronounced reduction (of 50%) in the incidence of tumors induced in rats when the diet was supplemented by 0.5 ppm Se supplied as selenite (Clayton and Bauman, 1949) Selenium has also been reported to be effective in animal trials against cancers of the breast, the colon, the lungs, the liver and the skin leading to reduction in tumor incidence of some 7–100% of the controls (Milner, 1985).

Evidence suggests the protective action of Se in carcinogenesis is not derived directly from its involvement in glutathione peroxidase. Recent reports on investigations in China lend credibility to the concept of a cancer preventative role for Se. It has been recognized for many years that diets high in fruits and vegetables exert some protective effect against cancer of the esophagus and stomach. This suggested a possible role of the minerals and vitamins in such foods. A large study was conducted in Linxiang province in rural north central China, a region where the incidence of esophageal cancer is among the highest in the world (Oldfield, 1993). It is known that dietary intake of several micronutrients in this area is low. In the trial, 29,584 adults (aged 40–69) were assigned to one of four daily mineral or vitamin treatments over a five-year period (March 1986 to May 1991). Over 2,000 deaths were reported during the treatment period with cancer of the esophagus or stomach being the leading cause of mortality (32%), followed by heart disease (25%). The most effective treatment, which resulted in a significantly lower cancer incidence, was a combination of Vitamin E (30 mgs tocopherol), betacarotene (15 mgs) and 50 g Se in the form of selenium yeast. None of the other dietary treatments showed a clear beneficial effect on cancer rates.

Although the results of this massive experiment are encouraging, many questions remained unanswered. No clear indication was given as to the specific roles of the three component parts of the treatment, although it was speculated that antioxidant properties were probably involved. In light of what is known about Se toxicity, safe levels of administration were identified. These would suggest that a maximum safe intake might be 600μg Se per day in the form of selenium yeast. These levels are similar to amounts recommended in other countries. For example, Swedish authorities cite a maximum acceptable daily intake of 500μg while the US Food and Nutrition Board recommends 50–200μg.

Levels of Se in human foods are of concern in regions where soils and cereal grains are deficient. For example in Finland, before decision to supplement all commercial fertilizers with Se, daily intake of Se in human diets was down to 20–55μg.

It is interesting to note that on the human side, just as on the animal side, the recommended daily intakes are without exception given without taking into consideration the chemical form of the Se in the diet. Possibly more available forms of Se will allow recommendations based on the sources employed for both human and animal foods.

A new era of natural Se sources is now with us. Not only are the organic sources more efficacious by some four- to six-fold, but by allowing a corresponding reduction in rate of use, any environmental concerns are reduced.

SELENIUM IN ANIMAL NUTRITION

Until about ten years ago, it was generally accepted that the dietary requirement of selenium for all animals except turkeys was met at 0.1 ppm. Turkeys were considered to need 0.2 ppm. In the US allowable levels were increased to 0.3 ppm in most species (for chickens, turkeys, ducks, swine, sheep and cattle) in 1987 after the scientific community presented information to the Food and Drug Administration (FDA) demonstrating the need for higher levels to combat frequent occurence of deficiency syndromes including muscular dystrophy, reduced fertility, mulberry heart in pigs and mastitis. The 1993 roll-back to the 0.1 ppm level has left many nutritionists, producers and veterinarians in the uncomfortable position of necessarily recommending inadequate dietary levels of Se. Within the EEC, the maximum content of 0.3 ppm Se in commercial feeds for all types of livestock animals still exists.

The advent of a selenium-enriching process for yeast and other microorganisms has given us the means to both more accurately supply Se and the potential to lower addition rate without sacrificing health. It has been demonstrated that when grown in the appropriate manner the Se is present in the form of a number of proteins including selenomethionine and selenocysteine. Selenium substitutes for sulfur in these amino acids; however seleno amino acids are used in tissue biosynthesis no differently than if the amino acid contained sulfur. As a result, the animal has a constant supply of Se as tissue proteins turn over. Additionally, by altering growth conditions and by feeding cereal-derived organic sources of selenium, the proportion of Se compounds in Se yeast can be controlled. The proportion of Se present as selenomethionine in Se yeast

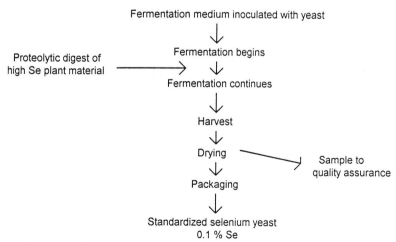

Figure 13 Production flow diagram for selenium yeast

can be controlled to consistently yield 40 to 45% in a recently-developed fermentation process (Figure 13).

The constant supply of bioavailable Se in muscle tissue protein may prove especially important to the sow and nursing litter. Though pig starter diets may remain at 0.3 ppm Se under the recent FDA ruling, all other diets are held to 0.1 ppm. Higher retention of the organic form of selenium, particularly in muscle tissue, may help reproducing animals maintain health and productivity in regions where Se deficiency syndromes are common. Studies are currently underway to examine effects of organic Se on Se concentration in sow's milk.

While pigs may benefit from the increased retention of Se in selenoproteins in muscle tissue, ruminants benefit first and foremost from the protection afforded Se through the rumen in the yeast cell. Pehrson has demonstrated that adding Se to diets fed lactating cows in the form of Se yeast increased milk Se content by a factor of six when compared with an equal amount of Se (0.4 ppm) from sodium selenite (Figure 14). He suggested that organic selenium compounds, such as selenium yeast, can be used sometimes at rates as low as 1/4 to 1/6 of the inorganic when added to ruminant diets. Presence of Se inside the yeast cell itself protects Se from reduction by rumen microbes. As the need to provide excess Se via inorganic forms is precluded by the "by-pass" function of the yeast cell, lower addition rates are possible. This, therefore, would reduce any potential negative influence of selenium on the environment. The beneficial effects of organic Se are not limited to animals but can also be seen in aquaculture. Pehrson (1993) summarized a series of experiments investigating organic Se conducted in Scandanavia (Table 16). Against the background of growing restrictions on use of sodium selenite, the advent of a new Se source with both less danger to human workers and improved bioavailable is timely.

To date, the most commonly used substance for Se supplementation is sodium selenite. However, selenite has many disadvantages, amongst them a tendency to be affected by oxidation. Pehrson (1993) suggested

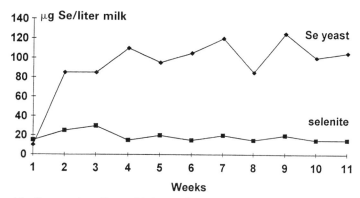

Figure 14 Comparative effects of 0.4 ppm dietary selenite and 0.4 ppm dietary Se from selenium yeast on Se content of milk. (Pehrson, personal communication)

Table 16. Summary of selenium yeast study results in Scandanavia.

Location	Parameter	Comparison	Results
University of Kuopio, 1988	Growth rate, tissue accretion	Se yeast supplement 0.5 ppm, control with 0.1 ppm from selenite. Initial weight 0.5 g	Better growth from 0.6 g to 12 g.
University of Kuopio, 1989	Growth rate, mortality	Se yeast supplement 0.5 ppm, control 0.5 ppm from selenite. Initial weight 0.1 g. Groups were divided into four individual tanks.	Significantly better growth rate. Improved survival during Costia infection.
Abo Akademi, 1989	Immunocompetence	Three groups, two with Se yeast supplement (0.5 ppm and 0-yeast). One Se yeast supplemented group was not vaccinated against vibriosis. Initial weight 50 g.	Improved survival during challenge with live *Vibrio anquillarus*, better growth rate.
University of Kuopio, 1990	Antibody production against *Vibrio anquillarus* following challenge.	Se yeast supplement 0.5 ppm, contol 0 yeast supplement. Initial weight 400 g. Both groups were fed trial feeds five months before challenge.	In the Se yeast group specific immune response was earlier and in higher proportion throughout the trial. Also, the specific response in Se yeast group was stronger (bigger concentration of circulating antibodies in blood).
University of Helsinki, 1990	Non-specific immune response against *Aeromas salmonicide*.	Se-yeast at 0.5 ppm, control 0.5 ppm selenite supplement. Initial weight 0.1 g. Weight after challenge 110 g (selenite group), 150 g (Se yeast group).	Long term Se yeast supplementation gave better blood O_2 carrying capacity, higher number of white blood cells and improved neutrophil activity.
University of Kuopio, 1989	Growth rate	Se from Se yeast versus selenite, 0.5 ppm. Initial weight 2 g, groups divided into four tanks	Faster growth rate. 14% higher growth (significant) in a 7 week trial.
University of Kuopio, 1987–88	Tissue absorption, mortality during infection	Se yeast supplements at 0.5 and 0.1 ppm. Initial weight 17.5 g	Good tissue uptake, especially skeletal muscle. Higher Se status in body, reduced mortality due to vibriosis.
University of Kuopio, 1988–89	Accumulation of Se from Se yeast in edible tissues	Se yeast supplementation for 5 months on several commercial farms with Se at 0.5 ppm. Control farms used 0.1 ppm from selenite...	Total Se in skeletal muscle ≈ 40% higher (1.21 in DM; range 0.96–1.57 ppm) compared with control. This higher level is comparable to wild fish selenium status in Finland.

Pehrson, 1993.

that organic Se in the form of Se yeast is the preferred source. He drew attention to the following:

- Organic Se compounds do not have the same peroxidative effects of selenite; therefore, will not tend to counteract themselves.
- Selenium yeast has a essentially the same composition of selenoproteins as most natural diets, whereas selenite is not a natural feed ingredient.
- Excretion of Se in milk and incorporation into tissues is four to six times higher after supplementation of organic Se compounds than after supplementation with similar levels of selenite (Figure 14).
- Improvements in fertility as measured by fewer services per conception occurred in dairy cows when inorganic Se was replaced by organic Se (Figure 15). Trials on other animals are underway.
- Selenium, stored in tissues after dietary supplementation with organic Se compound, can be used for glutathione peroxidase synthesis when body protein is mobilized.

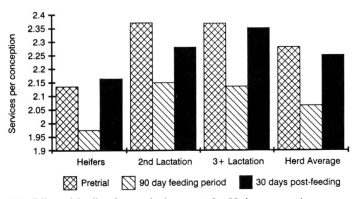

Figure 15 Effect of feeding Se as selenium yeast for 90 days on services per conception in a large commercial dairy herd. (Dildey, personal communication)

Chromium: a forgotten, but essential, nutrient

Nutritionists, like everyone else, most likely consider chromium (Cr) to be something associated with the bumpers of loud American cars from the Rock 'n Roll era. Perhaps it is the well-known toxicity of the hexavalent form of Cr used for chrome plating that has effectively caused the essential nature of the non-toxic trivalent Cr to be overlooked.

PHYSIOLOGICAL ROLE OF CR: GLUCOSE TOLERANCE FACTOR

In the trivalent form Cr is well known as an essential trace element for humans and laboratory animals. Numerous reviews report on the role of Cr in human nutrition, and, indeed, Cr was first determined to be an active constituent of the glucose tolerance factor isolated from yeast as early as 1959 (Schwartz and Mertz, 1959). This glucose tolerance factor appeared to work in conjunction with insulin in controlling glucose levels within the bloodstream. Though there may well be additional functions in physiology for Cr, at present it is only known to be essential to potentiate insulin in moving glucose into cells (Anderson and Mertz, 1977).

The chemical structure of the glucose tolerance factor, while it has not been completely identified, is thought to be a nicotinic acid-trivalent Cr-nicotinic acid axis with ligands of the amino acids glutamic acid, glycine and cysteine (Mertz et al., 1974; Figure 16). Chromium and insulin appear to work closely together in the shunting of glucose from the liver to peripheral tissues. Its dominant role seems to be to potentiate the action of insulin, both at the tissue and molecular levels. It is suggested that the glucose tolerence factor enhances the binding of insulin to its specific receptors possibly through initiating disulphide-bond linkages between insulin and cell membranes (Mooradian and Morley, 1987).

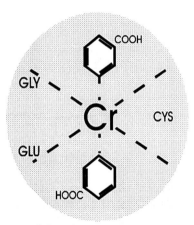

Figure 16 Possible structure of chromium in chromium yeast

SOURCES AND LEVELS OF CR IN PRACTICAL DIETS

Since inorganic Cr sources are only poorly available (<3%), those interested in supplementing practical diets have focused primarily on organic Cr ($\geq 10\times$ more available). Organic Cr sources include Cr yeast or Cr chelates or picolinate complexes.

It is of interest to know what diets or situations predispose an increased need for Cr. It is important to note that Cr is a nutrient; and as such, responses are only expected when Cr is deficient in the diet. Precise requirements for Cr are unknown. Adequacy of Cr levels in standard diet ingredients is difficult to evaluate as levels are low enough to challenge detection limits and sampling abilities. Beneficial responses in performance have suggested, however, that levels of Cr in most feed ingredients are inadequate for the combination of modern livestock diets, animals and rearing conditions. Suggested applications for supplemental Cr include rapidly growing animals/poultry, stress situations (eg. heat, transport) or high energy diets.

Effects of Cr in stressed or fattening animals
At present, research interests in Cr are focused on stressed animals and on fat deposition during the finishing phase in pigs. Mowat has suggested that the biggest potential for Cr supplementation is in the area of stress physiology (Mowat, 1993). He points out that periodic stress may be the norm rather than the exception in intensive rearing systems. His research indicates that when organic Cr was fed at a rate of 4 mg per head per day for the first three to five days after arrival at the feedlot, there was a significant increase in weight gain while morbidity was reduced to less than 1/3 of the control group (Chang and Mowat, 1992). Summarizing experiments with stressed feeder calves given Cr supplements in studies from 1989 to 1993, Mowat demonstrated an overall improvement in gain of 21% with the strongest responses coming from the animals with initially lowest performance (Figure 17). Morbidity was reduced by an average of 32%.

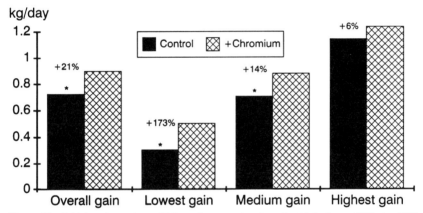

Figure 17 Initial gain and morbidity of stressed calves in trials from 1989 to 1993 (Mowat, personal communication)

Supplemental high Cr yeast also increased serum immunoglobulins and decreased serum cholesterol of growing steers (Moonsie-Shageer and Mowat, 1992). Decreasing serum cholesterol with Cr supplementation could have particular implications for the high-producing or metabolically stressed dairy cow. It is known that cortisol is antagonistic to milk production and that insulin availability may limit the onset of ovarian activity leading to first ovulation in cows.

Other researchers have reported beneficial effects of Cr in pigs. When organic Cr was used at 200 ppb in growing and finishing diets, increased loin eye area and percentage of muscling were reported (Page et al., 1993; Figure 18).

Recently, Gerber and Wenk (1993, personal communication) showed that while three sources of chromium all improved weight gain and

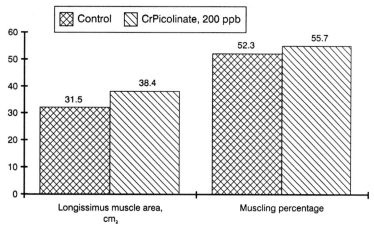

Figure 18 Effect of organic Cr on carcass characteristics of pigs (Adapted from Page et al., 1993)

feed efficiency, chromium in the form of chromium yeast resulted in the largest longissimus dorsii muscle area (Tables 17 and 18).

While further research is required to clarify the extent of improved milk production, improved reproduction and possible reduction in certain metabolic diseases in all animals, the role of organic chromium in the future appears assured. Though it is difficult to state a consensus on rates of use for various species, results of work to date suggest 400 ppb Cr for diets fed growing and finishing pigs and approximately 4 mg per head per day in diets for stressed cattle.

Table 17. Effects of the supplements on performance characteristics[1].

	Control	$CrCl_3$	Cr Yeast	Cr picolinate
Liveweight, kg				
Initial	27.5	27.6	27.3	27.2
End	106.5	107.3	107.1	105.0
Fattening period, days	105.3	100.4	103.0	102.3
End, corrected	99.8	101.2	102.0	98.2
Daily gain, g				
Early phase[2]	667	671	673	653
Final phase	829[a]	916[b]	875[ab]	878[ab]
Relative, %	100	111	108	106
Feed conversion				
Final phase	3.05[b]	2.73[a]	2.86[ab]	2.88[ab]
Relative, %	100	90	94	95
Entire period	2.75	2.58	2.65	2.66
Relative, %	100	94	96	97
Corrected	3.01[b]	2.79[a]	2.83[ab]	2.92[ab]
Relative, %	100	93	94	97

[1]Gerber and Wenk, 1993
[2] 27 to 60 kg

Table 18. Effects of the Cr supplements on carcass characteristics[1].

	Control	CrCl$_3$	Cr Yeast	Cr picolinate
Quality score[2]	2.1	2.2	2.3	2.4
Relative, %	100	105	110	114
Backfat thickness, cm				
Croupe (thinnest point)	1.6	1.5	1.7	1.6
Back	1.8	1.9	2.0	1.9
Longissimus dorsi				
Area, cm^2	49.5	51.2	53.8	48.6
per kg, relative %	100	102	103	100
per kg SG $_{corrected}$	100	102	106	100

[1]Gerber and Wenk, 1993
[2]Quality score: 2, normal; 3, high meat content.

Yeast cultures: the drive toward diet-specific strains

While the technique of using small amounts of live yeast to improve rumen function has been used for more than six decades, only recently has it been based on sound scientific concepts. A clear acceptance of yeast cultures as a dietary additive can be seen from the numerous positive reports in the scientific literature and papers presented at feed conferences around the world.

Dawson (1994) summarized results of yeast culture effects on milk production and liveweight gain reported in the literature. The range in liveweight gain response among trials was large and reflected both the variety of animals and diets involved; but was of the same order of other rumen modifiers such as ionophores. Furthermore, he demonstrated that the response to yeast culture, particularly the strain 1026 used in Yea-Sacc[1026], is related to effects on certain groups of rumen bacteria. Typically increases in total anaerobic bacteria and, in particular, those populations involved in cellulose digestion and lactic acid utilization, were observed. In attempting to build a model to explain the performance response to yeast culture, various researchers have concentrated on the ability of certain strains of the genus *Saccharomyces* to stimulate the growth and activity of specific groups of ruminal bacteria. These have been summarized in an overall model (Figure 19).

With the knowledge that Yea-Sacc[1026] stimulated growth and activity of the cellulolytic and lactic-utilizing rumen microbes, the development of yeast culture products that selectively stimulate other ruminal populations began. This has resulted in a new patented process whereby different strains can be screened for potential as rumen and(or) gut microbial modifiers. New species of yeast from which to make cultures have been selected from this screening procedure that examines effects on rumen parameters appropriate to specific rumen conditions or diet patterns. Of particular interest were the higher concentrate dairy formulas, beef fattening diets and high acid silages. While Yea-Sacc[1026] remains optimum for the mixed forage and concentrate diet (50–60%) fed the dairy animal, many fattening diets

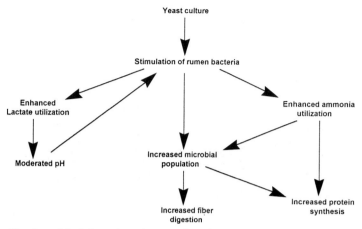

Figure 19 A model of the action of yeast culture in the rumen (Dawson, 1993)

contain little or no digestible fiber. Such diets support a different population of rumen microbes, promote a slower turnover rate and create a lower average pH environment. A desirable yeast culture supplement for this diet would have optimum ability to stimulate lactic acid-utilizing species while providing no advantage to species such as *S. bovis*. Yea-Sacc[8417] was selected after in vivo tests showed an improved ability to reduced lactic acid (Table 19).

Yeast culture made from strain 8417 selected by the screening process has now been subjected to a large number of validation studies under commercial production conditions. These field trials and controlled performance trials in high concentrate dairy and beef fattening programs have demonstrated that Yea-Sacc[8417] is useful in situations where potential for response from Yea-Sacc[1026] was limited. Beef steers initially weighing 873 pounds gained significantly faster (3.72 vs. 3.88 pounds per day) in a 95–day finishing trial utilizing a 17–day step-up period to an 87% concentrate diet (Figure 20; Birkelo, personal communication).

To further validate the initial in vitro data on the culture based on the new strain additional in vitro and in vivo studies were conducted in research herds and laboratory settings. In vivo studies and continuous culture work confirmed that anaerobic, cellulolytic and lactic acid-utilizing bacteria were indeed stimulated (Table 20). Additionally, ammonia concentrations were reduced, and there was a shift in the acetate to propionate ratio. In in vivo work with dairy cattle, an

Table 19. Comparative effects of two yeast cultures on rumen bacterial populations.

	Yea-Sacc[1026]	Yea-Sacc[8417]
Concentrations of anaerobes	Increased 58%	Increased 30%
Cellulolytic bacteria numbers	Increased 100%	Increased 51%
Lactate-utilizing bacteria numbers	Increased 42%	Increased 77%
pH	Little change	Slight increase
Ammonia concentration	No change	8% decrease

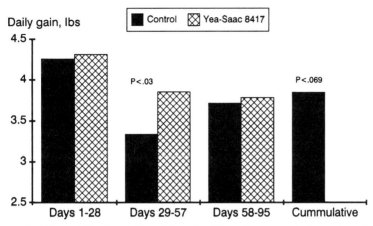

Figure 20 Effect of Yea-Sacc[8417] on daily gain of fattening steers during each period

Table 20. Summary of in vitro and in vivo responses to supplementation with Yea-Sacc[8417]

	Control	Yea-Sacc[8417]	P>F
In vitro studies in continuous culture[a]			
Anaerobic bacteria, billion CFU/ml	0.27	1.54	<0.05
Cellulolytic bacteria, million MPN/ml	28.5	43.5	<0.05
Lactate-utilizing bacteria, million CFU/ml	185	265	<0.07
Ammonia concentration, mg/dl	129.7	121.5	<0.01
Acetate:propionate ratio	0.39	3.29	<0.01
Performance of dairy cattle[b]			
Milk yield, kg/d	22.8	24.6	NA[c]
Performance of growing steers[d]			
Average daily gain, lb	2.73	2.96	<0.05
Dry matter intake, lb/d	15.25	15.95	0.06
Gain/feed	0.180	0.185	0.15

[a] Data from studies in rumen-simulating fermenter cultures fed a pelleted protein supplement with and without the yeast preparation.
[b] Performance data from a dairy trial conducted at Brookmount Farm in Northern Ireland. Milk production was from cows receiving a silage based diet and pelleted supplements which contained the yeast culture preparation. Twenty cows were assigned to each treatment group.
[c] NA indicates no statistical data were available.
[d] Performance data from growing steers fed corn silage with 10% of the diet supplied as one of four supplements. Two of the supplements contained Yea-Sacc[8417] while the other two supplements served controls. Forty-eight animals were assigned to each treatment group.

increase in milk yield of approximately 1.8 liters was observed. While the improvement was statistically non-significant, the trend was consistent. Diets used in this study were pelleted, again confirming that the yeast culture response was not lost during exposure to elevated temperatures.

For the growing steers, both dry matter intake and average daily weight gain improved significantly. The study clearly validated the concept that not all yeast strains work in the same way, and it is now possible to redefine appropriate applications of yeast culture preparations. The fact there are significant differences in the stimulatory effects of individual strains means yeast culture preparations can be designed which specifically stimulate the growth of lactic acid-producing bacteria in beef and dairy cattle fed high concentration rations. On the other hand, a specific strain has been found which stimulates cellulose-degrading bacteria and may be used for animals receiving poor quality forages or hay-based diets (Dawson, 1993).

On the analytic and(or) regulatory front, practical techniques have been developed whereby different yeast strains can be identified both in the concentrated product and subsequently in mixed final feed using DNA profiling.

Trace mineral proteinates in animal nutrition

Scientists have long known that trace minerals are required in small amounts for virtually every physiological process. From maintaining the rigidity of bones and teeth to the establishment of the actual structure of proteins and lipids, these trace minerals play key roles. Typically animals obtain trace minerals through a combination of those present in feedstuffs and those added via premixes, supplements, water, etc. Mineral supplementation is generally necessary to provide a constant supply of the minerals required to maximize animal growth and performance.

While very often the analysis of a given plant or cereal may reflect adequate mineral levels, the minerals may be either unavailable to the animal or availability may be variable. This variability may be caused by the presence of interfering substances such as phytic or oxalic acids in the plant tissue or by interactions which may occur among digesta constituents as they pass through the GI tract. With the availability of radiolabeling techniques, it has been possible to study mineral utilization within the body along with mineral balance. It is known that the absorption of mineral ions is dependent upon not just the level of the element ingested, but also the age of the animal, the pH of the GI tract and the state of the animal with respect to efficiency or adequacy of the element. Furthermore, the presence of antagonistic minerals or nutrients in the feed or any condition that increases the rate of passage of minerals through the intestine, for example digestive upset such as diarrhea, can lead to a deficiency. Few biological systems have so many variables as minerals absorption and utilization.

In recent years, scientists have recognized that if minerals are chelated to an amino acid or a peptide, then that mineral is better protected during passage through the acidic conditions of the stomach to the site of mineral absorption in the intestine. It is postulated that the amino acid or peptide may also carry the bonded mineral across the GI wall thereby enhancing its absorption despite presence of interfering substances.

The use of mineral Bioplexes or chelates has probably only received full scientific attention in the last few years, despite the fairly common use of these and other organic trace mineral complexes in commercial production. A recent survey (Feedstuffs, November 1993) estimated that 47.5% of all dairy farmers used zinc (Zn) proteinate compared with 50% using yeast culture. Farmers claim to observe fewer problems with feet in confined dairy and beef cattle, lower somatic cell counts in milk and generally improved performance in response to zinc proteinate despite the fact that few scientific data have existed to support these claims. The scientific evidence has begun to accumulate; and perhaps this can best be seen in the following examples.

- The role of a mineral proteinate (Bioplex Zinc) in mastitis incidence
- The role of a mineral proteinate (Bioplex Copper) in improving Cu availability.

ZINC BIOPLEX AND INCIDENCE OF MASTITIS

Zinc is well known for its role in the cellular immune response where it is a cofactor for one of the antioxidant enzymes. Though Zn is involved in over 200 reactions in physiology, a common symptom associated with deficiency is hyperkeratinization of the epithelium. Normal epithelial cells are the first defense against invasion by pathogenic microorganisms. It has been reported that nearly 20% of total body Zn is present in the skin. For this reason, the use of topical Zn mixtures to promote skin healing has been practiced for centuries. While oral Zn supplementation is often used in the control of foot rot for animals housed on concrete, many types of dermititis are also known to respond to Zn supplementation.

Adequate levels of Zn are crucial for any animal undergoing rapid tissue turnover such as the mammary glands for lactating dairy cows, the uterus for pregnant animals and the gastrointestinal tract for animals suffering from stress-related diarrhea. Of particular interest is the integrity of the teat canal epithelium, which is the first barrier against invasion by the organisms that cause mastitis. Zinc is required in the formation of keratin, the fibrous protein that lines the teat canal. It is known that lactating dairy cows have a higher incidence of mammary gland infections following the removal of keratin (26% compared with 8% on the control quarters). Additionally, since a single milking may remove up to 50% of the keratin, it is crucial that keratin be replaced on a daily basis. This requires sufficient bioavailable Zn.

A study at the University of Missouri was conducted to determine if supplying a portion of the dietary Zn in the form of a proteinate affected somatic cell count or the incidence of new mastitis infections. Forty lactating dairy cows given diets containing NRC-recommended levels of Zn were assigned to either control (Zn supplied as oxide) or Bioplex (50% oxide, 50% proteinate) diets. Production, somatic cell count (SCC) and milk culture data were collected for 112 days. Cows in the Bioplex group had significantly fewer new infections compared with the control (Spain et al., 1993; Table 21). Somatic cell counts were unaffected, however this herd had an initial SCC below 200,000 cells/ml. New infections in both groups were due to environmental pathogens. It

Table 21. Effect of Zn source on incidence of new infections[1].

	Zinc oxide	Bioplex Zn
Number of cultures	62	72
Number of quarters	248	304
Number cultured based on CMT[a]	72	63
Percent cultured	29.0	20.7
Positive cultures, n	14	18
New infections[b]	11	5

[1]Adapted from Spain et al., 1993
[a]Significantly different p=.106 (Yates corrected chi square)
[b]Significantly different p= .03

has been suggested that the response to Zn proteinate demonstrated in this study is related to the importance of Zn in immune response, perhaps for its role in replacing keratin lining the teat canal.

COPPER BIOPLEX: IMPROVED COPPER AVAILABILITY

In addition to providing bioavailable trace minerals to animals with increased nutrient demands, proteinated trace mineral forms aid in circumventing mineral interference problems common in livestock rearing. Often excesses of one mineral affect utilization of others. For example, excessive sulfur in ruminant feeds or water sources lowers copper utilization, especially if even low levels of molybdenum are present. Clark et al. (1993) found that commercial beef cows in a region known to be Cu deficient had hepatic Cu levels indicating low marginal to deficient status (Table 22). The herd was divided into three groups which received supplements containing either oxide, sulfate or proteinate (Bioplex) forms of Cu for one month. After one month the group on the supplement containing copper oxide remained low marginal or deficient while copper levels of the cows given the sulfate form fell in the marginal category. The group given copper proteinate demonstrated the largest increase in liver Cu and were classified as high marginal/low adequate. Average increases over the month were +9.45 mg, +33.18 mg and

Table 22. Effect of copper supplement form on liver Cu levels of beef cows after one month.

	Oxide	Sulfate	Cu Bioplex
No. cows	5	6	5
Initial			
Plasma Cu, ppm	0.96	0.89	0.90
Liver Mo, ppm	5.52	4.76	5.25
Liver Cu, ppm	24.82	23.66	29.29
28 days			
Liver Cu, ppm	34.27	56.84	79.33

+50.04 mg Cu for the oxide, sulfate and proteinate forms, respectively. Hair color changes in these animals were noted as well.

Copper and Zn at excessive levels may also affect mineral utilization as these two minerals compete for the same absorption site in the intestine. Therefore, pharmacological levels of Zn or copper sulfate used against microbial pathogens in pig starter diets leave mineral utilization unknown. In studies with rats Du et al. (1993) demonstrated that providing Cu in proteinate form resulted in higher hepatic Zn levels (Figure 21). This study suggested that the mineral proteinate is absorbed by a route different than that of the free ion; an advantage when absorbtion sites are flooded due to unintentional mineral imbalance.

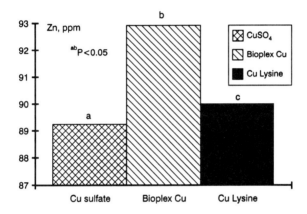

Figure 21   Effect of Cu source on hepatic Zn content

Further work by Du et al. (1993) also demonstrated that the organic ligand to which the trace mineral is chelated is important in determining bioavailability. Liver copper levels of Sprague-Dawley rats were measured after a growth period in which 5 ppm Cu fed as either copper sulfate, Cu proteinate or Cu-lysine were offered. The group given the proteinate had significantly higher target organ Cu levels (Table 23). This study illustrated the value of the peptide ligand in comparison to the single amino acid for copper supplementation.

Table 23. Effects of copper source in rat diets on copper and zinc content of organs.

	$CuSO_4$	Bioplex Cu	Cu Lysine
Cu content, ppm			
Liver	11.52[b]	12.55[a]	12.02[ab]
Spleen	4.67[b]	5.09[a]	4.97[a]
Heart	21.80[b]	22.66[a]	22.29[ab]
Kidney	29.25	29.85	29.39

[a,b] Means in a row with different superscripts differ ($P>0.05$)

MINERAL BIOPLEX QUALITY CONTROL

Bioplexes constitute a mixture of single amino acid chelates and larger proteinates (dipeptide proteinates, tripeptide proteinates, etc.) in order to gain maximum absorption of trace minerals across the gastrointestinal tract. Because research has demonstrated that mineral source and form are important for bioavailability and bioactivity, quality control in Bioplex production is paramount. A quality assurance method to ensure

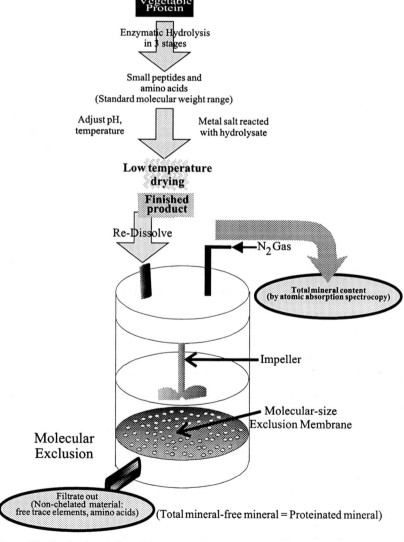

Figure 22 Production and quality control flow chart for Alltech trace mineral Bioplexes

chelation has been developed. This procedure is based upon the fact that the smallest chelate in the bioplexed product is perhaps 300 Daltons in size. It is quite straight forward then to pass a solution of Bioplex material through an ultrafiltration membrane with a molecular size "cut-off" of 300 Daltons. Chelated material is retained on the inlet side of the membrane while free amino acids and inorganic minerals pass through the membrane. Atomic absorption spectroscopy is then used on the retained material to determine the degree or the percentage chelation that has occurred (Figure 22). All batches are rejected or passed on this basis and it represents a rapid and accurate quality measure to ensure adequate chelation.

Natural toxin binders

A new area of investigation regarding both yeast culture and mannan-oligosaccharides is use of these supplements in mycotoxin-contaminated diets fed poultry. Feed grains contaminated with storage toxins, especially aflatoxin, are a major source of economic loss in many parts of the world. Field reports had indicated that Yea-Sacc[1026] improved performance of broilers in tropical climates when the diet unavoidably contained aflatoxin. Devegowda, in investigating this response, found that Yea-Sacc[1026] alleviated much of the ill effects of aflatoxin. Performance depression was reversed; and indications of liver damage and immune suppression were negated when 0.1% Yea-Sacc[1026] was added to diets containing 500 ppb aflatoxin (Table 24).

Explanation for the mode of action for yeast culture in reversing aflatoxicosis may rest with the binding capacity of the mannan-oligosaccharides on the yeast cell wall. In in vitro studies investigating ability of various combinations of gluco-mannan fractions to bind storage toxins, Trenholm (personal communication) found that both cell wall fractions and intact cells could bind zearalenone (Figure 23). Additionally, a greater percentage of the toxin was bound at pH 4 than at pH 9. A

Table 24. Effect of Yea-Sacc[1026] on body weight, feed conversion, mortality, total protein and HI titer level against Newcastle disease in broilers fed diets containing aflatoxin.

Aflatoxin, ppb	0	500	500
Yea-Sacc[1026], %	0	0	0.1
Body weight, g	1409[a]	912[b]	1308[c]
Feed:gain	2.27[a]	3.00[b]	2.36[a]
Mortality, %	3.87[a]	38.40[c]	3.30[a]
Liver, g/100g BW	2.65[a]	3.40[b]	2.87[a]
Bursa of Fabricius, g/100g BW	0.32[a]	0.20[b]	0.29[a]
Total protein, g/dl	2.68[a]	2.12[b]	2.56[a]
Gamma glutamyl transferase, IU/l	19.25[a]	15.25[b]	19.77[a]
HI titer, log 2 values	2.46[a]	1.74[b]	2.62[a]

[ab] Means in a row with different superscripts differ, P< 0.05

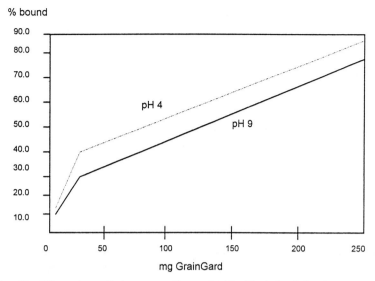

Figure 23 Effect of modified mannan-oligosaccharide (GrainGard) level on percentage of zearalenone bound in solution

technique for attenuating the binding capacity has been developed so that the mannan-oligosaccharide fraction can be effectively used when added to feed at rates as low as 1 to 2 kg/t. This modified oligosaccharide will soon be launched in a new commercial product called GrainGard.

Enzymes: emergence of substrate specific enzyme complexes

Enzymes have been used for many decades in the food processing and allied industries. Specific examples include the use of enzymes in the production of glucose and high fructose syrups from corn starch, as well as the use of pectinolytic enzymes in fruit juice processing and the use of proteases in cheese manufacture. Specific enzyme formulations have also been used by the medical community to supplement the endogenous digestive capability of patients suffering from certain digestive and related disorders. When such facts are considered, it seems somewhat surprising that the routine use of enzymes in animal feed processing has only recently become an established technology. It is only within the last decade that both the scientific and general agricultural communities have come to realize the range and magnitude of benefits that can accrue from the use of specific enzymes to achieve specific objectives in the animal feed industry. Today many excellent trial results have been reported pertaining to the use of enzymes for such purposes. The real benefits observed underline the present importance and future potential of enzyme use in animal feed.

Virtually all enzymes employed in this industry are hydrolases and are used as direct feed additives to achieve any or all of the following objectives:

- Supplementation of the host's endogenous enzymes, including proteases and amylases.
- Removal of antinutritional factors such as β-glucans and phytic acid.
- To render certain nutrients more available for absorption and to enhance the energy value of cheaper feed ingredients.
- Pretreatments of certain feed ingredients such as feathers or offal in order to render them more digestible.

THE EFFECT OF NON-STARCH POLYSACCHARIDES ON MONOGASTRIC ANIMALS: β–GLUCANASE AND PENTOSANASE

β-glucanase was perhaps the first enzyme used extensively in the feed industry. The presence of β-glucans in ruminant diets presents few problems because the rumen microbial populations produce enzymatic activities (β-glucanases) capable of degrading β-glucans, and hence destroying their antinutritive characteristics. Monogastrics such as pigs and poultry are devoid of endogenous activities capable of hydrolyzing β-glucans. Pigs are less susceptible to the adverse effects of these glucans as the longer retention times in the porcine digestive tract coupled to a proportionally greater dilution effect of the viscous NSP reduces their negative impact.

Inclusion of certain cereals such as barley in poultry diets is frequently regarded as being economically attractive. The presence of appreciable levels of β-glucan in barley had however traditionally precluded utilization of large quantities of this particular cereal in such feed. Some experimental data generated as early as the 1950s and '60s suggested that incorporation of microbial β-glucanase enzymes in barley-based rations could alleviate some of the adverse effects of this particular grain on animal performance. It was envisaged that a significant proportion of the β-glucanase activity retained its catalytic capability under the conditions encountered within the digestive tract. Decreased molecular weight of β-glucans promotes a significant reduction in viscosity. Therefore, relatively low levels of endo-glucanase activity could dramatically decrease its gelling ability and reduce the negative impact of such dietary constituents.

More recently numerous trials carried out have confirmed this, and it is now generally accepted that incorporation of suitable β-glucanase enzymes in the diet facilitates the incorporation of large quantities of barley in poultry feed (Table 25).

Incorporation of arabinoxylanase (pentosanase) in wheat-based and other diets known to contain pentosans can also facilitate removal of the antinutritional characteristics of these polysaccharides. As is the case with β-glucans, disruption of the backbone structure of pentosans promotes decreased viscosity.

Wheat contains variable quantities of pentosans (arabinoxylans) and, like the β-glucans of barley, ingestion of these non-starch polysaccharides

Table 25. Effect of β–glucanase on performance of broiler chicks fed a barley-based diet through 21 days.

Rate of Allzyme BG inclusion, kg/t	Total feed consumption, g	Total weight gain, g	Total feed: total gain
0 (Control)	618	361	1.72[a]
0.5	634	424	1.49[b]
1.0	630	423	1.49[b]
2.0	618	418	1.48[b]
4.0	635	431	1.47[b]

[ab] Means differ, $P < 0.05$.

results in increased viscosity of digesta, poor nutrient assimilation and wet litter problems. Inclusion of pentosanase preparations, such as Allzyme PT, in the diet can negate the deleterious effect of such antinutritional factors by conferring on the monogastric species the ability to digest dietary arabinoxylans. The effect of inclusion of Allzyme PT in the diets of broilers up to 21 days of age is presented in Table 26. As is clearly evident, inclusion of the enzyme in the diet promoted a statistically significant improvement in daily liveweight gain, feed weight and feed efficiency.

Table 26. Effect of pentosanase[1] on performance of broilers through 21 days.

Parameter	Control	Allzyme PT	SED	P>F
Daily gain, g	34.0	36.6	0.475	0.001
Feed usage, g	54.0	54.8	0.641	0.05
Feed efficiency[2]	0.634	0.657	0.003	0.001

[1] 1 kg/t
[2] Efficiency = body weight produced/feed used.

PHYTASE AND PHOSPHORUS POLLUTION

Supplementation of the diet with selected enzyme activities may promote a decrease in the overall pollutive effect of animal excreta. This is particularly true in the case of dietary phosphorus, a large proportion of which remains unassimilated by monogastrics.

In the region of 60–65% of the phosphorus present in cereal grains exists as phytic acid (myoinositol hexaphosphate) which, accordingly, represents the major storage form of phosphate in plants. However, in this form, the phosphate remains largely unavailable to monogastrics as these species are devoid of sufficient, suitable, endogenous phosphatase activity that is capable of liberating the phosphate groups from the phytate core structure. The animal's inability to degrade phytic acid has a number of important nutritional and environmental consequences. Phytic acid is considered antinutritional in that it chemically complexes a number of important minerals such as iron and zinc, preventing their assimilation by the animal. The lack of available phosphorus also forces feed compounders to include a source of inorganic phosphate (such as dicalcium phosphate) in the feed, with the result that a large proportion of total phosphate is excreted. It has been estimated that in the USA alone, 100 million tons of animal manure is produced annually, representing the liberation of somewhere in the region of 1 million tons of phosphorus into the environment each year. The potential pollutive effect of this in areas of intensive pig production is obvious. Many countries are currently enacting tough, new antipollution laws in an attempt to combat the adverse effect of animal waste on the environment. In Singapore, for example, pig production has been banned.

Several microbial species (in particular fungi) produce phytases (EC 3.1.3.8). The incorporation of suitable, microbially derived phytases in the diet can confer the ability to digest phytic acid on the recipient animals. This would have a threefold beneficial effect: the antinutritional properties of phytic acid would be destroyed; a lesser requirement of feed supplementation with inorganic phosphorus would exist; and reduced phosphate levels would be present in the feces.

Several trials have confirmed that the inclusion of phytase in animal feed promotes at least some of these effects (Table 27). However, despite the excitement generated by phytase within the industry, the enzyme is being used in comparatively few countries. This may be explained, in part, by the fact that most microbial species only produce low levels of phytase activity which, obviously, has an effect on the cost of the finished product. It seems likely that widespread utilization of phytase within the industry will only be made possible by the production of this enzyme from recombinant sources or by finding new ways to stimulate phytase production by intestinal microbes. Recent reports have also illustrated that the incorporation of transgenic tobacco seeds, which express the phytase gene, into poultry feed leads to enhanced phosphorus release from dietary phytate.

Table 27. Effect of phytase on phosphorus balance in pigs[1]

Parameter	Control	Phytase
Daily phosphorus intake, g/head	7.3	8.3
Phosphorus absorbed, g/day	3.5	4.5
Phosphorus absorbed, %	48.0	65.0
Phosphorus retained, %	47.0	60.0

[1]Increase in % phosphorus absorbed and retained due to phytase significant, $P < 0.05$.

Phytase belongs to a larger family of phosphohydrolase enzymes, of which acid phosphatases (orthophosphoric-monoester phosphohydrolases, EC 3.1.3.2) form a major sub-group. Several microbial species produce these enzymes, and while phytate is not the preferred substrate, many of these enzymes do display very significant phytate-hydrolyzing activity. The potential for use of such acid phosphatases capable of using phytate as substrate in the feed industry seems bright as such enzymes may well function efficiently at low pH values, thus facilitating the enzymatic conversion of phytate in acid regions of the GI tract. This in turn would maximize the level of phytate-derived phosphorus absorption in the proximal small intestine. Such enzymes are the subject of ongoing investigation in our research facilities. Initial studies indicate that a combination of these enzymes derived from non-recombinant sources plus stimulation of natural phytase production can have the desired effect. Interestingly, the combination of approaches allows levels of added phytase to be as low as 50,000 enzyme units per tonne compared with 500,000 to 1,000,000 units when phytase alone was added.

Galactosidase is yet another enzyme which merits increased attention with regard to its use in animal feed. This enzyme catalyzes the hydrolysis of terminal α–D-galactose residues in α–galactosides, including galactose oligosaccharides, galactomannans and galactolipids, and it also hydrolyzes α–D-fucosides. Galactosidase could therefore play a potentially important role in the destruction of certain antinutritional factors present in legumes such as soya, peas and beans (Table 28).

Table 28. Effect of treatment with α–galactosidase[1] on energy values of various legumes.

Raw material	Control	Allzyme G	Improvement, %
Soya	11.87	12.36	+4 %
Rape/pea mix	15.26	17.44	+14 %
Rape/pea mix	15.16	16.56	+9 %

[1]1kg/t

MATCHING THE ENZYME TO THE SUBSTRATE

In order to obtain maximum benefit from inclusion of enzymes in feed, it is necessary to ensure that the enzyme or enzyme cocktail added is chosen on the basis of feed composition, that is that the enzyme is matched to its substrate. In many instances, the choice of enzymes is rendered straightforward, i.e. the inclusion of β-glucanase in barley-containing diets destined for poultry. In other cases, more careful consideration is required. A judicious choice of enzyme cocktail rather than addition of a single enzyme may often be most effective and the proportion of each enzyme present in a cocktail should reflect the proportion of its substrate present in the feed. Proteases should be included in diets of particularly high protein content. The use of cellulase and hemicellulase enzymes in monogastric feed merits increased attention. Most diets will contain relatively large proportions of non-starch polysaccharides, against which monogastric animals produce no endogenous degradative enzyme activity. Supplementation of such diets with microbially-derived cellulase/hemicellulase preparations could promote direct degradation of such fibrous components. In addition to making some such carbohydrates biologically available, degradation of non-starch polysaccharides should facilitate more complete digestion by making nutrients available which were previously protected against digestive activity by such fibrous materials.

REGISTRATION OF ENZYMES

The increasing use of enzymes in animal feeds reflects the growing interest in these additives as natural alternatives to chemical products currently used to improve livestock production. Additionally, enzymes are expected to play an important role in the future as reducers of nitrogen and phosphorus pollution from intensive farming practices. Such a high profile has resulted in regulatory bodies undertaking a series of legislative initiatives to legally frame the use of enzymes in animal nutrition. The general directorate for agriculture within the European Union (EU) is no exception to this, and this body has perhaps passed the most stringent legislation noted worldwide to date. Under this legislation, enzymes are covered by the definition of additives stated in Directive 70/524/EEC concerning additives in feeds:

"Additives: Substances or preparations containing substances which, when incorporated in feedingstuffs, are likely to affect their characteristics or livestock production."

Accordingly, all enzymes much be registered with the EU by the submission of detailed dossiers, the guidelines for the compilation of which are detailed in another directive 87/153/EEC. The approval or registration of enzymes, like any other feed additive, is therefore based on three basic criteria: proof of the quality, efficacy and safety of the product. These criteria also apply to microbial products, including yeast culture.

Perhaps the single most important condition which must be satisfied by enzyme manufacturers is that the enzyme must be controllable (i.e., analyzable) in animal feeds. The difficulties associated with the development of such an assay may be the reason why only one dossier on an enzyme product has been submitted to date.

Assay of enzymes in feed
Thus far, no suitable method allowing detection of enzymes in finished feed has been widely discussed in the literature. Development of assay systems which facilitate direct detection and quantification of enzymes in finished feed is highly desirable for obvious reasons. Several strategies have been have been adopted in attempts to develop suitable in-feed assays. Most focus on extraction of the enzyme from feed followed by incubation of the enzyme extract with substrate, either artificial or natural, for an extended time period. While in most cases the enzyme is incubated with a suitable substrate in a test tube, an alternative approach involves incorporating the substrate into agar plates with subsequent application of the enzyme into wells cut in the agar. With time the enzyme diffuses outward, hydrolyzing the substrate which is effectively immobilized in the agar. Subsequent staining with a dye facilitates

visualization of the zones of hydrolysis promoted by the enzyme. The diameter of this zone of hydrolysis is proportional to the enzyme activity present.

Research at the European Biosciences Centre has shown this approach to enzyme assay to be feasible; and assays capable of detecting and quantifying a range of enzyme activities added to feed have been successfully developed. Such assay systems are likely to be introduced into the industry for routine analytical applications in the near future.

The figure below (Figure 24), along with the frontispiece, illustrate the agar plate-based assay. A feed containing enzyme added at a rate of 1 kg/t is analyzed by extracting the enzyme from the feed into buffer and subsequently loading the enzyme-bearing extract into the center well. Enzyme substrate is incorporated into the agar. A standard curve is constructed by taking samples of the feed devoid of enzyme activity and spiking it with enzyme such that these samples contain enzyme concentrations ranging between 0 and 2 kg/t. These samples can then be extracted in the same way as the commercial sample and the extracts loaded into the other wells in the agar plate. After incubation for an appropriate time period, zones of substrate hydrolysis are visualized by staining the agar plate with the dye congo red. The diameter of the various zones of hydrolysis are measured. The more enzyme activity present, the greater the diameter of the zone of hydrolysis.

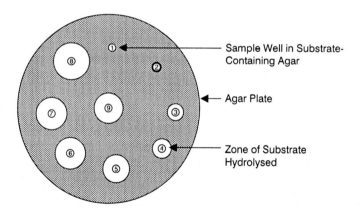

Figure 24 Determination of in-feed enzymes using the zone diameter of substrate hydrolysed. Well #1: buffer, Well #2: feed extract devoid of enzyme, Wells #3 to 8: spiked with various levels of enzyme. Well #9 was loaded with an extract obtained from a commercial feed sample to which enzyme had been added at 1 kg/t.

Peptides in animal nutrition

For a number of reasons, including the problem of nitrogen pollution from agricultural sources, researchers are examining alternatives for reducing the nitrogen content of the diet without sacrificing optimum nutritional standards. In poultry, almost 60% of the ingested nitrogen is excreted. The economic availability of a number synthetic amino acids e.g., methionine, lysine, threonine and tryptophan, has made possible a reduction in total nitrogen content by supplementation of pig and poultry diets with essential amino acids, production results have often been inconsistent, possibly due to several factors including other limiting amino acids or interference with release and absorption of other amino acids due to quality of the free amino acids in the feed as there are different transport sites for amino acids that are group specific exhibiting a degree of overlap.

Peptides produced during endogenous digestion of dietary protein or by any other means can be absorbed intact from the intestine. The significance of this in terms of animal nutrition is attested to by the fact that oral administration of glycine peptides results in higher plasma concentrations of glycine compared with equivalent amounts of free glycine. The uptake by carrier mechanisms for small peptides is distinct and independent of that for amino acids. The importance of this different absorption system from those utilized in amino acid uptake means that 1) there is no competion between amino acids and peptides for uptake, and 2) in certain clinical conditions of congenital defects in amino acid transport systems the same amino acids are absorbed when taken orally as dipeptides. Such scientific evidence indicates that dipeptides do not receive further digestion in the intestine and are absorbed more efficiently than the individual amino acids present in the peptides.

Conclusions

During the last 10 years we have seen great progress made in the fields of genetics, nutrition and biotechnology; and this has allowed the adaptation of livestock farming to the new agricultural structures and to the requirements of industrialized countries. In these 10 years, however, the ground rules have also changed. It is no longer simply a question of producing wholesome meat in the fastest way possible at the cheapest cost. Today's nutritionists instead must take into account the social implications of animal products including both the welfare of the animal and the environment. Ten years ago it probably would have been predicted that in 1994 genetic engineering would have played an enormous part in animal production with the application of new growth hormones and genetically modified rumen bacteria. What was not predicted, however, was the demand for natural products and a growing public skepticism of science. The winners have been those companies who have recognized this trend and moved towards systems which use natural products to achieve their objective.

The past 10 years have also brought to the floor a whole new vocabulary of biological aids. Today, when we talk about maximizing milk yield through the manipulation of rumen fermentation, we may be talking about using specific strains of yeast. If, on the other hand, the object is to improve milk protein, then the same yeast culture could help along with protected amino acids which will successfully bypass the rumen. If the emphasis is to improve the animal's defense against mastitis and to lower somatic cell counts, a focus would probably be given to tissue integrity and the role of bioplexed minerals.

Pollution concerns can now be addressed with the use of well-defined amino acids and at the same time the application of natural enzymes such as cellulases, proteinases and glucanases to minimize waste production. Concerns today about emission of gases such as methane and ammonia may be minimized by use of yeast cultures for the former and natural glycocomponents extracted from the yucca plant with the latter. Phosphorus pollution can also be minimized by either the use of the enzyme phytase or by stimulating those bacteria present in both the rumen and the hindgut which are either capable of breaking down phytin or better able to utilize inorganic phosphorus.

Today, and presummably tommorow, we cannot automatically rely on synthetic growth promoters, and a new understanding of both rumen and hindgut fermentation is occurring. Gut microflora can be changed towards the more benevolent microorganisms by the use of selected sugars which do not support pathogen growth and at the same time help block their colonization. The same oligosaccharides are now known to have powerful immune stimulation properties and with some modification can aid in overcoming the negative effects of feed toxins.

The whole focus, therefore, has been to emphasize the animal's own protection systems and to provide natural products which may supplement this protection. As we move into the future, a better understanding of the immune system and gut microbiology will make these techniques even more successful. The advent of new registration requirements has also focused the minds of the scientists on techniques that can be used to analyze these natural products present in complete feed. It has been said that the plant and microbial kingdom represent a virtually untapped source of new biomolecules. If the last 10 years have taught us anything, it must be the truth of this statement.

References

Anderson, R.A. and W. Mertz. 1977. Glucose tolerance factor: an essential dietary agent. Trends Biochem. Sci. 2:277.
Arce, J., E. Avila and C. Lopez-Coello. 1994. Effect of De-Odorase on performance and mortality due to ascites in heavy broilers. Proceedings Southern Poultry Science Association, 15th Annual Meeting, Atlanta, GA., January 17–18.
Ballou, C.E. 1970. J. Biol. Chem. 245:1197.
Chang, X. and D.N. Mowat. 1992. Supplemental chromium for stressed and growing feeder calves. J. Anim. Sci. 70:559.
Clark, T., Z. Xin, Z. Du and R. Hemken. 1993. A field trial comparing copper sulfate, copper proteinate and copper oxide as copper sources for beef cattle. J. Dairy Sci. 76(Suppl. 1):334.
Clayton C.C. and C.A. Bauman. 1949. Cancer. Research. 9:575–582.
Dawson, K.A. 1993. Current and future role of yeast culture in animal production: a review of research over the past seven years. In: Biotech-

nology in the Feed Industry. Proceedings of the 9th symposium. T.P. Lyons (Ed.) Alltech Technical Publications. Nicholasville, Kentucky.
Dawson, K.A. 1994. Manipulation of microorganisms in the digestive system: the role of oligosaccharides and diet-specific yeast cultures. Alltech Technical Symposium, California Nutrition Conference. May 12.
Drews. J. 1993. Biotechnology and the generation of economic value. In: the Biotechnology Report. Campden Publishing Ltd. London.
Du, Z., R. Hemken and T. Clark. 1993. Effects of copper chelates on growth and copper status in the rat. J. Dairy Sci. 76(Suppl. 1):306.
Hylands, P. 1993. Plant cell culture and drug discovery. *In:* Biotechnology Report, Campden Publishing, p. 99.
Mertz, W., E.W. Toepfer, E.E. Roginski and M.M. Polansky. 1974. Present knowledge of the role of chromium. Fed. Proc. 33:2275.
Milner, J.A. 1985. Selenium and carcinogensis. In: ACS Symposium Series, No. 277: Xenobiotic metabolism:nutritional effects. American Chemical Society, 267–282
Moonsie-Shageer, S. and D.N. Mowat. 1993. Levels of supplemental chromium on performance, serum constituents and immune status of stressed feeder calves. J. Anim. Sci. 71:232–238.
Mooradian, A.D. and J.E. Morley. 1987. Micronutrient status in diabetes mellitus. Amer. J. Clin. Nutr. 45:877.
Mowat, D. 1993. Organic chromium: a new nutrient for stressed animals. In: Biotechnology in the Feed Industry. Proceedings of the 9th symposium. T.P. Lyons (Ed.) Alltech Technical Publications. Nicholasville, Kentucky.
Mul, A.J., and F.G. Perry. 1994. The role oligosaccharides play in animal nutrition. Feed Manufacturer's Conference, University of Nottingham, Loughborough, Leics. Jan. 3–5.
Newman, K., K. Jacques and R. Buede. 1993. Effect of mannan-oligosaccharide supplementation on performance and fecal bacteria of Holstein calves. J. Anim. Sci. 71(Suppl. 1):271.
Newman, K., K. Jacques and R. Buede. 1993.
Newton, A.C., G.D. Lyon and T. Reglinski. 1993. Development of a new crop protection system using yeast extracts. Home Grown Cereals Association Project Report No. 1978. Great Britain.
Oldfield, J.E. 1993. Selenium and cancer prevention. In: Bulletin of the Selenium-Tellurium Development Association. November.
Onarheim, A.M. 1992. The glucan way to fish health. Fish Farming International. August.
Peat, S. W. Whelan and T. Edwards. 1958. J. Chem. Soc. p. 3862.
Pehrson, B. 1993. Selenium in nutriton with special reference to the biopotency of organic and inorganic selenium compounds. In: Biotechnology in the Feed Industry. Proceedings of the 9th symposium. T.P. Lyons (Ed.) Alltech Technical Publications. Nicholasville, Kentucky.
Parekh, R. 1993. Carbohydrate engineering in modern drug discovery. In: the Biotechnology Report. Campden Publishing Ltd. London p. 135.
Phaff, H.J. and C.P. Kurtzman. 1984. In: The Yeasts, a Taxonomic Study. Elsevier Biomedical Press, Amsterdam. pp. 252–262.

Page, T.G., L.L. Southern Ward and D.L. Thompson, 1993. Effect of chromium picolinate on growth, serum and carcass traits of growing-finishing pigs. J. Animal Science. 71:656–662.

Sharon, N. and H. Lis. 1993. Carbohydrates in cell recognition. Scientific American. January.

Spielvogel, B.F. 1993. Boron biochemistry: New organic-like compounds for life science applications. In: The Biotechnology Report, Campden Publishing, p. 109.

Spain, J. 1993. Effect of organic zinc supplementation on milk somatic cell count and incidence of mammary gland infections of lactating cows. J. Dairy Sci. 76(Suppl. 1):265.

Schwarz, K. and W. Mertz. 1959. Chromium (III) and the glucose tolerance factor. Arch. Biochem. Biophys. 85:292.

Spiegel, J., R. Rose, T. Karabell, V. Frankos and D. Schmitt. 1994. Safety and benefits of fructo-oligosaccharides as food ingredients. Food Technology. January.

ENZYMES: PRACTICAL APPLICATIONS FOR BIOLOGICAL CATALYSTS IN ANIMAL NUTRITION AND WASTE MANAGEMENT

PLANT POLYSACCHARIDES – THEIR PHYSIOCHEMICAL PROPERTIES AND NUTRITIONAL ROLES IN MONOGASTRIC ANIMALS

GEOFFREY ANNISON and MINGAN CHOCT
CSIRO Division of Human Nutrition, Glenthorne Laboratory, Majors Road, O'Halloran Hill, SA, Australia.

Introduction

The major components of animal feeds are the starch, protein, fat, and "fibre". In classical monogastric livestock nutrition the first three were considered to be nutrients whilst the "fibre" was thought to be essentially inert contributing little, if anything, to the animal. Research through the late 70s and 80s showed that some fibre components, namely the soluble non-starch polysaccharides (NSP), were by no means inert and indeed possessed anti-nutritive activity. This has been particularly well characterised by studies with broiler chickens where the soluble NSP were shown to depress the digestion of macronutrients which resulted in performance losses (Choct and Annison, 1990; 1992a;b). To date evidence has been gathered demonstrating that the soluble NSP of barley, rye and wheat exert anti-nutritive activity and it is likely that soluble NSP from other sources also have similar properties. Whilst the nutritional effects – ie reduced digestibility of nutrients – have been well characterised, the mechanism of the anti-nutritive activity is yet to be elucidated. It is clear, however, that the physicochemical properties of the NSP are responsible for their action – either directly, or through interactions with other factors. In order to develop techniques to counteract the anti-nutritive effects of soluble NSP, an understanding of their chemistry, physical properties and behaviour on ingestion by monogastrics is crucial.

Polysaccharide structure

Polysaccharides are macromolecular polymers of simple sugars or monosaccharides. The sugars are linked together by glycosidic bonds which are formed between the hemi-acetal (or hemi-ketal group) of one sugar and the hydroxyl group of another. Many different glycosidic bonds exist. They are distinguished by referring to the carbon atoms of each sugar which are involved in the bond and orientation of the oxygen in the bond (α– or β–). In starch the glucose molecules are joined mainly by

Plant polysaccharides – their physiochemical properties and nutritional roles

α–(1→4) bonds with a small number of α–(1→6) bonds. These bonds and the α–(1→2) link in sucrose, the β–(1→4) link between glucose and galactose and the α–(1→1) link of trehalose are cleaved by endogenous avian or mammalian enzymes. All other glycosidic bonds are resistant but they may be cleaved by microbially derived enzymes. Non-starch polysaccharides (NSP) have glycosidic bonds other than the α–(1→4), (1→6) bonds of starch. The importance of the bonds in determining the susceptibility to enzymes is illustrated by the resistance of cellulose (α β–(1→4) glucan) to starch degrading enzymes.

The great majority of NSP in feedstuffs are of plant origin. In modern formulations the variety present may be large which reflects the diverse ingredients currently available. Many different types of sugars may be present (Figure 1) and these may form many different structures. The polysaccharides may be relatively simple such as the cereal β-glucans (Figure 2) which are linear polymers of glucose with β–(1→3), (1→4) glycosidic links (Fincher and Stone, 1986). The other major cereal polysaccharides, the arabinoxylans, are more complex (Figure 3). These compounds are composed mainly of two sugars, arabinose and xylose, in a branched structure (Annison et al., 1992).

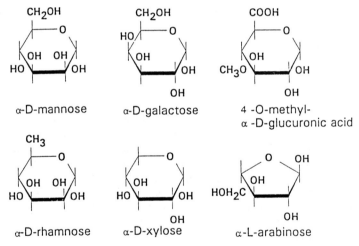

Figure 1. Sugars commonly found in plant non-starch polysaccharides

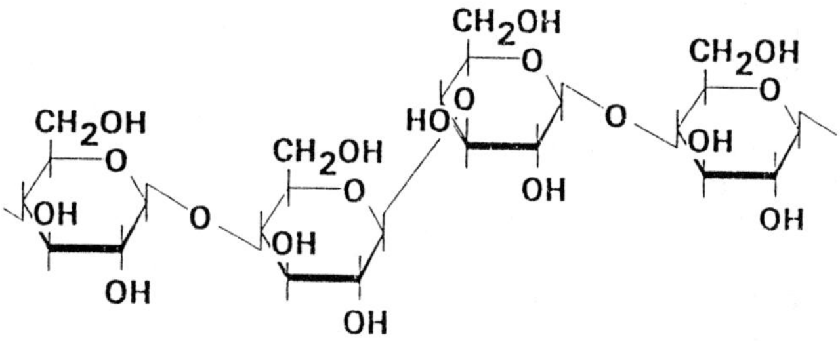

Figure 2. β–(1→3),(1→4)-D-glucan

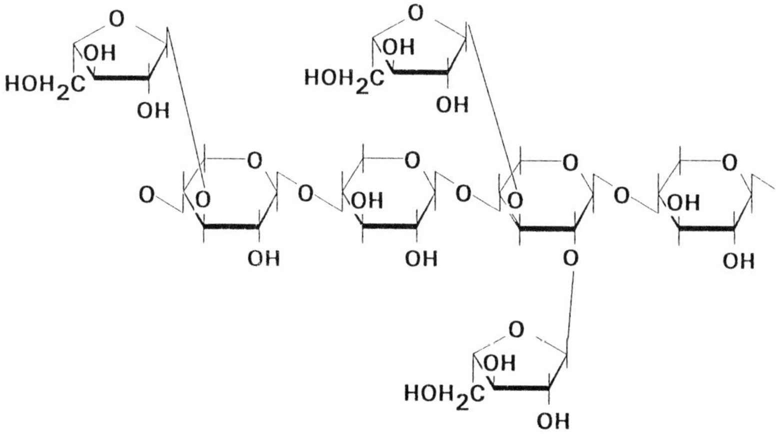

Figure 3. α–(1→2,3)–L-arabinosyl-β-(1→4)-D-xylan

Even more complex polysaccharides may be present if legumes are used in the ration. The main NSP of lupins is a highly complex branched structure containing a long β–(1→4)-D-galactose side chain attached to a pectin-like main chain of rhamnose and galacturonic acid linked by β–(1→4) and α–(1→2) bonds respectively. There are also side chains α–(1→5)-L-arabinose (Cheung, 1991).

Plant cell wall structure

The NSP found in feedstuffs are primarily components of plant cell walls. As such it must be appreciated they are not discrete materials but are in close association with other components which may mediate the way the NSP behave when ingested. The plant cell wall is a biphasic structure in which microfibrills of cellulose form a rigid skeleton which is embedded in a gel-like matrix composed of the non-cellulosic polysaccharides and glycoproteins (Fry, 1986). The cellulose microfibrills are highly ordered structures whilst the amorphous region is less ordered. Both types are deposited in the cell wall together, but as the plant tissue ages lignin is laid down encrusting the microfibrills (Selvendran et al., 1987). Whilst the nature of the cellulose microfibrills varies little between plants, the types and levels of the polysaccharides of the amorphous matrix can show considerable differences among species. In monocotyledonous plants such as the cereals, the cell wall polysaccharides are mainly arabinoxylans and β–glucans. In wheat and rye the arabinoxylans predominate whereas in barley the β–glucans are the main component. The cereals are virtually free of pectic substances (i.e. polysaccharides rich in uronic acid) which are found at high levels in dicotyledenous plants such as the pulses (Selvendran, 1984). It is of interest that most of the polysaccharides which comprise the cell walls are, after extraction, soluble, hydrophilic molecules; and yet in the intact cell wall they remain insoluble. This indicates that there must be considerable cross-linking between the polymers (Fry, 1986). Whilst the exact nature of the cell wall structure is yet to be elucidated it is clear that there are close associations between the polysaccharides, the proteins and lignin which may be a result of a variety of covalent links, ionic bonds or other associations (Fincher and Stone, 1986).

Chemical cross-linking involving NSP

Apart from being important to the structural integrity of the plant, the cross-linking between NSP and other components is likely to determine their nutritional activity and digestibility (Fry, 1986). The covalent bonding of cell wall polysaccharides to lignin has been shown to limit the digestibility of some forages in herbivores (Richards, 1976), and this will of course limit the polysaccharide's solubility when ingested by monogastrics.

PHENOLIC COUPLING

Some cell wall polysaccharides including the pectins of spinach and the arabinoxylans of wheat have the phenolic substituents p-coumarate and ferulate linked to them by ester bonds. Bridges between the polysaccharides can be formed by the oxidative coupling of these substituents. Evidence of this is provided by the isolation of diferulate molecules following the alkaline treatment of spinach (Fry, 1986) and from the observations (Iszydorczyk et al. 1990) that addition of hydrogen peroxide and peroxidase to solutions of isolated wheat arabinoxylans can result in the formation of gels which is interpreted as indicating a formation of cross-links between the polysaccharide molecules. It is also possible that this type of bond may exist between tyrosine residues of cell wall proteins and the polysaccharides (Fincher and Stone, 1986).

ESTER BONDS

Many plant polysaccharides are released by mild alkali treatment. This is interpreted as indicating the presence of ester linkages. Esters link the phenolic substances to polysaccharides and may also be present between the carboxyl groups of pectins and the hydroxyl groups of polysaccharides or other cell wall components.

CALCIUM ION BRIDGES

Treatment of plant material with strong chelating agents such as EDTA and CDTA releases pectin polysaccharides. The chelating agents bind Ca^{++} ions very strongly which strips them from binding sites within the cell wall structure and breaks the Ca^{++} ion bridges between the molecules. Other ionic interactions with other ions are also likely to contribute to cell wall structure.

OTHER INTERACTIONS

Due to the prepondence of hydroxyl groups on polysaccharides there is no doubt that hydrogen bonding plays a major role in the structure of plant cell walls. Intra-residue hydrogen bonding holds the cellulose molecule in a rigid, elongated structure which allows other cellulose molecules to become closely associated and held together by hydrogen bonds in the microfibrillar structure (Fincher and Stone, 1986).

Physical properties of NSP

Most of the nutritive activities of NSP which directly affect broiler performance have been ascribed directly to soluble polysaccharides. Insoluble NSP are by no means unimportant and they certainly have an effect on digesta passage and water holding. Detrimental effects, apart from nutrient dilution, have not been reported and in some cases addition of insoluble sources of NSP such as oat hulls can improve nutrient digestibility and broiler chicken performance (Rogel et al., 1987). When polysaccharides are in solution they exhibit many properties which may influence the digestive processes.

VISCOSITY

A majority of polysaccharides when dissolved in water give viscous solutions. Viscosity is dependent on a number of factors including the size of the molecule, whether it is branched or linear, the presence of charged groups and of course the concentration. The polysaccharides increase the viscosity at low concentrations by directly interacting with the water molecules. As the concentration increases the molecules of the polysaccharide themselves interact and become entangled in a network (Launay et al., 1986). This process can cause great increases in the viscosity and is dependent on the formation of junction zones between the polysaccharide molecules. Gel formation can occur when the interactions of the polysaccharide molecules becomes great. Increases in digesta viscosity associated with wheat arabinoxylan ingestion has been noted in broiler chickens (Choct and Annison, 1992a). Similar observations have been noted in studies on the β-glucans of barley (White et al., 1981; 1983).

SURFACE ACTIVITY

Polysaccharides can present charged (negative and less commonly positive) as well as weakly hydrophobic and weakly hydrophyllic surfaces. Therefore, when in solution they tend to associate with other surfaces. When ingested the polysaccharides may associate with the surfaces of food particles, the surfaces of lipid micelles or glycocalyx surfaces of the gut.

WATER HOLDING

Both soluble and insoluble NSP have water-holding activities. The insoluble forms such as cellulose and xylans behave like sponges whereas the soluble forms can entrap water molecules through the formation of networks. This becomes particularly marked at higher concentrations where gels may form. These effects can radically alter the physical properties of the digesta which in turn may alter the gut physiological activity (ie., through increased resistance to peristalsis).

BINDING OF IONS AND SMALL MOLECULES

Some NSP, such as the pectins, may have a high charge density at some pH values due to the presence of acidic groups. Apart from the association of cations with the negatively charged groups in some NSP, the stereo-structure of the polysaccharide allows chelation of ions to occur. Indeed, cations can form ionic bridges between NSP molecules profoundly influencing their viscosity and gel forming properties. Although there is little evidence for it, the possibility of cations acting as ion bridges between polysaccharides and small charged molecules in digesta should not be discounted. Small molecules may also be bound weakly to polysaccharides through both hydrophobic and hydrophyllic bond interactions.

OTHER EFFECTS

All of the properties of the NSP described may influence digestive processes. This has recently been reviewed by Annison (1993). It should be remembered, however, that the NSP become active in a complex mixture (namely the digesta) and other components may mediate their behaviour. It is known for example that polysaccharide–protein interactions can occur which affect the viscosity of solutions and the stability of oil–water emulsions.

Characterisation of the NSP anti-nutritive activity

In recent years the nutritional role of wheat arabinoxylans in broiler chicken diets has been characterised (Choct and Annison, 1992a;b; Annison, 1992; Annison, 1993). When wheat arabinoxylans are added to control diets a depression in apparent metabolisable energy occurs, which is a direct result of reduced ileal digestion of starch, protein and lipid (Table 1). If the wheat arabinoxylans are depolymerised with xylanase enzymes prior to addition to the feed, the anti-nutritive activity is ameliorated (Table 2). Similar improvements in the AME of wheat can be achieved by adding xylanases during feed formulation (Table 3). The xylanases cleave the wheat arabinoxylans to release lower molecular weight arabinoxylan fractions (Annison, 1992). Similar data have been

Table 1. Effect of adding graded levels of arabinoxylan to a control diet on apparent metabolisable energy (AME, MJ/kg DM), ileal starch digestibility, (ISD), ileal protein digestibility (IPD), and ileal lipid digestibility (ILD).

Diet	AME (MJ/kg DM)	ISD	IPD	ILD
Control	15.05	0.96	0.75	0.93
Arabinoxylan, 5 g/kg	15.00	0.96	0.75	0.93
Arabinoxylan, 10 g/kg	14.70	0.95	0.73	0.92
Arabinoxylan, 25 g/kg	13.34	0.92	0.69	0.76
Arabinoxylan, 40 g/kg	12.48	0.82	0.61	0.69
significant effects	$P<0.05$	$P<0.05$	$P<0.05$	$P<0.05$
regression on arabinoxylan level	$P<0.01$	$P<0.01$	$P<0.01$	$P<0.01$

(Adapted from Choct and Annison, 1992b)

Table 2. Effect of adding intact and partially depolymerised arabinoxylans (30g/kg) on the AME (MJ/kg DM) of sorghum. The effect on ileal starch digestibility (SD) and ileal digesta viscosity (IDV) is also shown.

Diet	AME (MJ/kg DM)	ISD	IDV[1]
Control	16.13[a]	0.98[a]	1.2[a]
Intact arabinoxylans, 30 g/kg	14.53[b]	0.92[b]	3.0[c]
Depolymerised arabinoxylans, 30 g/kg	15.74[a]	0.94[ab]	2.2[b]

[1] viscosity is expressed relative to the viscosity of water.
[a,b] values with unlike superscripts differ significantly at $P<0.05$.
(Adapted from Choct and Annison, 1992a).

Table 3. Effect of commercial feed enzymes on the apparent metabolisable energy (AME, MJ/kg DM), apparent pentosan digestibility (APD) and ileal starch digestibility (ISD) of a wheat-based broiler diet.

Diet	AME	APD	ISD
1. control, no enzyme	14.26[a]	0.26[a]	0.88[a]
2. enz. A, 1.5 g/kg	15.24[b]	0.43[b]	0.97[b]
3. enz. B, 1.0 g/kg	15.72[b]	0.39[b]	0.97[b]
4. enz. C, 1.0 g/kg	15.29[b]	0.37[b]	0.96[b]
5. enz. D, 1.0 g/kg	15.34[b]	0.30[a]	0.97[b]
6. enz. A, 0.75 g/kg	15.37[b]	0.44[b]	0.98[b]

[a,b] Values possessing unlike superscripts differ significantly ($P<0.05$). (Adapted from Annison, 1992)

gathered in experiments with rye and barley which also demonstrate that the non-starch polysaccharides of these cereals are anti-nutritive.

Many of the physical properties of the NSP are influenced greatly by the chain length of the molecules. If the main chains are cleaved by enzymes then the ability to form viscous solutions and gels is reduced. Indeed, when glycanases are added to wheat-based diets, apart from increases in energy metabolisability reductions in ileal digesta viscosity are observed (Choct et al., 1994). It should be remembered, however, that the cleaving of the NSP will also affect the surface activity and ion binding capacity and part of the anti-nutritive activity may be mediated by these factors. Attempting to examine the roles of these factors individually is very difficult experimentally as they are all affected greatly by the molecular weight of the polysaccharide.

Selecting NSP targets for feed enzymes

From the above data and many other studies as well as broad industry experience it is now well established that glycanase enzymes can improve the nutritive value of wheat, rye and barley based feeds for monogastrics. They do this by cleaving the soluble polysaccharides to low molecular weight fragments which have different physical properties (viscosity, etc.) compared with the native polysaccharide. This approach should also be of value when triticale and oats are included in diets as these ingredients contain considerable levels of arabinoxylans and β–glucans. It must be emphasised, however, that a systematic, scientific approach is required in assessing the potential benefit of glycanase use with feed ingredients. The following criteria must be met for an enzyme addition to feeds to improve production:

1) The ingredient must contain a NSP with anti-nutritive activity at a level high enough in formulated feeds to cause production problems. Assessing the NSP content of feed ingredients in order to determine whether they are at levels which may cause production problems is now possible with advanced analytical procedures. Routine "fibre" determinations (NDF, ADF etc) do not provide enough information as to the type of polysaccharides present.

2) The enzyme to be used must have a high activity for the substrate polysaccharide. As has already been discussed, the diversity of plant polysaccharide structures in feeds is great and glycanases are characterised by being highly specific enzymes. Thus it is crucial that enzymes are used which have demonstrated activity against the target NSP.

3) The enzyme must maintain activity through the feed preparation processes and into the animal to degrade the polysaccharide *in vivo*.

Recent work with rice bran illustrates how important establishing the presence of an appropriate substrate is in assessing a feedstuff as a possible target for enzymes. Rice bran production in Australia is over 70,000 tonnes per annum. Its composition and nutritive quality for poultry has recently been assessed by Warren and Farrell (1990a,b,c,d; 1991). The bran composition (g/kg) was 134–173 g/kg crude protein with a good amino acid balance, 204–234 g/kg lipid (as ether extract) and 105 g/kg ash. Neutral detergent fibre (NDF) was 256 g/kg (Warren and Farrel, 1990a). The nutritive value of the rice bran differed between cultivars but generally was tolerated well by broilers at dietary levels up to 200 g/kg above which growth was depressed. The growth depression was not associated with the lipid fraction of rice bran or a crude water soluble extract (Warren and Farrel, 1990b). In further experiments (Warren and Farrel, 1990c) the apparent metabolisable energy (AME) of the rice bran was estimated to be 9.6–10.9 MJ/kg in chickens, which was much lower than the values obtained using adult cockerels (14.7–15.0). This difference was explained by a reduced fat digestibility in the chickens compared with the adult cockerels.

Data on the levels of NSP in rice bran are not readily available in the literature. NDF and ADF determinations suggest milling techniques can greatly affect the levels of NSP which are the main components of "fibre" (Juliano, 1986). The crude fibre, ADF and NDF assays are now considered to be unsatisfactory techniques for "fibre" determinations as they fail to measure all soluble NSP (Mugford, 1993). The nature of the NSP is known, however, with pectins, β–(1→3, 1→4)-D-glucans, glucomannans and cellulose being present at low levels (<10%) and arabinoxylan being the major component (>60%) (Fincher and Stone, 1986). The arabinoxylan structure has not been fully characterised, but is probably very similar to the wheat arabinoxylan (Maningat and Juliano, 1982) which consist of a β–(1→4)-D-xylan main chain with α-L-arabinose residue substituted at the O2 and O3 positions (Annison et al., 1992). If they have a similar anti-nutritive activity to the wheat arabinoxylans, the presence of arabinoxylans in rice bran may limit its value in broiler chicken nutrition. In a recent study in Australia and New Zealand (Annison and Moughan, unpublished data) a rice bran NSP isolate was prepared using large scale procedures which were developed for the isolation of wheat arabinoxylans (Choct and Annison, 1990). This essentially consists of an incubation with α–amylase (to degrade starch) followed by an alkaline extraction and recovery of the arabinoxylan by ethanol precipitation. The rice bran NSP were added to a sorghum/casein diet at graded levels and fed to broiler chickens in a classical AME trial (see Choct and Annison, 1990). The data from the trial are shown in Table 4. As can be seen the rice bran NSP appeared to have no anti-nutritive effect. In fact, it appeared to increase the apparent metabolisable energy of the diet. Clearly, the nutritive activity of the rice bran NSP is quite different from the wheat arabinoxylan (compare

Table 4. Effect of adding rice bran NSP (RB-NSP) at graded levels on AME (MJ/kg DM) ileal starch digestibility (ISD) and ileal protein digestibility (IPD).

Diet	AME (MJ/kg DM)	ISD	IPD
Diet A control	13.26[a]	0.98	0.88
Diet B RB-NSP 20 g/kg	13.85[b]	0.98	0.88
Diet C RB-NSP 40 g/kg	14.26[c]	0.99	0.89
Diet D RB-NSP 60 g/kg	14.00[b]	0.97	0.88

[a, b, c] Values with different superscipt differ significantly (P<0.05) (Annison and Moughan, unpublished data).

Tables 1, 2 and 4). This may be explained by differences in the chemical and physical properties of the two arabinoxylans. The data in Table 5 indicate that although the rice bran NSP and wheat NSP preparation were predominantly arabinoxylans, the arabinose to xylose ratio in the rice bran NSP (1.23) was greater than in the wheat NSP (0.58). This means that the β-xylan main chain of the rice bran polysaccharide carries many more arabinose side chains than the wheat β-xylan main chain. An important factor in the formation of viscous polysaccharide solutions is the strength of the interaction between molecules. In solutions of arabinoxylans the strongest interactions occur between the unsubstituted sections of the xylan main chain which can come into close association with the formation of hydrogen bonds. The presence of the arabinose side chains inhibit these interactions. Solutions of the wheat NSP preparations were found to be much more viscous than rice bran NSP (Table 5), which is probably a direct reflection of the greater branched structure of the rice bran NSP. This difference in structure and physical properties is the most likely reason for the difference in the nutritive activity of the two NSP preparations.

Table 5. Chemical and physical properties of arabinoxylans from wheat and rice bran isolated by large scale procedures (for method see Choct and Annison, 1990).

Arabinoxylan	Carbohydrate[1] fraction	ara[2]	xyl[2]	man[2]	gal[2]	glc[2]	a/x[3]	vis[4] cP
Wheat	0.72	0.35	0.60	–	–	0.05	0.58	64
Rice bran	0.63	0.40	0.32	0.03	0.17	0.08	1.23	1.6

[1] Proportion of isolated material which is carbohydrate.
[2] ara = arabinose, xyl = xylose, man = mannose, gal = galactose, glc = glucose. The molar proportions of these sugars are shown.
[3] a/x = arabinose:xylose ratio.
[4] viscosity of a 1%(w/v) solution in 0.1M NaCl at 25°C.

These limited data are not sufficient to rule out the use of glycanases to improve the nutritive value of rice bran. It is possible that the isolation procedure was selective in the NSP fraction which was isolated. Further NSP components which were not extracted but which may be dissolved on ingestion and subsequently display anti-nutritive activity may be present. The studies, however, indicate establishing the presence of an appropriate substrate for enzymes is an important step in demonstrating their efficacy and benefit.

A further important factor which will influence the benefits of using feed enzymes is the variability in the levels and activity of anti-nutritive components of feed ingredients. To illustrate this it is appropriate to examine the low-AME wheat phenomenon in Australia. Two surveys (Mollah et al., 1983; Rogel et al., 1987) have shown that the AME of Australian wheats can vary considerably with 25% of samples having AME values below 13 MJ/kg DM. This is an ongoing problem for the

Australian broiler industry with recent results showing that the AME of wheats from the 1991/92 and 1992/93 harvests ranged from 9.52 to 14.5 MJ/kg DM. Recent studies in South Australia have identified low AME wheats and allowed further investigations of the low AME wheat phenomenon. The effects of adding a commercial enzyme preparation containing both xylanase and β–glucanase activity to diets prepared from a low-AME wheat and a normal AME wheat were compared. The AME of the wheats were assayed using the classical total collection method with 4-week-old broilers held in individual metabolism cages. At the end of the AME trial the birds were sacrificed and the small intestinal contents collected and their viscosity measured. The results are presented in Table 6. The glycanase supplementation greatly improved the nutritive value of the low-AME wheat, increasing its AME value by almost 3 MJ/kg DM. This was accompanied by a decrease in small intestinal digesta viscosity.

Table 6. Effect of glycanase supplementation on dry matter digestibility (DMD, %), apparent metabolisable energy (AME, MJ/kg DM), feed:gain ratio, (FCR, g/g) and small intestinal digesta viscosity (DV, cP) in broiler chickens.

Diet, (each n = 8)	DMD (%)	AME (MJ/kg DM)	FCR (g/g)	DV (cP)
Maize Control	81.01[a]	16.65[a]	1.96[a]	3.2[a]
Low AME Wheat	65.16[b]	12.02[b]	2.69[b]	20.3[b]
Normal Wheat	74.90[c]	14.52[c]	2.05[a]	10.4[a]
Low AME Wh. + Enz	76.26[c]	14.94[c]	2.01[a]	9.7[a]
Normal Wh. + Enz.	76.63[c]	14.83[c]	1.95[a]	5.7[a]

[a] Columns sharing superscripts are not significantly different (P<0.05).
(Choct et al., 1994)

The study above suggests that the energy digestibility of most wheats when fed to broiler chickens has the potential to be improved by the application of glycanase enzymes which cleave the NSP molecules and remove their ability to increase the viscosity of digesta. However the benefit to producers will clearly be greater when enzymes are used with low AME wheats and a cost–benefit analysis may reveal that in the case of normal or high-AME wheats enzyme supplementation is not appropriate. If so, clearly techniques are needed to allow the nutritive value of feed ingredients to be determined prior to feed formulation so that decisions can be made as to whether enzymes should or should not be used. Currently no convenient, routine laboratory techniques exist to predict the nutritive value of even the most common ingredients such as cereals, although some studies suggest that this is possible (Rotter et al. 1989, Annison, 1991). Thus although glycanase enzymes have come into common use it could be argued that the industry approach is still somewhat unsophisticated as there is little assessment of substrate levels and anti-nutritive activity in feed ingredients.

Assessing feed enzyme activity

In Australia there is still some resistance to the use of enzymes in the feed industries. This is because there is concern that the enzyme activity of supplements will be lost during steam pelleting (when temperatures may reach 90–95°C). To date there have been few attempts by feed enzyme manufacturers to provide data on the glycanase activity post-pelleting except by demonstrating production responses. This is because routine assays of very low levels of glycanase activity have yet to be established. The techniques for assaying the enzymes are, however, well established. The methods are based on incubating enzymes with substrate polysaccharides. There are three main approaches:

1) Follow the loss of viscosity of a solution of substrate polysaccharide. This may be very sensitive for some highly viscous polysaccharides, but the results are almost impossible to present in terms of International Units of activity.
2) Follow the generation of reducing sugar units. This method is relatively sophisticated and if used correctly the results can be presented in terms of International Units.
3) Follow the release of fragments soluble in organic solvents from a dyed polysaccharide substrate. This method is very convenient but less sensitive than method 2. It needs to be standardised against method 2 or incubations with known enzyme acitivities.

The main problem with assaying glycanase activity in enzyme-supplemented feeds is that the activities are very low and the feedstuffs themselves provide substrate to the incubation mixtures. Thus it is very difficult to characterise the enzyme activity using classical procedures because total substrate concentrations are unknown. It is possible, however, to detect the enzyme activity in feeds and with controlled conditions to develop standard procedures for assaying feed enzymes. This will allow the loss of glycanase activity through the pelleting process to be monitored. The results in Figure 4 (Annison, unpublished results) show the detection of β–glucanase activity in a feedstuff supplemented with a commercial feed glycanase. The feeds were assayed before and after commercial pelleting and the data demonstrate that a considerable amount of activity survives the pelleting process. The assay procedure which has been previously described (Annison, 1992) used a Remazol Brilliant Blue–β–glucan substrate (McCleary and Shemeer, 1987). The results are expressed in terms of an absorbance change rather than international units but the technique is amenable to standardisation which would make it more useful to industry. A similar technique has also been used to follow the destruction of enzyme activity at high temperatures (Annison, unpublished data). Three commercial glycanase preparations containing β–glucanase activity were mixed with ground wheat and held at 100°C. Samples were removed at intervals up to 20 minutes. The results shown in Figure 5 record a decline in β–glucanase activity, but even after 20 minutes at least 30% of the activity remains. It is likely that the loss of activity would be greater at higher moisture contents, but nevertheless these data and those in Figures 4 and 5 indicate

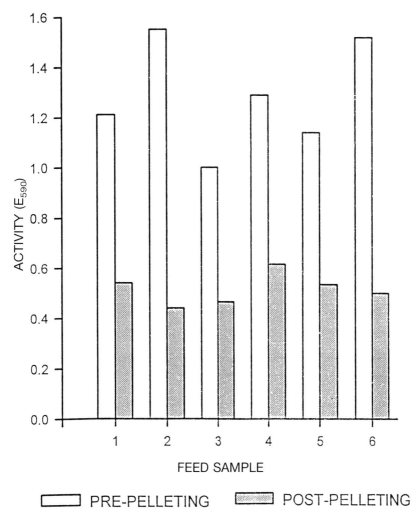

Figure 4. Effect of commercial steam pelleting on the β-glucanase activity in a commercial diet containing a glycanase supplement.

that the technology exists for assaying glycanase activity relatively easily in feedstuffs. Industry still lacks information linking glycanase activity in commercial preparations with its activity after feed formulation and pelleting and the expected biological response. Thus in promoting the use of feed enzyme supplements for use by industry it should be demonstrated that:

1) glycanase activity survives steam pelleting,
2) the activity of the glycanases in feed can be assayed reproducibly in the laboratory.
3) the assayed activity correlates with production responses when used against specified target substrates.

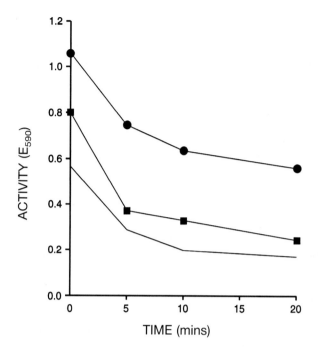

Figure 5. Loss of β–glucanase activity from a ground wheat/glycanase supplement mixture held at 100°C

With this approach both the users and suppliers of feed enzymes will benefit as production gains associated with the use of enzymes will be more predictable and more consistent.

Conclusions

The development of sophisticated plant NSP analysis procedures has allowed the nutritional roles of plant NSP in monogastrics to be established. This in turn has provided a scientific basis for the use of feed glycanases and allowed the development of sophisticated glycanase preparations with defined activities and target substrates. Thus in recent years the rationale for the use of feed glycanase enzymes in the intensive livestock industries has become clearer. Further work is required to characterise the type, levels and nutritive activity of the NSP found in feedstuffs. This information, coupled with standardised units of glycanase activity, will enhance the targeting of feed glycanase enzymes and lead to more predictable and reliable production responses following their use.

References

Annison G. 1991. Relationship between the levels of non-starch polysaccharides and the apparent metabolisable energy of wheats assayed in broiler chickens. J. Agric.Food Chem. 39:1252.
Annison, G. 1992. Commercial enzyme supplementation of wheat-based diets raises ileal glycanase activities and improves apparent metabolisable energy, starch and pentosan digestibilities in broiler chickens. Anim. Fd. Sci. Tech. 38:105.
Annison. G. 1993. The role of wheat non-starch polysaccharides in broiler nutrition. Aust. J. Agric. Res. 44:405.
Annison, G., M. Choct, and N.W. Cheetham. 1992. Analysis of wheat arabinoxylans from a large scale isolation. Carbohydr. Polym. 19:151
Choct, M. and G. Annison. 1990. Anti-nutritive activity of wheat pentosans in broiler diets. Brit. Poult. Sci. 30:811.
Choct, M. and G. Annison. 1992a. Anti-nutritive activity of wheat arabinoxylans: Role of viscosity and gut microflora. Brit. Poult. Sci. 33:821.
Choct, M. and G. Annison. 1992b. The inhibition of nutrient digestion by wheat pentosans. Brit. J. Nutr. 67:811.
Choct, M., R.J. Hughes, R.P. Trimble and G. Annison. 1994. The use of enzymes in low-AME wheat broiler diets: effects on bird performance and gut viscosity. Proc. Aust. Poult. Sci. Sym. 6:83.
Cheung, P.C.K. 1991. The carbohydrates of *Lupinus angustifolius*. A composite study of the seeds and structurual elucidation of the kernel cell-wall polysaccharides of *Lupinus angustifolius*. PhD Thesis. University of NSW., Sydney, Australia.
Fincher, G.B. and B.A. Stone. 1986. Cell walls and their components in cereal grain technology. In: Advances in Cereal Science and Technology. vol 8. Ed. Y. Pomeranz. Minnesota. AACC.
Fry, S. 1986. Cross-linking of matrix polymers in the growing cell walls of angiosperms. Ann. Rev. Plant. Physiol. 37:165.
Iszydorczyk, M.S., C.G. Biliarderis and W. Bushuk. 1990. Oxidative gelation studies of water-soluble pentosans from wheat. J. Cereal Sci. 11:153.
Juliano, B.O. 1985. Rice Bran, In: "Rice: Chemistry and Technology" (Ed). B.O. Juliano . (Minnesota, AACC).
Launay, B., J.L. Doublier G. Cuvelier. 1986. Flow properties of aqueous solutions and dispersions of polysaccharides. in "Functional properties of food macromolecules" eds. J.R. Mitchell and D.A. Ledward. Elsevier. London. New York.
Maningat, C.C. and B.O. Juliano. 1982. Compostion of cell wall preparations of rice bran and germ. Phytochem. 21:2509.
McCleary, B. V. and J. Shemeer. 1987 . Assay of malt β–glucanase using azo-barley glucan: an improved precipitant. J. Inst. Brew. 93:285.
Mollah, Y., W.L. Bryden, I.R. Wallis, D. Balnave and E.F. Annison. 1983. Studies on low metabolisable energy wheats for poultry in relation to the digestibility of starch. Brit. Poult. Sci. 24:81.
Mugford, D. 1993. Current methods for measurement of dietary fibre: choices and suitability. in "Dietary Fibre and Beyond" Eds. S.Samir

and G. Annison. Occasional Publications. Vol 1. pp19–36. Nutrition Society of Australia.

Rogel, A.M., E.F. Annison, W.L. Bryden and D. Balnave. 1987. The digestion of wheat starch in broiler chickens. Aust. J. Agric. Res. 38:639.

Rotter, B.A., R. R. Marquardt, W. Guenter, C. Biliaderis, and C. W. Newman. 1989. *In vitro* viscosity measurements of barley extracts as predictors of growth responses in chicks fed barley-based diets supplemented with a fungal enzyme preparation. Can. J. Anim. Sci. 69:431.

Richards, G.N. 1976. Search for factors other than "lignin shielding" in protection of cell-wall polysaccharides from digestion in the rumen. *In*: "Carbohydrates in Plants and Animals". Misc. papers 12, pp. 129–35. Landbauwhogeschool Wageningen. (Veeman and Zonen:Wageningnen.)

Selvendran, R.R., B.J.H. Stevens and M.S. Du Pont. 1987. Dietary fiber: Chemistry, analysis, and properties. Ad. Food Res. 31:117

Selvendran, R.R. 1984. The plant cell wall a source of dietary fiber: Chemistry and structure. Am. J. Clin. Nutr. 39:320.

Warren, B.E and D.J. Farrell. 1990a. The nutritive value of full-fat and defatted Australian rice bran. I. Chemical Composition. Anim. Fd. Sci. Tech. 27: 219–228.

Warren, B.E and D.J. Farrell. 1990b. The nutritive value of full-fat and defatted Australian rice bran. II. Growth studies with chickens, rats and pigs. Anim. Fd. Sci. Tech. 27:229.

Warren, B.E and D.J. Farrell. 1990c. The nutritive value of full-fat and defatted Australian rice bran. III. The apparent digestible energy content of defatted rice bran in rats and pigs and the metabolisability of energy and nutrients in defatted and full-fat bran in chickens and adult cockerels. Anim. Fd. Sci. Tech. 27:247.

Warren, B.E and D.J. Farrell. 1990d. The nutritive value of full-fat and defatted Australian rice bran. IV. Egg production of hens on diets with defatted rice bran. Anim. Fd. Sci. Tech. 27:259.

Warren, B.E and D.J. Farrell. 1991. The nutritive value of full-fat and defatted Australian rice bran. V. The apparent retention of minerals and apparent digestibility of amino acids in chickens and adult cockerels fitted with ileal cannulae. Anim. Fd. Sci. Tech. 34:323.

White, W.B., H.R. Bird, M.L. Sunde, N. Prentice, W.C. Burger and J.A. Martlett. 1981. The viscosity interaction of barley b-glucan with *Trichoderma viride* cellulase in the chick intestine. Poult. Sci. 62:853.

White, W.B., H.R. Bird, M.L. Sunde, J.A. Martlett, N. Prentice, and W.C. Burger. 1983. Viscosity of β-D-glucan as a factor in the enzymatic improvement of barley for chicks. Poult. Science 62:853.

THE USE OF ENZYMES IN DESIGNING A PERFECT PROTEIN SOURCE FOR ALL ANIMALS

S.L. WOODGATE

Beacon Research Ltd., Clipston, Leicestershire, UK

Summary

The term perfect protein is explained and the use of biotechnology to assist in the designing of perfect proteins for animal feeds is explored. The role of Beacon Research in applying the technology through to full scale production is reviewed in three separate application studies. The economic, nutritional and environmental benefits of manufacturing high quality, protein-based feed materials from by-products are highlighted for each application.

Introduction

A major objective of animal nutritionists today and in the future must be to effect improvements in animal production efficiency. The goals are clear to see: increased efficiency leads to improved animal productivity and decreased pollution from excreta. The reduction of agricultural pollution is absolutely vital in many areas of the world; and as Cole (1992) explained, research to find solutions to these problems must be a high priority for the biotechnology industry. One area where efficiency of the whole cycle is vitally important is in the safe recycling of animal by-products. As discussed in detail by Woodgate (1993), a significant proportion of the animal slaughtered for human consumption is unwanted by humans and therefore becomes a food waste. The term waste is unfortunate and misleading, so this material is preferably termed "animal by-product", as this describes the product and its potential uses much more accurately. It is the job of animal by-product processors or renderers, as they are also called, to maximise the economic return on the raw material in a safe and efficient manner.

Normally the maximum economic return for animal by-product would be to convert it to an animal feed ingredient such as meat and bone meal, animal fat, blood meal or feather meal. As long as the products of by-product processing are not tainted with the stigma that is sometimes attached to the practice of recycling, then the monetary values of the products will be in line with world commodity values for proteins and fats. The dominant factors in determining world protein and oil prices are soyabean meal and soya oil. Thus, whatever the costs to process the animal by-products by sterilisation and dehydration, their value is fixed by another commodity over which the renderers have little or no influence. Therefore, if all the other factors are equal, the only way for the renderer to "add value" is to produce something that is both unique and important to the animal feed industry which can be valued outside of the world commodity price formula.

Research has been conducted in many countries on the subject of improving the value of animal by-products. In the U.K., Prosper deMulder Ltd were at the forefront of this type of work until in 1990 when severe problems occured in the UK as a result of both the salmonella and bovine spongiform encephalopathy (BSE) crises. These problems resulted in this research and development work being shelved at that time. However, in 1992 Beacon Research, a company closely associated with Prosper deMulder, took up the challenge of attempting to upgrade and add value to by-product processing. It has since progressed the concept in many other countries around the world. The countries where most interest has been shown represent local situations where the concept of by-product "upgrading" fall broadly into two categories. Firstly, there are those countries that already have modern rendering systems and which produce commodity valued products, but who are trying to "add value". For example, Japan, Europe and North America fall into this category. The second type are those countries with immature or undeveloped by-product recycling facilities. These would include developing nations such as the former Eastern block countries and possibly some South American or Asian countries. To summarise therefore, it is the objective of Beacon Research to progress the development of added value processing schemes for both animal and vegetable by-products in any country where there is a requirement and commitment to achieve the goals discussed above. In the following sections, the major objectives of the work will be examined with reference to three application studies where the development work is discussed in detail. One of the major objectives of the upgrading programme is to manufacture products with an optimum blend of amino acids in the protein, ie., to produce a "perfect protein".

What is a perfect protein and why is it required?

From the animal's point of view, the perfect protein contains a range of amino acids that are both essential and balanced in a specific manner. For each species, and probably each age or physiological state, the amino acid balance of a perfect protein may vary. Therefore it is the job of the animal nutritionist to formulate the diet, both by species and age, as close to the perfect amino acid balance as possible. Of course an understanding of individual species requirements is a prerequisite to an attempt to supply the perfect protein. The most significant work in this area over the last twelve years has been the study of the essential amino acid requirement of the growing pig. Numerous studies have been completed by many workers in this field, for example, ARC (1981), Wang and Fuller (1989) and Chung and Baker (1992), to name only a few.

The suggestions for ideal amino acid balance by various authors are shown in Table 1, with the author's suggestion for a perfect protein balance shown as a reference example. This perfect protein can then be used in comparison with the amino acid balance of feed materials described later. For the pig in particular, the elucidation of the perfect amino acid balance has been very valuable in firstly maximising pig performance and secondly in minimising the wastage of non-essential amino acids in high "excess" protein pig diets. The move towards lower protein diets to minimise the polluting effect of nitrogenous pig manure has, in recent years, gathered momentum. In most countries, there will be even more pressure applied in the future to reduce this very insidious pollutant. Under these circumstances low protein diets containing perfectly balanced amino acids, will have a very significant part to play.

Table 1. Suggested "perfect" protein, amino acid profiles.

	ARC	Wang and Fuller	Chung and Baker	Suggested "perfect" protein
Lysine	100	100	100	100
Methionine + cystine	50	63	65	64
Threonine	60	72	67	68
Tryptophan	15	18	18	18

In other species, there has not been the same amount of research work as completed in the pig. However, in two very important farmed species, the dairy cow and the fish, there are data available which give strong indications about the ideal makeup of the amino acid composition of the perfect protein for these species. In the case of the dairy cow, as in all ruminants, the extent of rumen degradability of the protein source is particularly important. As is well documented, the requirement for amino acid at the small intestine is supplied by a mixture of rumen undegradable protein and microbial protein. Whereas the latter is difficult to manipulate in terms of amino acid balance, the former may be suitable for attempting to select sources of undegradable protein containing the correct amino acid balance. There is a requirement for a high proportion of essential amino acid within the total amino acid of the undegradable dietary protein (ARC, 1980) with amino acids such as leucine, isoleucine and valine also required in addition to lysine, threonine and methionine as precursors for de novo synthesis of milk protein. It is therefore entirely feasible that these essential amino acids may be supplied in the form of rumen undegradable amino acid from a perfect "dairy cow" protein source.

The farmed Atlantic salmon has been studied in detail, and here again there is considerable evidence to suggest that a perfect "fish" protein could be formulated from the available data. In his review article Cowey (1988) referred to the combination of the following amino acids in the proportions, lysine 100; methionine 30; threonine 45; tryptophan, 10; as possibly representing the most likely balance of amino acids to provide the growth requirement of the farmed salmon. Thus, defining the target amino acid balances for the relevant species in question allows progress to the next stage of the development programme.

What types of feed material are required to provide a perfect protein?

The world supply of raw materials for animal feeds is extremely varied with considerable differences among countries. Therefore, consideration will be given to only a few raw materials which are almost universally available. One obvious and very important consideration is that in manufacturing an individual perfect protein feed material, account should be given to the complete ration to which it will be added. The amino acid profiles, compared to lysine, of a range of raw materials are shown in Table 2; and of particular relevance are the wide range of amino acid profiles from only a very small number of feed materials.

Table 2. Amino acid profile of various feed materials.

Amino acid	Soya	Maize	Wheat	Fishmeal	Meat & bone	Blood meal	Feather meal
Lysine	100	100	100	100	100	100	100
Methionine + cystine	52	69	80	46	37	33	245
Threonine	62	135	80	55	47	53	258
Tryptophan	22	34	40	14	8	16	35

From the data shown in Table 2, it is apparent that in order to achieve the perfect amino acid profile in a compounded feed, an excess of total protein is currently required. Therefore, if a more appropriate source of balanced amino acid were available in a critical or key feed material, potential savings in total protein could be achieved while supplying adequate and balanced amino acids from the whole diet.

In much of the research work, digestibility of the amino acids in perfect protein has not been fully addressed, although the protein models of Wang and Fuller (1989) and Chung and Baker (1992), do take into account amino acid digestibilty. As new models are developed the importance of digestibility will be further recognised. It is therefore vital that when amino acids are delivered to the site of absorption, they must be highly digestible. This is not the case for many traditional sources of protein (amino acid), with conventionally processed blood meal and feather meal representing two good examples. Both provide amino acids with profiles shown in Table 2, and these data could be used in formulations to provide a ration with perfect protein profile. However, the digestibility of the individual amino acids from these materials may be quite variable, ranging from 50 to 80% for example. These data really confirm that it is inappropriate to consider only the amino acid content of a feed ingredient. The digestibility of the amino acids must also be taken into account.

Table 3 shows in vivo amino acid digestibility data for a range of feed materials (Eurolysine, 1988) These data illustrate both the range of data available and confirm the scope for improvement of some materials.

Table 3. In vivo digestibility of amino acids in feed materials.

Amino acid	Soya	Maize	Wheat	Fishmeal	Meat & bone	Feather meal
Lysine	80	68	71	87	81	51
Methionine	87	86	84	91	84	71
Cystine	79	74	80	62	60	72
Threonine	77	71	69	81	77	74
Tryptophan	78	72	81	74	64	60

From a world perspective it is clear that a wide range of feed materials exists with local availability depending on climatic or processing conditions. For each locality nutrient availability and animal requirement must be matched. A number of factors influence economics and availability, but the overall objective of any long-term feeding programme must be to maximise nutrient utilization and minimise waste and pollution generated.

Practical application of enzymes: progress toward perfect protein sources

CAN BIOTECHNOLOGY ASSIST IN DEVELOPMENT OF PERFECT PROTEIN RAW MATERIALS?

Feed processing using enzymes has been shown to make an effective contribution toward the objectives set out above, but progress has been difficult to achieve. The following studies will consider how biotechnology can enhance the value of certain by-products; and where appropriate, assist to produce a perfect protein material for the feed industry. In summary therefore, the main objective is to produce new or improved feed materials showing the following characteristics: Firstly, an increased supply of digestible amino acid at the digestion site of the target animal. Secondly, a balanced content of digestible amino acids representing as near as is possible, the ideal amino acid ratio or perfect protein. Thirdly, an overall reduction in pollution from excreta by supplying perfect protein to rations containing adequate, rather than excess, protein.

Bio-treated feather protein

In the UK (and many other developed countries), the majority of raw feathers are processed away from the poultry abattoir at dedicated feather processing sites. Under these circumstances the risk of contamination, particularly by metal, is extremely high. Therefore, traditional feather processing methods at dedicated sites have been developed that hydrolyse the raw feather, but can also cope with small particles of metal that may escape the metal detection systems. These traditional systems are characterised as steam pressure hydrolysis processes, operating at between 40 and 90 psi (276–622 kPa) for periods of time between 30 and 60 minutes. Following steam hydrolysis, the product is dried, usually under atmospheric conditions, in a separate drier. Currently, there are both batch and continuous systems in use around the world, with the major consideration for either type being logistical, as there are no significant advantages of either system from a nutritional point of view.

As discussed by Woodgate(1993), there are considerable nutritional drawbacks with feather meal made by the conventional steam hydrolysis process. The main problems of this process from a nutritional point of view are firstly. in order to make the majority of amino acids moderately digestible, a considerable amount of the volatile amino acids are lost due to the steam stripping effect of hydrolysis under pressure. Secondly, after hydrolysis the product is dried in such a manner as to cause further protein denaturation. There is an advantage, however, in that the conventional process produces a dense brown coloured hydrolysed feather meal looking not at all like feathers in their natural state. This standard product has two main markets, either as an undegradable protein source for cattle, mainly in concentrate rations, or as a protein source for lower quality feeds. For both of these markets, the priorities are: a) not to look like feathers, b) medium digestibility (65–75% pepsin digestibility), c) good bulk density (for transport in particular), d) high protein (~85% protein), and e) cheap price (~£2/unit protein compared with soya at ~£3/unit protein). However, due to its moderate nutritional quality and poor reputation, steam hydolysed feather meal is normally excluded from diets for monogastrics, pet foods, and many farmed fish feeds. Having set the scene, there appears to be significant scope for attempting to add value or upgrade feather, and in particular to try and penetrate those specialist markets discussed above.

A new biological process for feathers was described by Harvey (1992), with particular reference to a process developed by Poss Ltd. in Canada. In 1993, Woodgate described some preliminary data on a alternative system that was being developed in the UK by Prosper deMulder Ltd. The reasons for some divergence in the methods of processing in the UK compared with the Canadian situation are partly due to country differences, but mostly because of important logistical reasons some of which were descibed earlier. The Canadian operation is geared to the processor, (ie., the abattoir) with processing on site and recycling of the product (feather plus cereal) back to a growing unit via a feed mill possibly on the same site. The process described by Harvey (1992) uses an extruder at one point. In the integrated processing situation described above, contaminants such as metal hooks, chains and staples, etc. that could be capable of inflicting severe mechanical damage are most easily controlled. This fact is a distinct advantage. It may not appear very important to those outside the industry, but these stainless steel objects can cause considerable mechanical and financial damage when in the wrong place. For example, 1/4 inch pieces blocking an extruder can cause a major breakdown.

In the UK it was considered more appropriate to adapt a biotreatment process from the current system in operation (Woodgate, 1993). Therefore, while the objectives of the feather upgrading programme in Canada and UK are very similar, the method of achieving the added value is quite different. At this stage it is impossible to say that one process is better than the other, but it would be relevant to indicate where one or the other method would be most appropriate. Several major experiments using the UK system have been completed and the optimum process conditions outlined in Woodgate (1993) have been refined to gain confidence in the data. Several additional types of evaluation have been completed with the biotreated feather protein produced under the optimum conditions described in Table 4. Most notably, TME technique work has been completed at the Roslin Institute to evaluate both the TME and digestibility of amino acids using the method described by Sibbald and Slinger (1963). The nutritional data shown in Table 4 indicate that the enzyme treatment system conferred a very significant increase in TME and essential amino acid content of the biotreated feather protein when contrasted with the control (steam hydrolysed feather meal).

Table 4. The TME, protein and amino acid content of processed feather products produced under different conditions.

Description	Steam-hydrolysed feather meal	Biotreated feather protein
2 stage process conditions	30 mins 105°C + 30 minutes 165°C	45 mins 60°C (Enzyme) + 15 mins 120°C
TME, MJ/kg	13.8	15.9
Protein, %	84.5	84.2
Lysine, %	1.38	1.56
Methionine, %	0.52	0.61
Cystine, %	3.56	4.98
Threonine, %	2.70	3.25
Tryptophan, %	0.62	0.69

These results, though significant, do not alone provide the complete picture. It is essential that feather, as a source of protein, contain as much digestible amino acid as possible. This is the vital issue, as most modern feed formulations are based upon digestible amino acid rather than protein or amino acid levels alone. The very significant increase in digestible amino acids shown in Table 5 is very exciting, as it confirms the benefit of the new biotreatment process and the comparative wastefulness of the conventional steam hydrolysis process. It also indicates the potential way forward to achieve efficient and economic

Table 5. Amino acid digestibilities and the digestible amino acid content of processed feather products.

	Steam-hydrolysed feather meal		Biotreated feather meal	
	Digestibility of amino acids, %	Digestible amino acid	Digestibility of amino acids %	Digestible amino acid
Lysine	61.4	0.85	87.4	1.36
Methionine	70.4	0.37	90.2	0.55
Cystine	52.88	1.88	78.8	3.92
Threonine	65.2	1.76	83.5	2.71

production of a high value product from this bountiful raw material. The improved content and digestibility of amino acids in the enzyme-treated product also represents two significant pollution reduction effects of the enzyme process. Firstly, at the processing site, where there is a lower loss of sulphur and nitrogen as a result of the new process, and secondly at the animal level, where improvements of amino acid digestibility could represent as much as a 55% reduction in nitrogen pollution. These overall improvements are highlighted in Table 6 using the loss of cystine on processing and differences in cystine digestibility in the animal as markers of nitrogen and sulphur loss.

Table 6. The effect of processing on pollution potential: comparing steam hydrolysed feather meal with biotreated feather protein.

	Average loss of N and S as effluent or as excreta (as cystine), %			
	Steam hydrolysis	Biotreatment	Difference	% Reduction
In processing	49	29	20	41
In the animal	47	21	26	55

Further studies with enzyme-treated feather protein have also been completed in the rat. Availability of the sulphur amino acids in several test samples were compared using a rat growth bioassay. These data confirmed that the enzyme system is successful in producing a marked improvement in feed conversion when compared to the enzyme treatment control (Table 7). The biotreated feather protein was also significantly more efficient in promoting growth than standard hydrolysed feather meal.

Table 7. Rat growth bioassay: feed conversion ratios of feeds formulated to contain equal amounts of sulphur amino acid.

	Steam-hydrolysed feather meal	Biotreatment method	
		No enzyme	With enzyme
Feed conversion, Gain/diet consumed	0.270	0.240	0.341

Currently more trials are being completed using product manufactured under the same processing conditions as described earlier. These trials include feeding and growth studies in the rainbow trout, where feather protein may be a very valuable source of essential amino acid. Trials are also planned with the growing pig, where traditionally produced steam-hydrolysed feather meal has not featured at all in diets, due to its low nutritional value and poor image. Overall, the development work on upgrading or adding value to feather protein has been successful in several ways. Significant improvements in amino acid content and amino acid digestibility have been seen in biotreated feather protein. Reductions in pollution losses have been calculated for the new process. The process is currently being upgraded from pilot scale to production plant size, with a full evaluation of the economic benefits currently being undertaken. It is to be expected many will be slow to accept the value of feather meal derived by biotreatment. The reputation, looks and feel of "old" standard hydrolysed feather meal are poor to say the least. It will therefore be a major task to convert the sceptics into accepting the significant nutritional merits of the new biotreated feather protein.

Have the achievements described above produced a perfect protein source? The answer, probably has to be "no" when considering biotreated feather protein alone. However, the contribution of the new biotreated feather protein towards the total amount and balance of digestible amino acids in a diet has been very significantly enhanced when compared with the contribution from standard steam-hydrolysed feather meal. This in itself is an achievement that should be claimed as a phenomenal nutritional advancement.

Biotreated fish protein: improved utilisation of fish for feed production
Fishmeal is either manufactured from fish caught exclusively for producing fishmeal or is made from materials left over from fish caught for human consumption. The general method of manufacturing fishmeal is almost universal, but the type of product and its value will alter according to several important factors. These include the main type of species caught, the raw material's freshness and its specific composition ie., moisture, oil, protein and ash. The processing conditions, in particular the method of drying, may alter the nutritional value by potentially heat-denaturing the protein at low moisture levels during the final stages of drying.

An alternative method has been developed to manufacture an improved product from the same starting raw material. The main elements of this new process to manufacture a biotreated fish protein, illustrated in Figure 1, include the enzymic treatment of fish press solids followed by the removal of indigestible fishbones and scales. Stabilisation of the liquid product or evaporation of the water by a carefully controlled method will produce either a liquid or a solid product, respectively. Both liquid and solid products will have high nutritional values and can be used where appropriate in feed applications. A very significant advancement is that the processing equipment is not prohibitively expensive, thereby making the manufacture of such products a real economic possibility at last.

The use of enzymes in designing a perfect protein source for all animals

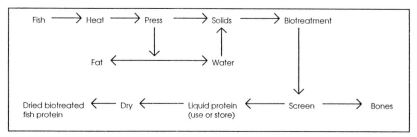

Figure 1. Biotreated fish protein process.

Table 8. Comparison of analyses of Biotreated fish protein with standard fish meal.

	Biotreated fish protein	Standard fish meal
% Product		
Protein	80	62
Oil	10	12
Ash	4	20
Digestible amino acid		
Lysine	5.6	4.2
Methionine + cystine	2.9	1.7
Threonine	3.1	2.1
Tryptophan	0.6	0.5

A typical analyses of the biotreated fish protein is contrasted with standard US fish meal in Table 8 to allow comparison of the traditional and new products. The most significant difference is of course the high protein content of the biotreated fish protein. However, it is the high protein content, combined with the very high digestibility that makes the biotreated fish protein able to fulfil a range of nutritional requirements at an economic advantage to other protein sources. Is this then, the product to fulfil the requirement of ideal amino acid profile or perfect protein? Here, the answer is a good deal more positive, and the relationship between the amino acids is significantly improved. The fact that the product contains highly digestible amino acids gives it a significant advantage over many other products when formulation is based upon the requirement for digestible amino acid and not just amino acid content. This type of product could be particularly useful for inclusion in dietetic pet food products where there is a very high requirement for easily digestible nutrients and in production feeds for animals or fish requiring very high nutrient digestibility.

Maximising utilisation of animal by-products by designing perfect protein products

The production of feed ingredients using biotechnology in the by-products industry was discussed in detail by Woodgate (1993). Progress has been made during 1993, notably in Japan, where more development work preceeded the installation of a production facility. This processing site is now able to manufacture a biotreated by-product, known as

Protagen, to exacting standards of both product safety and quality. In simplified form, the key elements of the process are:

a) Selection of raw materials to produce a final product with the desired amino acid balance.
b) Enzyme bioprocessing to solubilise the protein sources under optimum temperature and pH conditions.
c) Sterilisation of product under controlled temperature and time conditions.
d) Product drying. Method of drying, following sterilisation, must be optimised to prevent any protein denaturation at low moisture levels.

In summary, Protagen can be manufactured with varying levels of essential amino acid and with the required balance to create a more perfect protein to suit the final product. Therefore while Protagen can be targeted for several species, a specific example is shown in Table 9 of Protagen with an amino acid spectrum most suited for young pigs. This particular example shows a high protein level (70%) with a high lysine level and the well balanced ratio of methionine plus cystine, threonine and tryptophan particularly well suited for young growing pigs.

Table 9. A typical analysis of Protagen designed for inclusion in young pig grower feed.

Proximate analysis	%	Amino acid analysis		
		Amino acid	Content, % of product	Balance, Lysine=100
Protein	70	Lysine	4.5	100
Fat	20	Methionine + cystine	3.2	71
Ash	2	Threoine	3.1	68
Moisture	6	Tryptophan	0.8	18
Energy, DE MJ/kg	19.2			

The analysis in Table 9 also illustrates the high level of fat (~20%) normally present in Protagen. The high fat content boosts the digestible energy (DE) content to over 19 MJ/kg. This high energy value is useful in pig rations where it is very important to have adequate digestible energy present in high performance feeds. For other feed applications, for example as a forage supplement for ruminants, the high fat content together with the protein provide an excellent source of balanced energy and digestible undegraded protein. To indicate the effectiveness of this technology, some animal trial results where biotreated animal protein from the UK and Protagen from Japan were fed under practical conditions, are described in detail.

In Tables 10a and 10b the effect of substituting a biotreated animal protein (60% protein, 30% fat), for fish meal (66% protein, 10% fat), is shown. All diets contained equal levels of DE, protein, lysine, methionine and threonine. What then is the reason for the phenomenal increase in feed intake, mirrored by liveweight gain? The maintenance

The use of enzymes in designing a perfect protein source for all animals

Table 10a. Materials and methods summary.

Trial details	Diet formulation
6–30 kg liveweight	DE, 14.4 MJ/kg
Ad lib feeding	Protein, 22.5%
Records of feed intake	Lysine, 1.45% (100)
Records of liveweight	Methionine+Cystine, 0.85% (58)
	Threonine, 0.85% (58)

Table 10b. Production response of including biotreated animal protein (Protagen) in a pig grower ration as a replacement for fishmeal.

Protagen, %	0	2	4	6
Fishmeal, %	6	4	2	0
Feed intake, g/day	845	835	880	930
Liveweight gain, g/day	555	550	580	605
Feed:gain	1.52	1.52	1.52	1.53

of feed conversion over the period of increased intake indicates very efficient utilisation of nutrients at higher feed intakes, so improvement in feed efficiency cannot be the sole reason for improved intake. The major effect could therefore be due to a palatability effect. It is possible that enzymic hydrolysis within the production process resulted in formation of polypeptides which are known to be palatability enchancers.

A pig feeding trial, completed at a Japanese experimental farm, compared a range of protein supplements and evaluated their performance in terms of feed intake and liveweight gain. All proteins were accepted either as well as, or better than, the control (fishmeal and skim-milk). Protagen alone (at 6%) and a combination of 3% porcine plasma protein with 3% Protagen yielded the highest feed intake and liveweight gain over the trial period (Table 11). These findings again indicated a positive palatability effect of Protagen in this very vital early phase of the piglet's life. A further trial in Japan compared a standard pig weaner/grower diet containing 4% fishmeal, with a test ration with 4% Protagen replacing the fishmeal. Over a 21-day period, the Protagen-containing ration significantly out-performed

Table 11. Pig performance in Japanese feeds: comparing porcine plasma protein and spray dried blood meal with Protagen.

	Control (3% fishmeal, 3% skim milk)	6% Spray-dried blood meal	6% Porcine plasma protein	6% Protagen	3% Protagen, 3% porcine plasma protein
Intake, g/day	476	475	490	510	528
Liveweight gain, g/day	231	240	260	295	302

S.L. Woodgate

Table 12. Pig performance trial at Yebsin farm in Japan[1].

	Control	Test[2]
Feed intake g	657.1	826.6
Daily gain g/day	251.4	370.7
Feed conversion	2.61	2.24

[1]Piglets – 6kg liveweight. 30 animals per group. 21 day trial.
[2]Diets: 4% Protagen (test) replaces 4% fishmeal (control).
Both diets equal protein and energy

the control feed with a 26% increase in feed intake, 47.5% increase in live weight gain and a concomitant 15% improvement in feed conversion (Table 12).

Overall, these three pig trials indicated very positive effects which are not easily explained by reason of the biotreated by-product being a source of quality protein only. The additional palatability benefit accorded during the biotreatment process is of course particularly beneficial. This is of significant importance, especially in the young pig where early high intake is essential to give a platform for efficient growth performance. An phenomenal palatability effect has also been seen for Protagen included in complete diets for dogs and cats. Table 13 shows the positive effect on food intake for dogs when a complete feed included Protagen rather than an equivalent protein level from poultry meat meal. Again this positive effect is mostly easily explained by the production of palatability enchancing substrates during the biotreatment process.

In addition to the trials completed in monogastrics, some preliminary studies have been completed in ruminants. Rumen degradibility of a biotreated by-product using the *in sacco* technique has indicated that the protein fraction was approximately 72% undegradable at an average rumen outflow rate. Therefore, the supply of digestible undegraded protein (DUP) of a product with a 70% protein content could be as much as 45%, making this a very useful feed ingredient indeed. Some experimental data is also being generated in the farmed rainbow trout, as precursor to more detailed studies and performance trials. Here, there

Table 13. Dog food preference trial completed in Japan.

Materials and methods
    24 dogs used in preference test.
    Both diets equal protein and energy, in pelleted form.
    Each dog given both diets at same time and preference noted.
    Control diet contained 20% poultry meat meal
    Test diet contained 10% protagen, 10% poultry meat meal

Results
    Dogs which showed preference for one diet or other = 92%
    From above, those that showed preference for Control = 18%
    From above, those that showed preference for Test = 82%

Conclusion: Dogs show clear preference for diets including Protagen.

is great scope for a biotreated protein source to supply the high levels of digestible essential amino acids required by farmed fish.

The manufacture of biotreated animal proteins such as Protagen appears to offer the most potential for achieving perfect protein sources, both in terms of amino acid profile and in the excellent digestibility of the amino acids as supplied to the site of digestion.

Conclusions

There is scope for altering the amino acid profile to design perfect proteins for a range of species. The main limitation to this next step of designing species specific perfect protein sources is that with the exception of the growing pig, the amino acid profiles for other animals have not yet been confirmed as accurately. However, in the case of DUP supply to the ruminant or for protein requirement of the farmed fish, there are enough existing data to indicate the approximate balance required to achieve optimum performance. One additional and very important feature of biotreated by-products is their very positive effect on palatability which can complement and enhance the nutritional benefits. In all species intake is key. Therefore, ensuring that intake and liveweight gain goals are achieved in young pigs gives them an excellent boost, particularly in the very early period. In pet foods, acceptance in terms of palatability, together with good digestibility of the feed ingredients are essential. For ruminants, particularly in lactation, high intakes of the correct nutrients are required to provide the precursors for top quality milk production.

Three application studies have been described in which the benefits of using biotechnology to offer alternatives to existing standard technologies have been discussed. All biotreatment processes have yielded products with significantly higher levels of digestible amino acid compared with traditionally produced materials. However, it is still true that you cannot make something from nothing; so raw materials with deficiencies will not, in general, be able to be compensated for by processing alone.

Table 14. Simple pig formulation showing differences in protein level and perfect protein balance in finished feed with Protagen.

	Control	Protagen
Maize	70	75
Soya	30	15
Protagen	–	10
Protein, %	21	20
Lysine, %	1.1	1.1
Amino acid balance		
Lysine	100	100
Methionine + cystine	52	66
Threonine	62	68
Tryptophan	19	19

In complex compound feed formulations these deficiencies can be accommodated to a certain extent by adjusting the raw materials and possibly by adding synthetic amino acids, for example. This becomes both more difficult and costly the simpler the formulation used, eg. cereal plus protein source. It is therefore very important that protein feed materials are available that satisfy the requirement for perfect protein, particularly in simple formulations. A very simple feed formulation is described in Table 14 showing some of the features that have been discussed earlier. The lowering of the total protein content of the second diet, while maintaining the lysine level and potentially improving the perfect amino acid profile, is a positive and beneficial effect in terms of animal nutrition and pollution reduction.

References

Agricultural Research Council. 1980. The Nutrient Requirements of Ruminant Livestock. Commonwealth Agricultural Bureaux, Farnham Royal, Slough.
Agricultural Research Council. 1981. The Nutrient Requirements of Pigs. Commonwealth Agricultural Bureaux, Farnham Royal, Slough.
Chung, T.K. and D.H. Baker. 1992. Ideal amino acid pattern for 10–kilogram pigs. J. Anim. Sci. 70:3102.
Cole, D.J.A. 1990. Challenges facing pig production: Responses of the nutritionist. In: Biotechnology in the Feed Industry, T.P. Lyons (Ed). Alltech Technical Publications, Nicholasville, KY. p.183–214.
Cowey, C.B. 1988. The nutrition of fish: The developing scene. Nutrition Research Reviews.1:255.
Eurolysine. 1988. Apparent ileal digestibility co-efficients of crude protein and essential amino acids in feedstuffs for pigs. Eurolysine Information Bulletin No 15
Harvey, J.T. 1992. Changing waste protein from a waste disposal problem to a valuable protein source: A role for enzymes in processing offal, feathers and dead birds. In: Biotechnology in the Feed Industry, T.P. Lyons (Ed.) Alltech Technical Publications, Nicholasville, KY. p.109–119.
Sibbald, I.R. and S.J. Slinger. 1963. A biological assay for metabolisable energy in poultry feed ingredients. Poultry Sci. 42:313.
Wang, T.C. and M.F. Fuller. 1989. The optimum dietary amino acid pattern for growing pigs. 1.Experiments by amino acid deletion. Bri J. Nutr. 62:77–89.
Woodgate, S.L. 1993. Animal By-Products – The case for recycling:possibilities for profitable nutritional upgrading. In: Biotechnology in the Feed Industry, E.P. Lyons (Ed.) Alltech Technical Publications, Nicholasville, KY, p.395–408.

MANIPULATION OF FIBRE DEGRADATION: AN OLD THEME REVISITED

ANDREW CHESSON

Rowett Research Institute, Bucksburn, Aberdeen, UK

Introduction

Many authors have speculated about the possibilities of manipulating the rumen fermentation of fibre by the introduction of new or genetically modified organisms or their gene products (Forsberg et al., 1986; 1993; Russell and Wilson, 1988). However, the realisation of these possibilities generally has foundered on difficulties surrounding the selection of appropriate enzyme(s) for manipulation and the inability of introduced organisms to survive rumen conditions (Wallace, 1992).

It now seems likely that much of the fibre-degrading capacity of rumen bacteria is located in cellulosomes (Lamed et al., 1987), highly organised multienzyme complexes of a type first described for *Clostridium thermocellum* (Béguin et al., 1992). Cellulosomes are organised on the bacterial cell surface and may be released to become associated with the substrate or retained to contribute to the binding of the organism to the fibre surface.

Insufficient is known about the internal organisation of these specialised structures to consider manipulation by introduction of new or "improved" catalytic activities. Although the genes of some of the proteins likely to be contributing to cellulosomes of rumen bacteria have been isolated and sequenced (Flint et al., 1993; Paradis et al., 1993), structural attributes other than the catalytic centre have received little attention. As a result it is not known how the synergistic interaction of some enzymes, notably the enzymes of the cellulase complex, is engineered within the cellulosome. In the absence of such essential information it seems unlikely that any modified protein, which would have to retain a structure capable of interlocking with other proteins of the complex, could be constructed and introduced without disrupting the activity of the cellulosome as a whole.

The value of introducing gene products to the rumen directly as free enzyme has met with some scepticism not least because it has yet to be satisfactorily demonstrated that added enzymes survive proteolytic attack and that the normally-functioning rumen ecosystem is deficient in specific enzyme activities. It is difficult to see how a highly evolved ecosystem like the rumen could have major deficiencies of a type capable of being rectified by the simple addition of an existing gene product, albeit from a different source. Nonetheless, there is limited evidence that addition of enzyme or related products can increase the rate, if not the extent, of fibre digestion in ruminants and thus a potential for improving fibre digestion with existing products apparently does exist.

Cell-free extracts of cultures of *Aspergillus* spp. have been shown to increase the rate of fibre digestion in the rumen by up to 5% (Figure 1); an effect accompanied by an increase in both the numbers of cellulolytic bacteria present and microbial protein outflow from the rumen (Wiedmeier et al., 1987; Gomez-Alarcon et al., 1990; Wallace and Newbold, 1992). These effects are destroyed by autoclaving the extracts but not by gamma irradiation (Newbold et al., 1991). The hydrolysis of hay by a mixture of *Aspergillus oryzae* fermentation product and an extract of rumen bacteria was more extensive than the sum of the individual effects (Newbold, 1992). This apparent synergism was interpreted to mean that the *Aspergillus* extract contained enzyme activities which were, in some way, complementary to those produced by the rumen organisms. Since the amount of saponifiable phenolic acids remaining in the fibre fractions of Brome grass and switchgrass undergoing digestion by rumen bacteria *in vitro* was reduced when cultures were supplemented with an *Aspergillus* fermentation product, it has been suggested that esterase(s) may be the key enzymes provided by the fungus (Varel et al., 1993). However, phenolic acid esterases are produced by rumen organisms (McDermid et al., 1990) and the *Aspergillus* preparation can only be supplementing an existing potential.

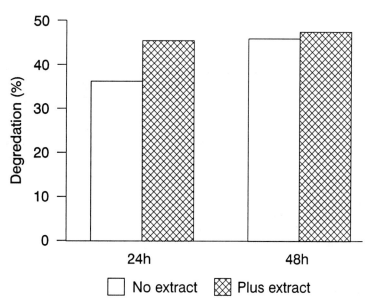

Figure 1. Effect of an *Aspergillus* fermentation extract on the loss of dry matter from a forage plus concentrate diet measured in a Rusitec after 24 and 48 hours. (from Newbold et al., 1991)

Whatever the effects of *Aspergillus* culture extracts, it is evident that they are realised only during the initial stages of fibre degradation and are relatively short-lived. Differences in the fibre degradation detectable after 12h incubation tend not to be evident at 24h and thereafter (Varel et al., 1993). This might suggest that the additional activities supplied by the fermentation extract act more to promote colonisation of the fibre than its actual degradation.

It can be argued that in the normally-functioning rumen it is the nature of the fibre source which determines both the maximum rate and extent of its degradation and not the size or composition of the microbial flora. The latter develops in response to the availability of nutrients. The partial identification of the physical and chemical constraints to degradation has allowed strategies to be developed to alleviate their effects and promote the digestion of fibre both in ruminant and non-ruminant species. The approach to date, however, has been predominantly chemical and has involved non-selective treatments with alkaline or oxidative reagents. Substantial progress has been made in the last five years in understanding the molecular architecture of plant cell walls and, in particular, the nature of the interactions between its component polymers. At the same time the molecular biology of rumen and related organisms has developed to a point when gene transfer is becoming commonplace. This would seem an opportune time in the light of a greater understanding of the nature of the substrate, the availability of recombinant techniques for organisms able to survive gut conditions and the option of producing transgenic plants and animals, to re-examine the possibilities for manipulating the metabolism of fibre in livestock.

Dictates of the fibre component

The primary walls of dicotyledonous plants, including the forage legumes, and many monocotyledenous plants such as *Allium* spp. are now known to consist essentially of two discrete but intertwining polysaccharide networks. One network is formed of cellulose microfibrils with xyloglucan acting as the cross-linking agent between individual microfibrils and the other solely of pectic polysaccharides. The pectic network, contrary to previous views, is not an amorphous or gel-like material but is capable alone of resisting turgor pressure and providing a skeletal framework for the cell. It has been shown that cells in culture adapted to DCB (2,6–dichlorobenzonitrile), a potent inhibitor of cellulose and xyloglucan biosynthesis, function normally with a wall consisting only of pectic substances (Shedletzky et al., 1990; 1992).

Walls of the Poaceae (Gramineae), the family that includes the grasses and cereals, differ considerably from the cell walls of other higher plants. In these walls the cellulose microfibrils are interlocked by glucuronoarabinoxylans with xyloglucan occurring only in small or trace amounts. The pectic network, although present, represents a far smaller fraction of the wall and does not contribute to its skeletal function. The extent to which the xylan backbone is substituted with glucuronic acid, arabinose and other side-chains varies considerably depending on the age, phylogenic origin and part of the plant examined. Substitution is highest in young primary walls and in the walls of seed grains and least in the secondary layers of older lignified walls. Xylan substitution constitutes steric hindrance to intermolecular hydrogen bonding, increases water solubility and may give rise to anti-nutritional effects in non-ruminants, particularly poultry fed high levels of wheat grain (Annison, 1991; Wiseman et al., 1993).

The effectiveness of enzyme supplementation for the disruption of gel forming polysaccharides leached from the cell walls of cereal-based diets has been well documented (see, for example Chesson, 1993). More recent approaches aimed at removing the need for constant supplementation have investigated the introduction of selected bacteria able to degrade cereal polysaccharides (Johnsson and Hemmingsson, 1991) and increasing the digestive enzyme repertoire of the host species (Hall et al., 1993). The successful expression of an endo-glucanase from *Cl. thermocellum* in the exocrine pancreas of transgenic mice demonstrates the feasibility of the latter approach.

PHYSICAL PROPERTIES

The dual cellulose–xyloglucan and pectic networks can be visualised by the use of deep-etched, freeze fracture electron microscopy'(McCann et al., 1990). From such electron micrographs it is possible to measure pore size in the wall and to obtain, for *Allium* cell walls for example, a mean diameter of approximately 10 nm. Given that the freeze-fracture techniques strips all bound water from the cell wall, this value is in reasonable agreement with the 5 nm value obtained with methods applied to hydrated samples involving polymer (Gogarten, 1988) or gold particle probes (O'Driscol et al., 1993) and with physical chemical methods (Figure 2). Treatment of the *Allium* wall to remove pectic polysaccharide before examination by freeze fracture increased the mean porosity to approximately 20 nm. Since removal of the cellulose–xyloglucan network in DCB-treated tomato cells did not alter wall porosity it seems likely that the pectic network is the major determinant of porosity in the walls of dicotyledonous and many monocotyledonous plants. In contrast, evidence from DCB-adapted barley cells suggests that, in the Poaceae, it is the extent of cross-linking between glucuronoarabinoxylan chains which determines wall porosity (Delmer et al., 1992).

Globular proteins of 17 kD molecular weight have an effective diameter of approximately 5 nm. The vast majority of enzymes produced by rumen and other organisms have molecular weights well in excess of this figure. It is thus well nigh impossible for such enzymes, even in monomeric form, to freely diffuse into a cell wall even if it is assumed that enzymic proteins act in an inert manner. This seems unlikely given the charged nature of both the protein and plant cell wall surface. The inevitable conclusion is that enzymes do not penetrate cell walls and that fibre digestion occurs only by erosion of available surfaces. The rate of erosion is constant, and for primary cell walls whose digestion normally goes to completion

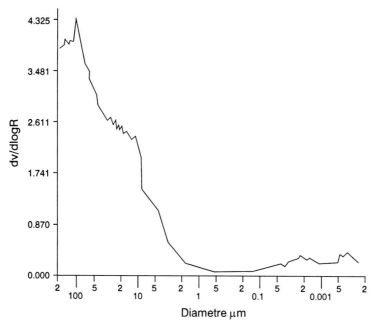

Figure 2. Pore size and distribution in a hammer-milled wheat and rapeseed meal mix measured by mercury intrusion porosimetry. Two populations of cell wall pores are evident, the first around 0.02 μm (20 nm) from the rape seed and the second at 0.004 μm (4 nm) from the wheat. Other larger "pores" (>10 μm) represent the cell lumen and interstitial spaces between particles.

in the rumen, the time to complete degradation is determined simply by the diameter (usually ~ 2–300 nm) of the wall. The porosity of tissue is reduced by the presence of lignin and by the increasing hydrophobicity of the wall surface. Superficial attack of lignified walls occurs at a rate which, in the case of isolated walls, appears to be constant and independent of age and cell type (Lomez et al., 1993). However, the amount of degradation is determined by the nature and extent of the cross-linking between wall polymers.

The value of cellulosomes to a rumen microorganism operating in a aqueous environment is evident when the nature of fibre is taken into consideration. If enzymes are active only at cell wall surfaces and cannot diffuse into the substrate, then erosion of the plant surface is most effectively co-ordinated by the physical organisation of multiple catalytic sites. Cellulosomes of *Cl. thermocellum* are known to contain both glucanase and xylanase activities. Both activities possess highly conserved regions which bind to a scaffold protein which in turn shows a high affinity for cellulose (Béguin et al., 1992). Genes for a number of bifunctional enzymes, each containing two catalytic domains (glucanase + glucanases, glucanase + xylanase, xylanase + xylanase), have been isolated from rumen microorganisms (Zhang and Flint, 1992; Flint et al., 1993; Paradis et al., 1993). These genes also contain a common highly conserved region which does not have an identifiable catalytic function. It is possible to speculate that this represents a binding domain for a protein functioning as a scaffold in a manner analogous to *Cl. thermocellum*. The enzyme activities identified suggest that cellulosomes have evolved principally to attack the xylan/cellulose-rich secondary-thickened walls of forages and that primary walls, with their more amorphous form of cellulose, are degraded by less organised and possibly soluble enzymes.

The importance of the pectic network in determining wall porosity and restricting attack to the surface of the cell wall might suggest that galacturonan should be a primary target for disruption. The application of endo-polygalacturonse (EC 3.2.1.15) or endo-pectic acid lyase (EC 4.2.2.2) would be expected to significantly increase wall porosity of dicotyledonous plants. However unless there was also present a population of soluble low molecular weight (< 30 kD) enzymes able to take advantage of the increased porosity, little would be gained. The high molecular weight (~ 2 mD) cellulosomes would still be restricted to the surface by the cellulose–xyloglucan network. Targeting the pectic fraction of the wall, therefore, may be more appropriate to the design of supplementary enzymes for non-ruminants where all fibre-degrading activities are added in soluble form, than for the promotion of cell wall degradation in the rumen. A similar argument can be made for the addition of endo-xylanases (EC 3.2.1.8) to graminaceous feeds.

CHEMICAL PROPERTIES

The inherent capacity of the rumen to disrupt and degrade available cell wall polysaccharides is high, the rate of primary cell wall degradation being only slightly less than that of starch (Chesson et al., 1986). In animals fed high roughage, whether legume or grass based, increasing the extent of polysaccharide availability is likely to provide more substantial gains than simply attempting to increase the rate of degradation. It is well recognised that the overriding constraint to the digestion of roughage polysaccharide is the presence of lignin and the nature of its association with the other wall components. Identifying the specific structures cross-linking lignin into the wall would allow, in the medium term, the appropriate the microbial/enzymatic or chemical treatment to be designed. In the longer term it may be more effective to engineer a change to the plant itself.

There are relatively few molecules in the plant cell wall which are bifunctional in nature and able to interact covalently both with polysaccharide and phenolic polymers. Phenolic acids (*p*-coumaric acid and ferulic acid) with two functional groups, the phenolic hydroxyl and the acid group, and have been implicated in a number of putative bridge structures in members of the Poaceae. Ester-linked dehydrodiferulate (Fry and Miller, 1989) and cyclobutane-type dimers (Ford and Hartley, 1990) are thought to bridge between polysaccharides, and ferulic acid and dehydrodiferulate between polysaccharide and lignin (Lam et al., 1992) (Figure 3). Ferulic acid, in particular appears to play a unique role in the control of lignification. Microsomal preparations of plants

Ferulic acid monomers

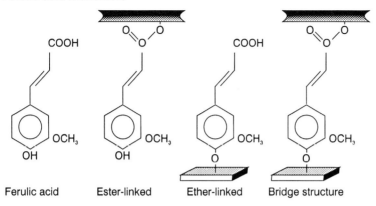

Dimeric structures

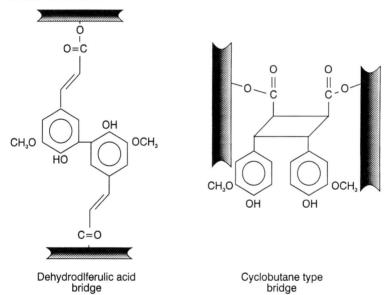

Figure 3. Cross links between polymers formed by ferulic acid in Poaceae cell walls. Cell wall polysaccharide, usually glucuronoarabinoxylan, is represented by a rod and lignin by a block.

have been shown to be able to transfer [^{14}C]-labelled ferulate from its CoA-thioester to a polysaccharide receptor (Kohler and Kauss, 1992). It is probable that all ferulate is ester-linked through the arabinose side-chains to glucuronoarabinoxylans before they are exported into the developing cell wall. Once *in situ* within the wall ferulate units appear able to act as initiation sites for lignin biosynthesis, becoming ether-linked to other phenylpropanoid units. Half or more of the ferulic

Table 1. Ester-linked, ether-linked and total p-coumaric (PCA) and ferulic acid (FA) content (g/kg) of cereal straws[1].

Linkage	Phenolic acid	Barley	Maize	Wheat
Ester-linked	PCA	5.5	22.4	6.3
	FA	4.4	5.9	4.6
Ether-linked	PCA	0.2	0.6	0.5
	FA	4.5	4.8	5.3
Total	PCA	5.7	23.0	6.8
	FA	8.9	10.7	9.9

[1](Data from Provan et al., 1994)

acid in the walls of cereal straws appears to function as a cross-linking agent retained in the wall by both ester and ether bonds (Table 1).

The ester bond formed between arabinoxylan and ferulic acid is labile to dilute alkali. As a result, up to 60% of phenolic material in graminaceous cell walls can be solubilised by treatment with dilute alkali at ambient temperatures with the consequent well known improvements to degradability. The same bond is also attacked by specific esterases (ferulic acid esterases), suggested as a possible key activity in the *Aspergillus* preparations. In theory, treatment of grass/straw based diets with ferulic acid esterase should produce the same production response seen with the weaker alkalis such as ammonia. In practice this does not happen because of the difference in the ability of the enzyme compared with the chemical to penetrate the wall. Selecting or engineering a low molecular weight esterase able to migrate through pores of 5 nm could provide an effective means of upgrading most grass and straw-based diets for ruminants. A similar, but less selective effect could also be achieved with a low molecular weight arabinofuranosidase (EC 3.2.1.55), removing the sugar unit to which the bulk of ferulate units are attached (Hartley et al., 1990). If such enzymes were coupled with a low-molecular weight xylanase, the internal structure of the wall would be sufficiently disrupted to produce the swelling and hydration of the wall evident after alkaline treatments. To a limited extent the need for low molecular weight enzymes can be offset by time. Extended treatment of lignified residues by ensiling with xylanase will also increase the extent of degradation compared with controls. The slow breakdown of glucuronoarabinoxylan engineered by such treatments is sufficient to promote the solubility of lignin, leaving rumen organisms greater access to the remaining carbohydrate (Figure 4). The net gain in the degradability of the ensiled material, however, may be counteracted by a loss in digestible carbohydrate caused by run off and fermentation during the ensiling process.

Ferulate is found associated with the pectic fraction in plants other than grasses and may play a small part in the lignification of the primary wall. However, treatment of dicots under conditions which liberate ether-linked ferulate from cereals fails to release phenolic acids in anything other than trace amounts. This lack of ether-bound ferulate coupled with the known insolubility of dicot lignin in alkali,

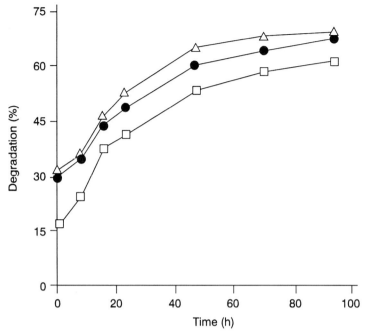

Figure 4. Effect on digestibility of ensiling barley straw for a period of 30 days in the presence and absence of xylanase prepared from a strain of *Aspergillus awamori*. (□) Original untreated straw; (●) straw ensiled for 30 days and (Δ) straw ensiled for 30 days in the presence of added xylanase (from Smith and Orskov, 1989).

confirms that the bonding between lignin and cell wall polysaccharides is substantially different from that found in the Poaceae. Treatments effective in increasing the extent of fibre degradation for one group are unlikely to be effective in the other. The specific nature of the structures linking cell wall polysaccharide to lignin are not known for forage legumes and other dicots. It is possible that tyrosyl residues in wall protein act in the same manner as ferulate. More probable is that direct coupling between carbohydrate and lignin to form benzyl ether linkages occurs during lignin biogenesis by the addition of the carbohydrate hydroxyl or carboxyl function of uronic acids to quinonemethide intermediates (Brunow et al., 1989). Such bonds would not be amenable to enzymatic attack, particularly under rumen conditions.

Complete or partial destruction of lignin by oxidative means substantially improves the degradability of all lignified feedstuffs. Breakdown of lignin by enzymatic means, however, is not a practical proposition in the context of the rumen and has proved ineffective as a pretreatment (Khazaal et al., 1993). Lignin peroxidase and monooxygenases may have a role in the preparation of silage when used in conjunction with other cell wall degrading activities able to promote the solubilisation of lignin. Any accumulation of soluble phenolics would, however, result in a chain polymerisation reaction rather than extended breakdown. Attempts to improve degradability by composting fibrous by-products with selected lignolytic organisms also have proved unsuccessful except under carefully controlled and prohibitively expensive laboratory conditions.

Engineering fibre

Modification of lignin in terms of its nature, the quantity deposited in the maturing plant and its association with cell wall polysaccharides would be expected to have a profound effect on digestibility. However, while the steps leading to the production of the lignin precursors (p-coumaryl, coniferyl and sinapyl alcohols) are well documented, relatively little is know about the biogenesis of lignin itself. The polymerisation stage takes place outside the cell and within the cell wall and consequently is beyond the normal regulatory signals associated with metabolic pathways. It seems unlikely that lignin biogenesis is controlled simply by the flux of the precursors. Several other controlling factors are likely to be operative, including the location and specificity of the peroxidase/laccase system involved in radical generation, the production of hydrogen peroxide, the location of glycosidases possibly responsible for liberating the aglycone from precursors supplied as glycosides, and the pre-programming of initiation sites. The details of these possible control mechanisms and their relative importance is unknown, making it difficult to select a target gene. Since a prerequisite for any transgenic experiment is a good knowledge of the biochemistry of the metabolic pathway to be manipulated, most experimental work to date has focused on the enzymes of the cinnamic acid pathway and the supply of lignin precursors.

TARGET ENZYMES OF THE CINNAMIC ACID PATHWAY

The gateway to the cinnamic acid pathway is controlled by phenylalanine ammonia lyase (PAL), the enzyme responsible for the production of cinnamic acid by deamination of phenylalanine. Since the pathway branches at several points thereafter and gives rise to a number of biologically important phenolic compounds in addition to lignin, direct interference with this gateway enzyme would be expected to produce multiple effects. This was confirmed when a heterologous bean PAL was introduced into tobacco. The transgenic plants produced, paradoxically, showed a reduced expression of PAL and, while the lignin content was reduced, a number of other aberrant traits were expressed. These included stunted growth and reduced viability (Elkind et al., 1990). Downregulation of PAL and the second enzyme in the pathway, cinnamate 4–hydrolase, clearly would give rise to too many undesirable effects to be useful.

Only the enzymes cinnamoyl-CoA reductase (CCR) and cinnamyl alcohol dehydrogenase (CAD) function after all possible branch points in the pathway and thus are specific to lignin biosynthesis. Evidence that down-regulation of these activities can reduce lignin content and improve digestibility is provided by the brown midrib mutants which have been found or induced in a number of forage species including maize, sorghum and pearl millet (Cherney et al., 1991). These mutant forms, which are characterised by a reddish-brown coloration in lignified feed tissue indicative of modification to lignin, have been shown to be deficient in one or more enzymes of the cinnamic acid pathway. In addition to reduced levels of CAD and CCR, O-methyltransferase (OMT) activity responsible for the production of ferulic and sinapic acids from p-coumarate also appears to be affected in some plants (maize bm_3, sorghum bmr_{12}, bmr_{18}). Transgenic tobacco plants expressing CAD antisense RNA, in which CAD activity is virtually suppressed, also show the reddish-brown pigmentation in the xylem typical of the brown midrib mutants. Although the total amount of lignin laid down differed little from normal plants, its composition and structure was modified and the degradability of fibre prepared from the transgenic plants improved (Halpin et al., 1993). Tobacco plants expressing antisense RNA for OMT also had a reduced lignin content compared with normal plants and the lignin contained a lower amount of syringyl monomers (Dwivedi et al., 1993). The effect on digestibility was not determined, but by extrapolation from results obtained with the brown rib mutants it would be surprising if fibre digestibility was not improved.

Transformation of crop plants such as maize and wheat is a relatively recent success and still is not routine. Instead, tobacco plants, which are far easier to transform, have been used to more rapidly establish that the activities chosen for manipulation were appropriate and to demonstrate feasibility. It has now been shown that down-regulating selected enzymes in the cinnamic acid pathway can alter the supply of lignin precursors to the cell wall and influence both the amount and the composition of the lignin produced. This done, the way is open for the introduction of antisense genes into elite crops to improve their degradability characteristics.

POLYMERISATION OF LIGNIN MONOMERS

The polymerisation stage in the biogenesis of lignin has been rarely targeted. The little work that has been attempted has simply demonstrated the difficulties involved in this approach. Peroxidase is an obvious target enzyme since its presence is central to the initiation and maintenance of polymer formation. However this enzyme has multiple functions within the plant and exists in isoforms of varying specificity (Frias et al., 1991). Down-regulation of a peroxidase thought to be important in suberin formation in tomato by expression of its antisense gene successfully removed all detectable activity of the isoform but had no effect on suberin composition or deposition (Sherf et al., 1993). It seems likely that, as the authors suggest, its function simply was replaced by one of the other peroxidase isoforms present in tomato. This exemplifies the difficulties of attempting to manipulate genes that occur in high copy numbers and in multiple forms. In another experiment a ten-fold overproduction of peroxidase in tobacco by the introduction of a chimeric peroxidase under the control of a constitutive promotor was successfully achieved. However, while causing a small increase in lignin content, this had the unexpected and unfortunate effect of introducing severe wilting at flowering (Langrimini et al., 1990).

Down-regulating the transferase responsible for the introduction of ferulic acid units into glucuronoarabinoxylans in the Poaceae and into pectic polysaccharides in other plants is an option that has yet to be investigated. It has the obvious attraction that its effect would be selective, that the total amount and type of lignin synthesised would be unchanged but that the association of lignin with polysaccharide could be substantially reduced. There is the danger, however, that in the absence of specific initiation sites for lignin formation, a more general and direct coupling with carbohydrate via benzyl ether bonds may substitute. If this were the case digestibility might not be improved and could be reduced.

Options for the future

Ruminants: Except in areas of the world in which concentrate feeding is difficult because of price or shortage there is little real pressure to use or further develop chemical pretreatments for roughage. Of the alternative biological pretreatments, most effort has gone to the use of lignolytic fungi. This has proved unattractive for a number of reasons not least of which are cost and effectiveness. If there is scope for biological pretreaments it is likely to be in the area of silage additives able to improve the dry matter digestibility of the finished product. Target structures for enzymatic disruption would be lignin itself or the units involved in cross-linking lignin to structural polysaccharide. Delivery could be as the free enzyme(s) or as a genetically modified silage inoculant.

Modest benefits in the digestibility of fibre can be obtained with existing rumen fungal probiotics. There is obvious mileage in determining the precise mode of action of the *Aspergillus*-based enzyme products with the aim of optimising effects and improving the consistency of performance. The introduction of other activities with the intention of influencing more substantially the rate and extent of fibre degradation would seem to present fundamental problems. Low molecular weight enzymes capable of good penetration into the cell wall would introduce a factor not currently present in the rumen ecosystem. However, since the same or better results could be permanently obtained by the engineering of crop plants, this would seem the better long term investment.

Non-ruminants: Soluble fibre leached from the walls of cereal grains may present problems in young animals, particularly at high inclusion levels in the diet. Supplementary enzymes have been successfully applied to correct such problems and there would seem little pressure to find alternative solutions. In fact the need for enzyme supplementation is likely to increase as processing conditions become more severe and the use of extruders and expanders becomes commonplace. Permanent solutions involving the expression of polysaccharidase activity by the host, although attractive, is very unlikely to provide an early alternative.

The use of enzymes to promote fibre degradation and ensure release of any entrapped nutrients could be made more effective by the use of selected low molecular weight activities. Targeting either glucurnoarabinoxylan or the pectic network could substantially improve the effectiveness of existing enzyme preparations. Ultimately however, as for ruminants, genetic modifications to the feed itself are likely to provide the most effective solution to many of the existing problems.

Acknowledgement

Support for this work was provided by the Agriculture and Fisheries Department of the Scottish Office (SOAFD).

References

Annison, G. 1991. Relationship between levels of soluble nonstarch polysaccharides and the apparent metabolizable energy of wheats assayed in broiler chickens. J. Agric. Food Chem. 39: 1252.
Beguin, P., J. Millet, and J-P. Aubert. 1992. Cellulose degradation by *Clostridium thermocellum*: from manure to molecular biology. FEMS Microbiol. Lett. 100: 523.
Brunow, G., J. Sipila, and T. Makela. 1989. On the mechanism of formation of non-cyclic benzyl ethers during lignin biosynthesis. Part 1. The reactivity of β–O-4–quinone methides with phenols and alcohols. Holzforschung 43: 55.
Cherney, J.H., D.J.R. Cherney, D.E. Akin, and J.D. Axtell. 1991. Potential of brown-midrib, low-lignin mutants for improving forage quality. Adv. Argron. 46: 157.
Chesson, A. 1993. Feed enzymes. Anim. Feed Sci. Technol. 45:65.
Chesson, A., C.S. Stewart, K. Dalgarno, and T.P. King. 1986. Degradation of isolated grass mesophyll, epidermis and fibre cell walls in the rumen and by cellulolytic rumen bacteria in axenic culture. J. Appl. Bacteriol. 60: 327.
Delmer, D.P., E. Shedletzky, M. Shmuel, T. Trainin, and S. Kalman. 1992. DCB-induced cell modifications: a comparison between dicot and monocot. Page 13 in Abstracts of the Sixth Cell Wall Meeting, University Press, Nijmegen, The Netherlands.
Dwivedi, U.N., J. Yu, R.S.S. Datia, W.H. Campbell, G.K. Podila, and V.L. Chiang. 1993. Expression of lignin-specific aspen O-methyltransferase antisense gene in tobacco and woody species to reduce lignin content. CTAPI 7th International Symposium on Wood and Pulping Chemistry Proceedings 2:633.
Elkind, Y., R. Edwards, M. Mavandad, S.A. Hendrick, R.A. Dixon, and C.J. Lamb. 1990. Abnormal plant development on down-regulation of phenylpropanoid biosynthesis in transgenic tobacco containing a heterologous phenylalanine ammonia-lyase gene. Proc. Natl. Acad. Sci. USA. 87: 9057.

Flint, H.J., J. Martin, C.A. McPherson, A.S. Daniel, and J-X. Zhang. 1993. A bifunctional enzyme, with separate xylanase and β(1,3–1,4)-glucanase domains, encoded by the *xynD* gene of *Ruminoccus flavefaciens*. J. Bacteriol. 175:2943.
Ford, C.W. and R.D. Hartley. 1990. Cyclodimers of *p*-coumaric and ferulic acids in the cell walls of tropical grasses. J. Sci. Food Agric. 50: 29.
Forsberg, C.W., B. Crosby, and D.V. Thomas. 1986. Potential for manipulation of the rumen fermentation through use of recombinant DNA techniques. J. Anim. Sci. 63: 310.
Forsberg, C.W., K-J. Cheng, P.J. Krell, and J.P. Phillips. 1993. Establishment of rumen microbial gene pools and their manipulation to benefit fibre digestion by domestic animals. Pages 281–316 Proceedings of the VII World Conference on Animal Production, Edmonton, Canada.
Frias, I., J.M. Siverio, C. Gonzalez, J.M. Trujillo and J.A. Perez. 1991. Purification of a new peroxidase catalysing the formation of lignin-type compounds. Biochem. J. 273:109.
Fry, S.C. and Miller, J.G. 1989. Towards a working model of the growing cell wall. Phenolic cross-linking reactions in the primary wall of dicots. *In:* Plant cell polymers, biogenesis and biodegradation (ACS Sym. Ser. 339). Lewis, N.G. and M.G. Paice (ed.) American Chemical society, Washington.
Gogarten, J.P. 1988. Physical properties of the cell wall of photo-autotropic suspension cells from *Chenopodium rubrum* L. Planta 174: 333.
Gomez-Alarcon, R.A., C. Dudas, and J.T. Huber. 1990. Influence of cultures of *Aspergillus oryzae* on rumen and total tract digestibility of dietary components. J. Dairy Sci. 73: 703.
Hall, J., S. Ali, M.A. Surani, G.P. Hazlewwood, A.J. Clark, J.P. Simons, B.H. Hirst and H.J Gilbert. 1993. Manipulation of the repertoire of digestive enzymes secreted into the gastrointestinal tract of transgenic mice. Bio/technology 11: 376.
Hartley, R.D., W.H. Morrison, D.S. Himmelsbach, and W.S. Borneman. 1990. Cross-linking of cell wall phenolic arabinoxylans in graminaceous plants. Phytochemistry 29: 3705.
Jonsson, E. and S. Hemmingsson. 1991. Establishment in the piglet of lactobacilli capable of degrading mixed-linked β–glucans. J. Appl. Bacteriol. 70: 512.
Khazaal, K.A., E. Owen, E. Dodson, A.P. Palmer, and P. Harvey. 1993. Treatment of barley straw with ligninase: effect of activity and fate of enzyme shortly after being added to straw. Anim. Feed Sci. Technol. 41: 15.
Kohler, A. and Kauss, H. 1992. Biosynthesis of hydroxycinnamate esters of cell wall polysaccharides in endomembranes from parsley cells. Page 99 in Abstracts of the Sixth Cell Wall Meeting, University Press, Nijmegen, The Netherlands.
Lagrimini, L.M., S.Bradford, and S. Rothstein. 1990. Peroxidase-induced wilting in transgenic tobacco plants. The Plant Cell 2:7.

Lam, T.B.T., K. Iiyama, and B.A. Stone. 1992. Cinnamic acid bridges between cell wall polymers in wheat and phalaris internodes. Phytochemistry 31: 2655.
Lamed, R., J. Naimark, E. Morgenstern, and E.A. Bayer. 1987. Specialised cell surface structures in cellulolytic bacteria. J. Bacteriol. 169: 3792.
Lopez, S., S.D. Murison, A.J. Travis, and A. Chesson. 1993. Degradability of parenchyma and sclerenchyma cell walls isolated at different developmental stages from a newly extended maize internode. Acta Bot. Neerl. 42: 165.
Newbold, C.J. 1992. Probiotics: a new generation of rumen modifiers. Med Fac. Landbouww. Univ. Gent 57/4b, 1925.
McCann, M.C., B. Wells, and K. Roberts. 1990. Direct visualization of cross-links in the primary plant cell wall. J. Cell Sci. 96:323.
McDermid, K.G., C.R. MacKenzie, and C.W. Forsberg. 1990. Esterase activities of *Fibrobacter succinogenes* subsp. *succinogenes* S85. Appl. Environ. Microbiol. 56: 127.
Newbold, C.J., R. Brock, and R. J. Wallace. 1991. Influence of autoclaved or irradiated *Aspergillus oryzae* fermentation extract on rementation in the rumen simulation technique (Rusitec). J. Agric. Sci., Camb. 116: 159.
O'Driscoll, D., S.M. Read, and M.W. Steer. 1993. Determination of cell-wall porosity by microscopy: walls of cultured cells and pollen tubes. Acta Bot. Neerl. 42: 237.
Orskov, E.R. and D.C. Smith. 1990. Enzyme treatment to improve nutritive value of straws. Paper 20B Proceedings of the 2nd International Conference Straw – opportunities and innovation Vol.2. PIRA; Leatherhead, UK.
Paradis, F.W., H. Zhu, P.J. Krell, J.P. Philips and C.W. Forsberg. 1993. The xynC gene from *Fibrobacter succinogenes* S85 codes for a xylanase with two similar catalytic domains. J. Bacteriol. 175: 7666.
Provan, G.J., L. Scobbie, and A. Chesson. 1994. Determination of phenolic acids in plant cell walls by microwave digestion. J. Sci. Food Agric. 64:63.
Russell, J.B. and D.B. Wilson. 1988. Potential opportunities and problems for genetically-altered rumen microorganisms. J. Nutr. 118: 271.
Shedletzky, E., M. Shmuel, D.P. Delmer and D.T.A. Lamport. 1990. Adaption and growth of tomato cells on the herbicide 2,6–dichlorobenzonitrile (DCB) leads to production of unique cell walls virtually lacking a cellulose-xyloglucan network. Plant Physiol. 94:980.
Shedletzky, E., M. Shmuel, T. Trainin, S. Kalman, and D.P. Delmer. 1992. Cell wall structure in cells adapted to growth on the cellulose-synthesis inhibitor 2,6–dichlorobenzonitrile (DCB): a comparison between two dicotyledonous plants and a graminaceous monocot. Plant Physiol. 100: 120.
Sherf, B.A., A.M. Bajar, and P.E. Kolattukudy. 1993. Abolution of an inducible highly anionic peroxidase activity in transgenic tomato. Plant Physiol. 101: 201.
Varel, V.H., K.K. Kreikemeier, H-J. G. Jung, and R.D. Hatfield. 1993. In vitro stimulation of forage fiber degradation by ruminal

microorganisms with *Aspergillus oryzae* fermentation extract. Appl. Environ. Microbiol. 59, 3171.

Wallace, R.J. 1992. Rumen microbiology, biotechnology and ruminant nutrition: the application of reseach findings to a complex microbial ecosystem. FEMS Microbiol. Lett. 100: 529.

Wallace, R.J., and C.J. Newbold. 1992. Probiotics for ruminants. *In:* Probiotics – the scientific basis (R. Fuller, ed.). pp.317–353. Chapman & Hall, London.

Wiedmeier, R.D., M.J. Arambel, and J.L. Walters. 1987. Effect of yeast culture and *Aspergillus oryzae* fermentation extract on ruminal characteristics and nutrient digestibility. J. Dairy Sci. 70:2063.

Wiseman, J., N. Nichol, and G. Norton. 1993. Biochemical and nutritional characterisation of wheat varieties for performance. Aspects Appl. Biol. 34: 329.

Zhang, J-X. and H.J. Flint. 1993. A bifunctional xylanase encoded by the *xynA* gene of the rumen cellulolytic bacterium *Ruminococcus flavefaciens* 17 comprises two dissimilar domains linked by an asparagine/glutamine-rich sequence. Mol. Microbiol. 6: 1013.

THE UNITED STATES MARKET FOR FEED ENZYMES: WHAT OPPORTUNITIES EXIST?

C.W. NEWMAN
Montana Agricultural Experiment Station, Montana State University
Bozeman, Montana USA

Introduction: An historical perspective

The science of enzymology began over 150 years ago with the discovery that starch could be converted into fermentable sugars by an extract from malt. As technology in the biological sciences grew through the ensuing years, major breakthroughs in enzymology occurred throughout the scientific community. Enzymes were defined as protein catalysts that initiated and controlled the rate of biological reactions which changed substrates into secondary products. Early industrial and agricultural applications of enzymes were slow to develop due in part to a lack of understanding of enzyme kinetics and the requirements for minimum and maximum activities in applied situations. As general knowledge in enzymology evolved and purification procedures were developed, it became possible to target specific substrates. From a crude beginning where mixtures of unknown composition were added to feeds, it became possible to add relatively pure enzymes or mixtures of enzymes to achieve a planned response.

Nearly 70 years ago Sumner isolated and purified the enzyme urease. With this accomplishment, concepts of utilizing these specialized proteins began to evolve in industrial and agricultural laboratories. Some of the early use of enzymes in feedstuffs was in all probability accidental. Feeding of partially sprouted barley and rye was found to increase the nutritive value of these grains in the early 1950s. Wheat millrun, a byproduct of the flour milling industry, was reported to be significantly improved in "metabolizable energy" for poultry when soaked in water for 24 hours and later dried (Davis et al., 1959). In all probability, the sprouting and wetting processes activated beneficial endogenous enzymes and(or) stimulated the production of enzymes from microbes present in these feedstuffs. One of the first reports of supplementation of enzymes in animal diets was that of Jensen et al. (1957). This study showed that supplementation of barley-based poultry diets with a crude mixture of amylase and protease enzymes gave a significant improvement in the performance of the birds as well as an improvement in litter quality. It was later found that this crude mixture contained β–glucanase activity, which in view of present knowledge, was probably the beneficial active factor (Rickes et al., 1962). Other pioneering work in enzyme supplementation was that of Nelson et al. (1968), who demonstrated the effectiveness of a microbially produced phytase for increasing the utilization of phosphorus from plant sources by poultry.

Goals, substrates and applications

GOALS

Although a great amount of research has been accomplished, it is only recently that the potential of utilizing enzymes to enhance animal production has been truly appreciated. Goals of enzyme supplementation of animal diets as outlined by Inborr (1989) are to remove or destroy anti-nutritive factors, enhance overall digestibility, render certain nutrients biologically available and reduce the pollution impact of animal excreta. Since the early beginnings, applied research on the use of enzymes in diets for poultry and other monogastric animals has greatly intensified.

SUBSTRATES

Cereal grains
A three-year average (1989–90–91) of feed grains fed livestock and poultry showed that corn represented 85% of the total fed. Grain sorghum and the combined use of oats and barley make up 6 and 5% of the total, respectively. Wheat represents <3% and rye only 0.1% of the total (USDA, 1992). This is reflected somewhat in the feed potential for swine and poultry in the USA (Tables 1 and 2). Feed potential is greatest in the areas where corn is the predominant grain. This is especially true for swine but also holds true for poultry. Never the less, there is considerable tonnage of feed in the West that is largely composed of small grains. Improvement in the utilization of nutrients in these feeds could enhance their usage in the West and perhaps encourage greater use in other regions as well in other countries.

Table 1. Feed potential for swine in the U. S. and by region, 1991 (000 tons)[a].

Region	Sows	Pigs	Total	% of Total
U.S.	13,237.0	38,040.8	51,277.6	
N. Atlantic	214.5	643.7	858.2	1.67
N. Central	10,462.1	29,853.7	40,315.8	78.62
S. Atlantic	1,370.3	4,180.4	5,550.7	10.82
S. Central	893.2	2,489.8	3,383.0	6.60
West	296.7	873.2	1,169.9	2.28

[a]Schoeff and Castaldo, 1992

Table 2. Feed potential for poultry in the U. S. and by region, 1991 (000 Tons)[a].

Region	Layers	Broiler	Turkey	Total	% of total
U.S.	11,472	24,554	9,263	45,289	
N. Atlantic	1,404	471	298	2,173	4.80
N. Central	3,572	881	3,517	7,970	17.60
S. Atlantic	2,334	9,606	2,841	14,781	32.64
S. Central	2,245	11,754	780	14,779	32.63
West	1,917	1,194	1,151	4,262	9.41
Other	–	648	676	1,324	2.92

[a]aSchoeff and Castaldo, 1992

Carbohydrates
More than 80% of the components in cereal grains are carbohydrates which are primarily starch and nonstarch polysaccharide (NSP) with a small portion (1 to 3%) of free sugars (Henry, 1985). Of the carbohydrate in cereal grains, 70 to 90% is composed of starch, with nonstarch polysaccharides (NSP) making up from 10 to 30% of the total. Although grains make a significant contribution of protein to poultry and swine diets, the major nutrient furnished is energy. Dietary energy is primarily derived from starch with smaller amounts coming from lipids, non-essential amino acids, free sugars and NSP.

Two forms of starch exist in cereal grains, amylose and amylopectin. Amylose is a linear chain of glucose units linked via α–1,4 glycosidic bonds. Amylopectin, on the other hand, is highly branched with successive glucose units linked via α–1,4 glycosidic bonds with branching points linked with α–1,6 glycosidic bonds. Under most conditions in which cereal grains are fed to poultry and swine, cereal grain starch is effectively and efficiently utilized. On the other hand, most of the NSP in cereal grains is not digested by monogastrics due to the absence of the necessary enzymes in the gastrointestinal tract. That portion of NSP that is digested furnishes only a small percentage of the total energy released. This is accomplished primarily in the cecum and large intestine by microbial fermentation. In many instances, the soluble portion of NSP in cereal grains increases the viscosity of the digesta in the small intestine to the extent that nutrient absorption is restricted. This not only affects the absorption of basic nutrients such as glucose, fat and protein (Fengler and Marquardt, 1988; Wang et al., 1992), but the utilization of certain minerals, including calcium, phosphorus and zinc, is also reduced in the presence of soluble NSP (Gordon, 1990). Under these conditions, soluble NSP are classified as antinutrients.

Cellulose, the predominant NSP constituent in corn, is primarily found in the aleurone. In small grain cereals, i.e., barley, oats, wheat, NSP components are more varied than in corn. The NSP of these grains are composed of cellulose, mixed linked (1–3),(1–4)-β–glucans, commonly referred to as β–glucans, and arabinoxylans (pentosans). Covered or hulled barley and oats are the only major feed grains that contain significant amounts of cellulose. Most of the cellulose in barley and oats is in the hulls, although small amounts are found in the aleurone and endosperm cell walls as is the case with wheat, rye, and triticale. Naked or hulless barley is similar to wheat and rye in cellulose content. Although not a carbohydrate, lignin is generally classified as an NSP because of its relationship with cellulose. β–Glucans are the major NSP in the endosperm and cell walls of barley and oats. In barley, β–glucans are found in both the aleurone and endosperm cell walls but in greater concentrations in the latter. The concentration of β–glucans in barley, as well as their molecular weights, have been shown to vary with genotype (Bengtsson et al., 1990; Xue et al., 1991). In contrast to barley, the β–glucans in oats are concentrated in the outer portion of the kernel with considerably less in the endosperm cell walls (Bacic and Stone, 1981). About one-third of the cell wall NSP in barley is arabinoxylan. Only small amounts of β–glucans are found in wheat, rye and triticale, in which arabinoxylans are the major NSP (Mares and Stone, 1973; Henry, 1985). As with β–glucans in barley and oats, the arabinoxylans of wheat, rye and triticale are located in the aleurone and endosperm cell walls (D'Appolonia and MacArthur, 1976). Two types of pentosans have been described in rye (Bengtsson and Åman, 1990). Pentosan 1 is characterized by single arabinose units linked β–(1–3) to the primary xylan β–(1–4) chain. A second fraction, Pentosan 2, was isolated that contained sequential xylose residues (4–5) that were doubly branched at the –2 and –3 positions.

Phytic acid
Myoinositol 1,2,3,4,5,6–hexa*kis*phosphate, commonly called phytate, comprises 1.0 to 1.5% of the content of cereal grains and typically represents from 50 to 80% of total seed phosphorus (Raboy, 1990). Phytate is principally deposited as discrete globular inclusions in single-membrane storage microbodies referred to as protein bodies (Pernollet, 1978; Lott, 1984). In wheat and barley, the protein bodies contain a proteinaceous matrix, which surrounds phytate-rich globoid crystals (Jacobsen et al., 1971; Raboy, 1990). Most of the phytate (approximately 90%) in wheat (and barley possibly) is found in the aleurone, with about 10% occurring in the germ (embryo and scutellum). In corn, the reverse occurs, with nearly 90% of the phytate localized in the germ and 10% in the aleurone (O'Dell et al., 1972).

Applications
The widespread use of enzymes in animal feeds on a commercial basis has only recently occurred. β–Glucanase use in barley based poultry diets is perhaps the most well known and best documented application of enzymes in agriculture. More recently, pentosanases for poultry rations based on wheat, rye and triticale have shown some success for commercial application. β–glucanases and pentosanases are the two major categories of enzyme supplements that have been most extensively researched and are currently used commercially in poultry feeding systems with the intent to improve nutrient utilization, litter quality and(or) egg cleanliness. Phytase has intrigued animal nutritionists for many years. Recent advances in the production of phytase has made

the commercial use of this enzyme realistic. Poultry research reports on the specific effects of other enzyme categories such as cellulases, pectinases, proteases, lipases and amylases are limited. Most enzyme additives for animal feeds, however, are crude preparations and generally exhibit activity towards a range of substrates (Campbell and Bedford, 1992). Commercial enzyme products are often blends of two or more of the enzyme groups listed above and are referred to as "enzyme cocktails" (H. Graham, *personal communication*).

There is little scientific evidence to support the use of α–amylases in nonruminant diets, even though the concept of improving starch digestibility for genetically improved, fast growing animals is intriguing. The pioneering studies of Willingham et al. (1959) showed that crystalline α–amylase was ineffective in improving barley diets for poultry and that the beneficial effects he reported were due to β–glucanase present in the crude enzyme mixture. Amylase secretion in the small intestine of poultry is obviously at such a level that starches are well digested and utilized under most conditions (Moran, 1982). It has been postulated by a number of authors (Campbell and Bedford, 1992) that very young birds or pigs could benefit most from amylases and other enzyme supplements in the feeds. Such is the case with β–glucanase and pentosanases, but scientific evidence with other enzyme systems to support this contention is limited (Campbell and Bedford, 1992).

Equally intriguing as the idea of using amylases, is the concept of using cellulase to enhance energy levels in feedstuffs high in insoluble fiber, such as covered (hulled) barley, oats or by-products of the brewing and distilling industries. The process of enzymatic hydrolysis of cellulose is extremely complex, however, involving numerous different cellulase activities. Additionally, cellulose is rarely found in pure form in nature (cotton being the exception), especially in feedstuffs. In feedstuffs, cellulose is generally intimately associated with other polymers such as lignin and pentosans. Lignin encrustation renders access to cellulose by the enzyme difficult if not impossible (Sears and Walsh, 1993).

Currently, phytase is the only enzyme that has the potential to dramatically improve nutrient utilization in poultry feeds at the same magnitude as that of β–glucanases and pentosanases. Given a maximum effect on phytate, the resulting improvements could exceed that of these two carbohydrases. Whereas the problems of β–glucans are limited to barley and oats, and pentosans to rye and wheat, phytate is universally present in plant material. Phytate in feedstuffs represents a major source of phosphorus for meeting the requirements for growth and bone development in animals, but as such is almost entirely unavailable to monogastrics.

Phytate is an antinutrient in that it irreversibly chelates divalent cations and interferes with amino acid absorption in the gastrointestinal tact of birds as well as other monogastrics. This is sufficient cause to attempt to remove it from feedstuffs. Additionally, the fecal excretion of phytate phosphorus and chelated minerals is a major source of soil and water pollution when wastes are applied to farm land. Given these three reasons, improved utilization of plant phosphorus, removal of an antinutrient and the reduction of pollution, the successful utilization of phytase could surpass the overall benefits of any other single or multiple enzyme system used in feeding regimes for poultry and other monogastric animals.

Pertinent research

GLUCANASE

Chicks
Supplementation of barley diets with β–glucanase has been shown in numerous trials to be effective in improving the growth rate and feed efficiency of poultry. The greatest benefits have been shown in the young broiler chick (Elwinger and Saterby, 1987), although feeding trials to market weight have also demonstrated benefits in older birds (Classen et al., 1988). In the early work with carbohydrases, the enzymes utilized were crude mixtures. Rickes et al. (1962) obtained a purified β–glucanase from the enzyme mixture fed earlier by Jensen et al. (1957). Rickes et al. (1962) concluded that the β–glucanase was responsible for the improvement in the performance of the birds. Gohl et al. (1978) reported that β–glucanase or water treatment (mixing with warm water for 2 h followed by drying) did not significantly influence the nutritional value of medium viscosity barley. However, when applied to high-viscosity barley, β–glucanase or water treatment improved litter quality as well as performance of the birds. Hesselmann et al. (1981) showed that β–glucanase supplementation as a dry powder in the feed or in the drinking water of broiler chicks improved feed consumption, weight gain, and feed efficiency up to 21 d of age. Dry matter of the excreta was increased and cage cleanliness was improved where the enzyme was consumed. These and later studies (Hesselmann et al., 1982; 1986) confirmed the earlier findings of Rickes et al. (1962) that β–glucanase was the active enzyme in improving the growth rate and feed efficiency of broiler chicks.

Absolute viscosity and β–glucan levels of barley are affected by both genotype and environment (Aastrup, 1979; Hesselman and Thomke, 1982; Hockett et al., 1987; Newman and Newman, 1987;1988). Barleys having higher levels of β–glucans always show greater response to β–glucanase supplementation as measured by improved chick performance (Newman and Newman, 1987; 1988; Classen et al., 1988; Campbell et al., 1989). Although studied much less extensively than barley, oats appear to behave similarly in regard to β–glucan content and β–glucanase supplementation (Elwinger and Saterby, 1987; Pettersson et al., 1987; Cave et al., 1990).

Several studies reported improvement in the absorption of fat, starch, nitrogen and amino acids by chicks fed enzyme-treated barley (Classen et al., 1985; Hesselman and Åman, 1986; Edney et al., 1989; Rotter et al., 1989; Wang et al., 1992). The improvement in nutrient digestibility in barley is believed to be due to the reduction of digesta viscosity by disruption of the β–glucan molecule. Complete conversion of β–glucan to glucose by β–glucanase would theoretically increase the metabolizable energy of barley or oats, however, most researchers conclude that the major effect is due to the reduced digesta viscosity.

Swine
Responses to β–glucanase supplementation have been low and inconsistent with swine. The available data suggest that β–glucans are not as detrimental to swine performance as with chicks. Bhatty et al. (1979) reported that a high-viscosity hulless barley which significantly retarded growth of chicks, produced good growth in pigs. Even with crude enzyme mixtures, inconsistent response has been reported. Newman et al. (1980, 1983) showed positive growth and feed efficiency response to a bacterial diastase with a hulless waxy barley but not with nonwaxy hulled or hulless barleys. It is now known that waxy barleys contain higher levels of β–glucans (Ullrich et al., 1986; Xue et al., 1991; Han and Froseth, 1992) and one report suggests waxy barley β–glucans have higher molecular weights compared with that of β–glucan in nonwaxy barley (Bengtsson et al., 1990). It may be surmised that differences in barley β–glucan levels and molecular weights predispose different responses in swine as was shown in poultry (Newman and Newman, 1987).

Bedford et al. (1992) found a positive response to β–glucanase with swine fed barley; however Thacker et al. (1988, 1989, 1992), reported little effect of a β–glucanase on swine which was beneficial in poultry diets. In a study in our laboratory with young pigs weighing about 12 kg initially, supplemental β–glucanase improved rate of gain and feed conversion on a barley diet by 11 and 10%, respectively (Newman et al., 1993). Ileal digestibility of β–glucan in pigs seems to increase with age (Graham et al., 1988) but as suggested in our poultry studies (Newman and Newman, 1987), is also influenced by source and solubility of the β–glucans (Bach et al., 1991). High digestibility values (95.7–97.1%) for β–glucans have been reported with older swine (Graham et al., 1986; Graham et al., 1989). In a recent study with young pigs fed a diet composed of 35% barley, 35% wheat and 22% soybean meal, a mixture of xylanase, amylase and β–glucanase did not influence weight gain or feed efficiency despite a significant increase in the digestibility of soluble NSP (Inborr et al., 1993).

PENTOSANASE

Chicks
Halpin et al. (1936) concluded that rye was unsuitable for poultry because of reduced feed consumption and poor growth which was accompanied by sticky droppings. Similar results were reported by Wieringa (1967) and Moran et al. (1969). It was then shown that nutrient utilization was depressed in chicks fed rye-based diets (Misir and Marquardt, 1978a,b,c,d; Marquardt et al., 1979; Lee and Campbell, 1983) which severely depressed chick performance. Studies by Marquardt et al. (1979) and Antoniou et al. (1980) revealed that the depression of nutrient digestion, especially that of saturated fat, was due to a nonspecific antinutritional factor in rye. This factor was found to be concentrated

in a water-extractable fraction and it was hypothesized to be a water soluble portion of the pentosans (Fernandez et al., 1973; Antoniou and Marquardt, 1981). Fractionation studies by Antoniou et al. (1981) indicated that the fraction causing the nutrient depression was water soluble and was in fact rich in pentosans. The in vitro and in vivo studies of Fengler and Marquardt (1988a,b) confirmed these findings.

The causative factor of sticky droppings in poultry consuming barley was determined to be β–glucans which causes excessive losses of fat (Wang et al., 1992). The similar problem with sticky droppings from poultry fed rye diets was reported to be eliminated by supplemental β–glucanase and xylanase (Pettersson and Åman, 1989). These studies and that of GrootWassink et al. (1989) who fed a crude arabinoxylanase preparation to broiler chicks, confirmed that the antinutritive factor in rye grain was a water soluble pentosan. Further, these reports confirm the efficacy of pentosanase enzymes in improving the nutritive value of poultry diets based on rye.

The pentosans of wheat have also been implicated as antinutritive factors for poultry. Choct and Annison (1990) reported that the addition of isolated arabinoxylans to broiler chick diets caused a depression in apparent metabolizable energy (AME) and growth. It has also been demonstrated that glycanase (β–glucanase + xylanase) supplementation of wheat-based broiler chick diets is beneficial (Inborr and Graham, 1991) indicating that wheat NSP are deleterious to broiler chick performance. A recent study showed that a strong negative correlation exists between Australian wheat AME values and the level of water soluble NSP, which are predominately arabinoxylans (Annison, 1991). A later report by this author provided further evidence that the cell wall material of wheat possesses antinutritive activity which may be reduced by supplementation of diets with glycanase preparations (Annison, 1992). In this study, supplemental enzymes raised the AME of the wheat from 14.3 MJ/kg to 15.2 to 15.8 MJ/kg. In current studies at Montana State University, we have not been able to show any beneficial effects on broiler chick performance due to xylanase supplementation in diets prepared from soft white wheat or hard red winter or spring wheats. It is conceivable that the lack of response to the enzyme in our studies compared with the results reported by European and Australian researchers is due to differences in solubility of the pentosans in these wheats and the European and Australian wheats. It may be possible at some future date to predict the nutritive value of wheat based on viscosity extracts which are directly related to intestinal viscosity (Choct and Annison, 1992) as with barley (Campbell et al., 1989, Rotter et al., 1989; Wang et al., 1992).

Swine
Pentosanase has not been effective in improving the performance of pigs fed rye as compared to that of barley fed pigs. Thacker et al. (1991, 1992a) fed pentosanase in rye-based diets at very high levels and were not able to show any improvement in pig weight gain. Bedford et al. (1992) suggested that solubilization of pentosans in rye with pentosanase tended to increase the viscosity of the small intestine contents. Such an effect could in all possibility be detrimental to animal performance. The

improvement in digestibility of soluble NSP in a diet mixture of wheat and barley (Inborr et al., 1993) may have been due in part to xylanase as well as β–glucanase.

PHYTASE

Endogenous phytase
Low levels of active phytase occur in the gastrointestinal tracts of humans, poultry and other animals. Additionally, wheat, rye, triticale, and their byproducts, and to a lesser extent barley, are fairly rich sources of phytase whereas, oats, corn and soybean meal contained little or no phytase (McCance and Widdowson, 1944; Møllgaard, 1946; Bos, 1990). It has been documented that pigs fed wheat or barley based diets require less supplemental phosphorus than those fed corn- or grain sorghum-based diets to maximize performance and bone mineralization (Cromwell et al., 1972a, 1974). These results could have been influenced by the higher total level of phosphorus in these grains compared with that in corn and grain sorghum. Other researchers, however, have reported that the availability of wheat or triticale phosphorus is higher than that for corn for pigs (Pontillart et al., 1984, 1987) and poultry (Sauveur, 1989). A review of literature in 1967 on the utilization of phytate phosphorus by poultry (Nelson, 1967) cited widely varying views of researchers up to that date on the availability of plant phosphorus. The preponderance of data presented by Nelson (1967), however, suggest that it is questionable whether any portion of the phytate phosphorus should be considered available for utilization by poultry. Recent findings in our laboratory at Montana State University, which will be discussed later, concur with this conclusion.

Chicks
Nelson et al. (1968) were the first to report an improvement in the availability of phytate phosphorus in chicks due to supplemental phytase. The enzyme, produced by a culture of *Aspergillus ficuum* strain NRRL 3135, was added to liquid soybean meal and incubated at 50°C for 24 h. When the treated dried soybean meal was fed to day-old chicks, a considerable increase in bone ash was observed compared with controls receiving no inorganic phosphorus. In a follow-up study, Nelson et al. (1971) demonstrated the beneficial effects of supplemental phytase added directly to chick diets on the in vivo utilization of phytate phosphorus. The addition of phytase produced by *Aspergillus ficuum* to diets as a dry powder produced an increase in percentage bone ash and increased rate of gain in White Leghorn cockerels. Chicks utilized phosphorus from the phytate as well as supplemental phosphate from sodium orthophosphate or β–tricalcium phosphate. Simons et al. (1990) confirmed the findings of Nelson et al. (1968, 1971) in a series of experiments with broiler chicks fed diets based on corn and grain sorghum supplemented with soybean and sunflower meals. Growth performance of supplemented chicks was equivalent to that of birds fed adequate inorganic phosphorus. Calcium and phosphorus availability was also improved by the phytase

supplement. Growth rate and feed conversion ratio of the broilers were dependent on the level of supplemental phytase. An additional benefit of the phytase was a decrease in mortality in birds fed treated diets. Other researchers have shown significant improvements in phosphorus retention (Simons and Versteegh, 1990) and decreased phosphorus excretion (Saylor et al., 1991) in chicks fed phytase supplemented diets. In contrast, Perney et al. (1993) reported that the addition of dietary phytase to low (0.21 –0.32%) phosphorus diets in two experiments did not significantly improve weight gain, feed intake or feed conversion of broiler chicks. Chicks fed the 0.21% phosphorus diet with or without phytase exhibited signs of rickets. Supplemental phytase, however, did result in increased levels of plasma inorganic phosphorus and toe ash. These authors suggested that the relative lack of response to phytase in these studies compared with that reported by Simons et al. (1990) was possibly due to the use of higher dietary calcium and lower levels of phytase.

The location of phytate in the aleurone layer of wheat kernels suggests that the use of fiber degrading enzymes such as β–glucanase and pentosanase in combination with phytase could possibly enhance the activity of the latter or endogenous phytases in the gastrointestinal tracts of monogastric animals. In our research at Montana State University, however, the use of xylanase alone or in combination with phosphatase and protease has not resulted in any significant improvement in the utilization of phytate phosphorus by broiler chicks up to six weeks of age. In some instances in our studies, it appears that the use of these enzymes increases the excretion of minerals such as calcium. It is possible that the enzymes are releasing the phytate from cellular structures and allowing greater access to minerals for chelation.

Swine
It is generally accepted among nutritionists that microbial phytase has the potential to enhance phosphorus availability when added to the diets of swine as has been shown with poultry. Recent studies with pigs have clearly demonstrated this to be true (Simons et al., 1990; Jongbloed et al., 1990; Ketaren et al., 1991; Mroz et al., 1991; Lei et al., 1991; Young et al., 1993; Cromwell et al., 1993; Power and Kahn, 1993). These studies showed one or more of the following improvements: increased overall phosphorus digestibility, improved phytate phosphorus availability, increased growth rate, increased feed efficiency, improved ileal protein digestibility and protein deposition, increased bone strength and decreased fecal phosphorus. Such reports are extremely encouraging for the deployment of commercial phytase to the modern swine industry regardless of geographical location.

Conclusions

Despite the lack of consistency in research reports, the use of enzymes in swine and poultry feeds to improve the utilization of nutrients has great potential for the feed and allied industries as well as livestock and poultry producers. The successful application of β–glucanase supplements in barley-based broiler chick diets has greatly benefitted the poultry industry in Western North America as well as barley growers in these regions. The use of barley in poultry diets has been severely restricted prior to the commercial availability of β–glucanases. In effect, the enzyme industry has produced an alternate market for barley. Pentosanase has not made the major impact as seen with β–glucanase, however, recent reports where the two enzymes are used together with barley and wheat are very promising. Perhaps phytase has the greatest potential of all the current commercial enzymes for successful application in swine and poultry diets throughout the world as well as in the USA because of the universal occurrence of phytate in cereal gains and plant protein supplements. Cromwell and Coffee (1991) estimated that poultry and swine in the US excrete 200 and 120 thousand tons of phosphorus, respectively, on an annual basis. A large percentage of this is phytate phosphorus which represents a tremendous waste of a natural resource not to mention the negative effect on the environment which can be partially if not totally eliminated through the use of phytase.

The use of new techniques in biotechnology to produce enzymes that are more adaptable to gastrointestinal environments, less thermolabile, and that can be directed at specific target substrates should be the theme of future enzyme research. Utilization of active supplemental enzyme packages in modern feeds not only will improve animal performance and increase profits of the producer, but will significantly reduce soil and water pollution. While reducing cost of production, with a resulting increase in net returns to the farmer, enzymes can assist in making agricultural production more compatible with the environment of this planet for current and future generations of mankind.

References

Aastrup, S. 1979. The relationship between the viscosity of an acid flour-extract of barley and its beta-glucan content. Carlsberg Res. Commun. 44:289–304.

Annison, G., 1991. Relationship between the levels of soluble non-starch polysaccharides and the apparent metabolizable energy of wheats assayed in broiler chickens. J. Agric. Food Chem., 39:1252–1256.

Annison, G. 1992. Commercial enzyme supplementation of wheat based diets raises ileal glycanase activities and improves apparent metabolizable energy, starch and pentosan digestibilities in broiler chickens. Anim. Feed. Sci. Technol. 38:105–121.

Antoniou, T., R.R. Marquardt, and R. Misir. 1980. The utilization of rye by growing chicks as influenced by calcium, vitamin D-3, and fat type and level. Poult. Sci. 59:758–769.

Antoniou, T., R. Marquardt, and P.E. Cansfield. 1981. Isolation, partial characterization, and antinutritional activity of a factor (pentosans) in rye grain. J. Agric. Food Chem. 29: 1240–1247.

Antoniou, T.C. and R.R. Marquardt. 1981. Influence of rye pentosans on the growth of chicks. Poult. Sci. 60:1898–1904.

Bach Knudsen, K.E., B. Borg Jensen, J.O. Andersen, and I. Hansen. 1991. Gastrointestinal implications in pigs of wheat and oat fractions. 2. Microbial activity in the gastrointestinal tact. Br. J. Nutr. 65:233–248.

Bacic, A. and B.A. Stone. 1981. Isolation and ultrastructure of aleurone cell walls from wheat and barley. Aust. J. Plant Physiol. 8:453–474.

Bengtsson, S., P. Åman, H. Graham, C.W. Newman, and R.K. Newman. 1990. Chemical studies on mixed link beta-glucans in hulless barley cultivars giving different hypocholesterolemic responses in chickens. J. Sci. Food Agric. 52: 435–445.

Bengtsson, S. and P. Åman. 1990. Isolation and chemical characterization of water soluble arabinozylan in rye grain. Carbohydr. Poly. 12:267–271.

Bhatty, R.S., G.I. Christison and B.G. Rossnagel. 1979. Energy, and protein digestibilities of hulled and hulless barley determined by swine feeding Can. J. Anim. Sci. 59:585–588.

Bos, K.D. 1990. Chemical background of phosphorus compound and phytase in livestock feed. Proc. Symp. Livestock Feed and Environment: The Manure Problem: Tackling through Pig and Poultry Feed. Lelystad, The Netherlands.

Campbell, G.L., B.G. Rossnagel, H.L. Classen, and P.A. Thacker. 1989. Genotypic and environmental differences in extract viscosity of barley and their relationship to its nutritive value for broiler chickens. Anim. Feed Sci. Technol. 26:221–230.

Campbell, G.L. and M.R. Bedford. 1992. Enzyme applications for monogastric feeds: A review. Can. J. Anim. Sci. 72:449–466.

Cave, N.A., P.J. Wood, and V.D. Burrows. 1990. The nutritive value of naked oats for broiler chicks as affected by dietary addition of oat germ, enzyme, antibiotic, bile salt and fat-soluble vitamins. Can. J. Anim. Sci. 70: 623–633.

Choct, M. and G. Annison. 1990. Anti-nutritive activity of wheat pentosans in broiler diets. Br. Poult. Sci., 30:811–821.

Choct, M. and G. Annison. 1992. Anti-nutritive effect of wheat pentosans in broiler diets: Roles of viscosity and gut microflora. Br. Poult. Sci. 33:821–834.

Classen, H.L., G.L. Campbell, and J.W.D. GrootWassink. 1988. Improved feeding value of Saskatchewan-grown barley for broiler chickens with dietary enzyme supplementation. Can. J. Anim. Sci. 68:1253–1259.

Cromwell, G.L., V.W. Hays, and J.R. Overfield. 1972a. Effects of phosphorus levels in corn, milo and wheat based diets on performance and bone strength of pigs. J. Anim. Sci. 35:1103.

Cromwell, G.L., V.W. Hays, and J.R. Overfield. 1974. Effects of phosphorus levels in corn, wheat and barley diets on performance and bone strength of swine. J. Anim. Sci. 39:180.

Cromwell, G.L., Stahly, R.D. Coffey, H.J. Monegue, and J.H. Randolph, 1994. Efficacy of phytase in improving the bioavailability of phosphorus in soybean meal and corn-soybean meal diets for pigs. J. Anim. Sci. (at press).

Cromwell, G.L. and R.D. Coffey. 1991. Phosphorus – a key essential nutrient, yet a possible major pollutant – its central role in animal

nutrition. In: Biotechnology in the feed industry T.P. Lyons (ed.). Alltech Technical Publication, Nicholasville, KY.
D'Appolonia, B.L. and L.A. MacArthur. 1976. Comparison of bran and endosperm pentosans in immature and mature wheat. Cereal chem. 53:711–718.
Davis, G.T., A.F. Beeckler, K.J. Goering, D. Beardsley, E. Guenther, and T.W. Wilcox. 1959. Use of Montana products in poultry feeds. Proc. Montana Nutr. Conf., Feb 9–10, 1959. pp 20–24.
Edney, M.J., G.L. Campbell, and H.L. Classen. 1989. The effect of β-glucanase supplementation on nutrient digestibility and growth in broilers given diets containing barley, oat groats or wheat. Anim. Feed Sci. Technol., 25:193–200.
Elwinger, K. and B. Saterby. 1987. The use of β–glucanase in practical broiler diets containing barley and oats. Effect of enzyme level, type and quality of grain. Swedish J. Agri. Res. 17:133–140.
Fengler, A.I. and R.R. Marquardt. 1988a. Water-soluble pentosans from rye: I. Isolation, partial purification, and characterization. Cereal Chem. 65:291–297.
Fengler, A.I. and R.R. Marquardt. 1988b. Water-soluble pentosans from rye: II. Effects on rate of dialysis and on the retention of nutrients by the chick. Cereal Chem. 65:298–302.
Fernandez, R., E. Lucas, and J. McGinnis. 1973. Fractionation of a chick growth depressing factor from rye. Poult. Sci. 52:2252–2259.
Gohl, B., S. Alden, K. Elwinger, and S. Thomke. 1978. Influence of β–glucanase on feeding value of barley for poultry and moisture content of excreta. Br. Poult. Sci. 19:41–47.
Gordon, D.T. 1990. Total dietary fiber and mineral absorption. In: Dietary Fiber (D. Kritchevsky, C. Bonfield, and J. W., Anderson, ed.) pp. 105–128. Plenum Press, NY.
Graham, H., J.G. Fadel, C.W. Newman, and R.K. Newman. 1989. Effect of pelleting and beta-glucanase supplementation on the ileal and fecal digestibility of a barley based diet in the pig. J. Anim. Sci. 67:1293–1298.
Graham, H., K. Hesselman, E. Jonsson, and P. Åman. 1986. Influence of beta-glucanase supplementation on digestion of a barley-based diet in the pig gastrointestinal tract. Nutr. Rep. Int. 34:1089–1096.
GrootWassink, J.W.D., G.G. Campbell, and H.H. Classen. 1989. Fractionation of crude pentosanase (arabinoxylanase) for improvement of the nutritional value of rye diets for broiler chickens. J. Sci. Food Agric. 46:389–400.
Halpin, J.G., C.E. Holmes, and E.B. Hart. 1936. Rye as a feed for poultry. Poult. Sci. 15: 3–8.
Han, M.S. and J.A. Froseth. 1992. Composition of pearling fractions of barleys with normal and waxy starch. Proc. W. Sec. Amer. Soc. Anim. Sci. 43:155–158.
Henry, R.J. 1985. A comparison of the non-starch carbohydrates in cereal grains. J. Sci. Food Agric. 36:1243–1253.
Hesselman, K. and P. Åman. 1986. The effect of β–glucanase on the utilization of starch and nitrogen by broiler chickens fed on barley of low- or high-viscosity. Anim. Feed Sci. Tech. 15:83–93.

Hesselman, K., K. Elwinger, and S. Thomke. 1982. Influence of increasing levels of β–glucanase on the productive value of barley diets for broiler chickens. Anim. Feed Sci. Tech. 7:351–358.

Hesselman, K., K. Elwinger, M. Nilsson, and S. Thomke. 1981. The effect of β-glucanase supplementation, stage of ripeness and storage treatment of barley in diets fed to broiler chickens. Poult. Sci. 60:2664–2671.

Hesselman, K. and S. Thomke. 1982. Influence of some factors on development of viscosity in the water extract of barley. Swed. J. Agric. Res. 12:17–22.

Hockett, E.A., C.F. McGuire, C.W. Newman, and N. Prentice. 1987. The relationship of barley beta-glucan content to agronomic and quality characteristics. In: (S. Yasuda and T. Konishi, ed.). Barley genetics V (Proceedings of the 5th International Barley Genetics Symposium), Sanyo Press, Okayama, Japan, pp. 851–60.

Inborr, J. 1989. The use of supplementary enzymes in pig nutrition. Proc. of the 25th Annual Nutrition Conference for Feed Manufacturers, Toronto, Ontario. pp. 32–44.

Inborr, J. and H. Graham. 1991. Effect of enzyme supplementation of wheat-based diets on performance of broiler chickens. Proc. Aust. Poult. Sci. Sym., pp. 50–55.

Inborr, J., M. Schmitz, and F. Ahrens. 1993. Effect of adding fibre and starch degrading enzymes to a barley/wheat based diet on performance and nutrient digestibility in different segments of the small intestine of early weaned pigs. Anim. Feed Sci. Tech. 44:113–127.

Jacobsen, J.V., R.B. Knox, and N.A. Pyliotis. 1971. The structure and composition of aleurone grains in the barley aleurone layer. Planta 101:189.

Jensen, L.S., R.E. Fry, J.B. Allred, and J. McGinnis. 1957. Improvement in the nutritional value of barley for chicks by enzyme supplementation. Poultry Sci. 36:919–921.

Jongbloed, A.W., P.A. Kemme, and Z. Mroz. 1990. The effect of *Aspergillus niger* phytase in diets for pigs on concentration and apparent digestibility of dry matter, total phosphorus and inositol phosphates in different sections of the alimentary tract. Report no. 221. Research Institute for Livestock Feeding and Nutrition, Lelystad, The Netherlands.

Ketaren, P.P., E.S. Batterham, and D.J. Farrell. 1991. Recent advances in the use of phytase enzyme in diets for growing pigs. In: Recent Advances in Animal Nutrition in Australia. (D.J. Farrell, ed.).

Lee, B.D. and L.D. Campbell. 1983. Influence of rye and dietary salt level on water and sodium metabolism in intact and colostomized roosters. Poult. Sci. 62:472–479.

Lei, X.G., P.K. Ku, E.R. Miller, and M.T. Yokoyama. 1991. Improvement of phytate phosphorus utilization by a microbial phytase in weanling pigs. J. Anim. Sci. 69(suppl.1):374.

Lott, J.N.A. 1984. Accumulation of seed reserves of phosphorus and other minerals. In: Seed Physiology (D.R. Murray, Ed.) pp 139–166, Academic Press, NY.

Mares, D.J. and B.A. Stone. 1973. Studies on wheat endosperm. I. Chemical compositions and ultrastructure of the cell walls. Aust. J. Biol. Sci. 26:793–812.

Marquardt, R.R., A.T. Ward, and R. Misir. 1979. The retention of nutrients by chicks fed rye diets supplemented with amino acids and penicillin. Poult. Sci. 58:631–640.

McCance, R.A. and E.M. Widdowson. 1942. Mineral metabolism of healthy adults on white and brown bread dietaries. J. Physiol. Sci. 101:44–85.

Misir, R. and R.R. Marquardt. 1978a. Factors affecting rye (*Secale cereale* L.) utilization in growing chicks. I. The influence of rye level, ergot and penicillin supplementation. Can. J. Anim. Sci. 58:691–701.

Misir, R. and R.R. Marquardt. 1978b. Factors affecting rye (*Secale cereale* L.) utilization in growing chicks. II. The influence of protein type, protein level and penicillin. Can. J. Anim. Sci. 58:703–715.

Misir, R. and R.R. Marquardt. 1978c. Factors affecting rye (*Secale cereale* L.) utilization in growing chicks. III. The influence of milling fractions. Can. J. Anim. Sci. 58:717–730.

Misir, R. and R.R. Marquardt. 1978d. Factors affecting rye (*Secale cereale* L.) utilization in growing chicks. IV. The influence of autoclave treatment, pelleting, water extraction and penicillin supplementation. Can. J. Anim. Sci. 58:731–742.

Møllgaard, H. 1946. On phytic acid, its importance in metabolism and its enzymic cleavage in bread supplemented with calcium. Biochem. J. 40:589–603.

Moran, E.T. 1982. Starch digestion in fowl. Poult. Sci. 61:1257–1267.

Moran, E.T., S.P. Lall, and J.D. Summers. 1969. The feeding value of rye for the growing chick. Effect of enzyme supplements, antibiotics, autoclaving and geographical area of production. Poult. Sci. 48:939–949.

Mroz, A., A.W. Jongbloed, P.A. Kemme, and N.P. Lenis. 1991. Ileal and overall digestibility of nitrogen and amino acids in a diet for pigs as influenced by *Aspergillus niger* phytase and feeding frequency or levels. 6th Int. Symp. Protein Metabolism and Nutrition, Herning DK.

Nelson, T.S., 1967. The utilization of phytate phosphorus by poultry – A review. Poult. Sci. 46:862–871.

Nelson, T.S., T.R. Shieh, R.J. Wodzinski, and J.H. Ware, 1971. Effect of supplemental phytase on the utilization of phytate phosphorus by chicks. J. Nutr. 101:1289–1294.

Nelson, T.S., L.W. Ferrara, and N.L. Storer. 1968. Phytate phosphorus content of feed ingredients derived from plants. Poult. Sci. 47:1372.

Newman, C.W., R.F. Eslick, J.W. Pepper, and A.M. El Negoumy. 1980. Performance of pigs fed hulless and covered barleys supplemented with or without a bacterial diastaste. Nutr. Rep. Int. 22:833–837.

Newman, C.W., R.F. Eslick, and A.M. El Negoumy. 1983. Bacterial diastaste effect on the feed value of two hulless barleys for pigs. Nutr. Rep Int. 28:139–146.

Newman, C.W., K.S. Bryan, and R.K. Newman. (1993). Improvement of pig performance with supplemental enzymes and microbial fermentation products. J. Anim. Sci. (Suppl. 1) 71:163.

Newman, R.K. and C.W. Newman. 1987. β–Glucanase effect on the performance of broiler chicks fed covered and hull-less barley isotypes having normal and waxy starch. Nutr. Rep. Int. 36:693–699.

Newman, R.K. and C.W. Newman. 1988. Nutritive value of a new hulless barley cultivar in broiler chick diets. Poul. Sci. 67:1573–1579.

O'Dell, B.L., A.R. de Boland, and S.R. Koirtyohann. 1972. Distribution of phytate and nutritionally important elements among the morphological components of cereal grains. J. Agric. Food Chem. 20:718–721.

Perney, K.M., A.H. Cantor, M.L. Straw, and K.L. Herkelman. 1993. The effect of dietary phytase on growth performance and phosphorus utilization of broiler chicks. Poul. Sci. 72:2106–2114.

Pernollet, J.C. 1978. Protein bodies of seeds: ultrastructure, biochemistry, biosynthesis and degradation. Phytochem. 17:1473–1480.

Petterson, D and P. Åman. 1988. Effects of enzyme supplementation of diets based on wheat, rye or triticale on their productive value for broiler diets. Anim. Feed Sci. Technol. 20: 313–324.

Petterson, D., H. Graham, and P. Åman, 1990. Enzyme supplementation of broiler chicken diets based on cereals with endosperm cell walls rich in arabinoxylans or mixed-linked β–glucans. Anim. prod. 51:201–207.

Petterson D. and P. Åman. 1989. Enzyme supplementation of a poultry diet containing rye and wheat. Br. J. Nutr. 62:139–149.

Pettersson, D., H. Graham, and P. Åman. 1987. The productive value of whole and dehulled oats in broiler chicken diets and influence of beta-glucanase supplementation. Nutr. Rep. Int. 36:743–750.

Pointillart, A., N. Fontaine, and M. Thomasset. 1984. Phytate phosphorus utilization and intestinal phosphatases in pigs fed low phosphorus: wheat or corn diets. Nutr. Rep. Int. 29:473–483.

Pointillart, A., A. Fourdin, and N. Fontaine. 1987. Importance of cereal phytase activity for phytate phosphorus utilization by growing pigs fed diets containing triticale or corn. J. Nutr. 117:907–913.

Power, R. and N. Kahn, 1993. Phytase: The limitations to its universal use and how biotechnology is responding. In: Biotechnology in the Feed Industry. (T. P. Lyons, Ed.) pp 355–368. Alltech Publications, Nicholasville, KY.

Raboy, V. 1990. 5. Biochemistry and genetics of phytic acid synthesis. In: Inositol Metabolism in Plants, pp 55–76. Wiley-Liss, Inc.

Rickes, E.L., E.A. Ham, E.A. Moscatelli, and W.H. Ott, 1962. The isolation and biological properties of beta-glucanase from *B. subtilis*. Arch. Biochem. Biophys. 96:371–375.

Rotter, B.A., R.R. Marquardt, W. Guenter, C. Biliaderis, and C.W. Newman. 1989. In vitro viscosity measurements of barley extracts as predictors of growth responses in chicks fed barley-based diets supplemented with a fungal enzyme preparation. Can. J. Anim. Sci., 69:431–439.

Sauveur, B., 1989. Phosphore phytique et phytases dans l'alimentation des volailes. INRA Prod. Anim. 2:5:343.

Saylor, W.W., A. Bartnikowski, and T.L. Spencer. 1991. Improved performance of broiler chicks fed diets containing phytase. Poult. Sci. 70(suppl. 1):104 (Abstr.)

Schoeff, R.A., and D.J. Castaldo. 1992. Market data: 1992. Feed Management, Oct. pp 6–24.

Schwarz, G. and P.P. Hoppe. 1992. Phytase enzyme to curb pollution from pigs and poultry. Feed Magazine 1/92, pp 22–26.

Sears, A. and G. Walsh, 1993. Industrial enzyme applications: Using these concepts to match animal, enzyme and substrate in feed industry applications. In: Biotechnology in the Feed Industry. (T. P. Lyons, ed.) pp 373–394. Alltech Technical Publications, Nicholasville, KY.

Simons, P.C.M. and H.A.J. Vereestgh. 1990. Phytase in feed reduces phosphorus excretion. Poultry – Misset. June/July:15–17.

Simons, P.C.M., H.A.J. Versteegh, A.W. Jongbloed, P.A. Kemme, P. Slump, K.D. Bos, M.G.E. Wolters, R.F. Beudeker, and G.J. Verschoor, 1990. Improvement of phosphorus availability by microbial phytase in broilers and pigs. Brit. J. Nutr. 64:525–540.

Southgate, D.A.T. 1987. Minerals, trace elements and potential hazards. Amer. J. Clin. Nutr. 45:1256–1266.

Thacker, P.A., G.L. Campbell, and J.W.D. GrootWassink. 1988. The effect of beta-glucanase supplementation of the performance of pigs fed hulless barley. Nutr. Rep. Int. 38:91:99.

Thacker, P.A., G.L. Campbell, and J.W.D. GrootWassink. 1989. The effect of sodium bentonite on the performance of pigs fed barley-based diets supplemented with beta-glucanase. Nutr. Rep. Int. 40:613–619.

Thacker, P.A., G.L. Campbell, and J.W.D. GrootWassink. 1992. The effect of salinomycin and enzyme supplementation on nutrient digestibility and the performance of pigs fed barley- or rye-based diets. Can. J. Anim. Sci. 72:117–125.

Ullrich, S.E., J.A. Clancy, R.F. Eslick, and R.C.M. Lance. 1986. β–glucan content and viscosity of extracts from barley. J. Cereal Sci. 4:279–285.

U.S.D.A. 1992. Agricultural Statistics. United States government Printing office, Washington D. C.

Wang, L., R.K. Newman, C.W. Newman, P.J. Hofer, and A.I. Fengler. 1992. Barley β–glucans alter intestinal viscosity and reduce plasma cholesterol concentration in chicks. J. Nutr. 12:2292–2297.

Wieringa, G.W. 1967. On the occurrence of growth inhibiting substances in rye. Publ. 156. Institute for storage and processing of agriculture produce: Wageningen, The Netherlands.

Willingham, H.E., L.S. Jensen, and J. McGinnis, 1959. Studies on the role of enzyme supplements and water treatment for improving the nutritional value of barley. Poultry Sci. 38:539–544.

Xue, Q., R.K. Newman, C.W. Newman, and C.F. McGuire. 1991. Waxy gene effect on β–glucan, dietary fiber content and viscosity of barleys. Cereal Res. Comm. 19:399–404.

Young L.G., M. Leunissen, and J.L. Atkinson. Addition of microbial phytase to diets of young pigs. J. Anim. Sci. 71:2147–2150.

REGISTRATION OF ENZYMES AND BIOLOGICAL PRODUCTS AROUND THE WORLD: CURRENT ANALYTICAL METHODS FOR FEED ENZYMES AND FUTURE DEVELOPMENTS

RONAN POWER and GARY WALSH

European Biosciences Research Centre of Excellence, The National University of Ireland, Galway, Ireland.

Summary

Enzymes represent a valuable tool to the animal feed industry as a natural means of improving feed utilization and controlling pollution through reducing animal wastes. This is reflected by the rapidly increasing use of enzymes by the industry, particularly in pig and poultry applications. However, this high profile status has resulted in a drive by regulatory agencies worldwide to legislate for the control of such products in terms of quality, safety and efficacy. One of the more important consequences of such legislation for manufacturers is that they have to devise accurate and routine methodologies for the measurement of enzyme activity in feed. The difficulties associated with this task are significant and may be compared to similar problems associated with the detection of vitamins in feed, the analytical procedures for which are still questionable after almost 25 years of development. However, in-feed analysis of enzyme preparations is the topic of much concentrated research and considerable progress has already been made in this area.

Introduction

The potential benefit of using enzymes as feed additives was demonstrated over 35 years ago by researchers at Washington State University who found that soaking cereal grains in water prior to feeding led to significant improvements in liveweight gain and litter quality in broiler chicks. These improvements were correctly attributed to the release of endogenous cereal enzymes during the soaking process and the researchers noted that the effect could be mimicked by the addition of crude microbial amylase preparations to such grains. However, it was not until the 1980s that the use of enzymes in monogastric animal diets became widespread and it is only in the past 5–6 years that intensive adoption of this technology has been noted. This is perhaps best illustrated by figures from the United Kingdom which show that the percentage of broiler feed receiving enzyme treatment rose from almost zero in 1988 to 95% in 1993. One result of this veritable explosion in enzyme utilization within the animal feed industry has been the intensification of efforts by the relevant regulatory authorities around the world to control the use of these additives and to ensure that spurious or ineffective preparations are not allowed in the marketplace.

Representative legislation worldwide

THE US AND CANADA

To obtain an impression of the variability of worldwide registration requirements for microbially-derived products such as enzymes destined for use in animal feedingstuffs, one could choose the United States, Canada and the European Union as good examples. The U.S. is one of the least stringent countries in terms of requirements for enzyme authorization or registration and does not actually have specific legislation for enzymes in animal nutrition. Various enzymes have been used for different applications since 1959 and enjoy a unique position as they benefit from the GRAS (Generally Recognized As Safe) listing system agreed upon between the industry and regulatory authorities. However, the FDA does intend to draw up a list of enzymes approved for use in animal feed which will be based primarily on safety criteria that must be provided by the manufacturer.

Enzyme preparations in Canada are considered as feed ingredients and are listed in the Feed Regulation (P.C. 1983.214) under the "Fermentation Products" group. In 1993 this group included 35 products for which a corresponding description is given. As in the U.S., the labelling requirements for these products include a guarantee of enzyme activity and additional guarantees for minimum crude protein, maximum crude protein from non-protein nitrogen, maximum crude fiber and maximum moisture. Furthermore, enzyme preparations must be stated to be free of anti-microbial activity and not to be a source of viable microbial cells. In terms of registering novel products not already included in the "Fermentation Products" group, the required information is much more detailed. Companies must submit to the Canadian authorities details of the product formulation, enzyme activity guarantees and appropriate methods to quantitatively detect the enzyme preparation in complete feedingstuffs. Additionally, information on the safety and efficacy of the product is required.

EUROPE

The European Union (formerly the European Community) has passed quite strict legislation governing the use of enzymes and microorganisms in animal feedingstuffs. Under this new legislation, enzymes and microorganisms become covered by the definition of feed additives stated in Directive 70/524/EEC concerning additives in feedingstuffs. This definition is as follows:

"Additives: Substances or preparations containing substances which, when incorporated into feedingstuffs, are likely to affect their characteristics or livestock production".

As such, these preparations must be registered in the same manner as any other feed additive used in the E.U. in that a detailed dossier on the preparation must be submitted to the relevant Directorate for consideration. The guidelines for the compilation of such registration dossiers are detailed in another Directive (87/153/EEC) and are very stringent in their demands for the proof of quality, safety and efficacy of the additive. Once a dossier has been completed, it can begin the long process of assessment depicted in Figure 1. In addition to the modifications which have been made to the existing Directives to allow the inclusion of enzymes and microorganisms, a provisional "Identification Note" scheme has been adopted to allow the marketing and use of these products at member state level while full dossiers are being compiled. This provisional measure is aimed at gaining basic information about the 100 or so enzyme and microorganism products currently being used in the E.U. in an effort to ensure that the products are of a suitable quality and that they do not represent a danger to human or animal health (on the basis of the available information).

Furthermore, strict labelling requirements will be enforced throughout the E.U. concerning the indication of enzyme activity on the label of the additive or the label of premixtures and feedingstuffs containing the

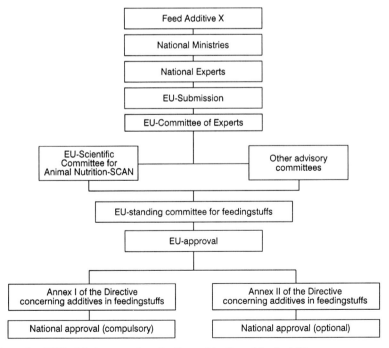

Figure 1. EU legislative procedure for the registration of feed additives

additive. The specific name of the active constituent(s) according to its (their) enzymatic activity(ies), the identification number(s) according to the International Union of Biochemistry and the activity units (activity units per g or activity units per ml). Furthermore, products containing microorganisms must declare the identification of the strain(s), the depositing number(s) of the strain(s) and the number of colony-forming units (CFU/g).

From this brief overview of current or impending legislation in representative countries, therefore, it can be seen that manufacturers of fermentation products such as enzymes are faced with the rather difficult task of furnishing the required information to the registration authorities within a given time limit. From a technical standpoint, one of the most difficult challenges will be the provision of routine methods of analysis for such additives in complete feedingstuffs. The scope of this problem is one which merits some attention and represents the topic for the rest of this paper.

Assay of enzyme activity

Direct detection of enzyme activity in animal feed is desirable from more than just a regulatory point of view. Suitable methods allowing direct detection and quantification of enzymes in feedstuffs would allow more rigorous quality assurance on finished feeds. This is desirable not only from the point of view of the compounder but also from the point of view of the consumer. Furthermore, consumer groups could then verify that the enzyme activities included in the feed were indeed present in the quantities specified. Direct detection of enzyme activity in finished feed formulations would also allow proper evaluation of the effect of processes such as pelleting on enzyme activity.

Increased awareness of the importance of detecting and quantifying enzyme activity in feed highlights a critical consideration which has not often been sufficiently emphasised in the past. Enzymes are biological catalysts and achieve their effect by virtue of their biological activity measured in units. As the effectiveness of enzymes is a factor of activity, it is the total activity incorporated rather than the weight of enzyme incorporated in the feed which dictates how successful inclusion of the enzyme will be in achieving pre-defined goals.

Enzyme activity is normally measured or "assayed" by one of two methods: (a) monitoring the disappearance of the enzyme's substrate over a given time, or; (b) the production of product may be assessed over a given time, as is indicated in Figure 2. This latter approach is the one usually adopted with units of activity per gram of enzyme often being expressed as the amount of product formed per unit time (i.e. often μmoles of product/minute) at a defined temperature.

Assay of concentrated enzyme preparations which are destined for subsequent inclusion in feed is generally uncomplicated and is normally accomplished using the enzyme's natural substrate. The high levels of activity associated with such enzyme preparations coupled with the fact

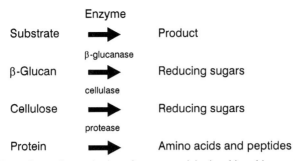

Figure 2. Detection and quantitation of enzyme activity is achieved by monitoring either disappearance of substrate or more often appearance of product, over a given time period and under defined conditions of pH and temperature. Reducing sugars can be detected and quantified by reaction with dinitrosalicylic acid (DNS). Amino acid and peptides are usually quantified by measurement of absorbance at 280 nm

that generally no substances are present which can interfere with the assay procedure renders such assay procedures straightforward. Assays of enzyme activity subsequent to the enzyme's incorporation into feedstuffs has proven more technically challenging for a number of reasons.

IN FEED ASSAYS: TECHNICAL DIFFICULTIES

Most enzyme activities achieve their intended effect even when added to feed at low levels. Inclusion of enzyme preparations at levels of 1 kg/tonne feed obviously has a considerable dilutive effect on enzyme activity. This may be even more pronounced if an enzyme cocktail consisting of two or more activities is added at the same overall level. In order to assay the in-feed enzyme, the enzyme must be extracted from the feed which is usually achieved by agitation in the presence of up to 10 volumes of buffer. This effectively represents an additional dilution of enzyme activity. Detection of low levels of activity thus require sensitive assay systems and prolonged assay incubation times.

Another difficulty in relation to development of in-feed assays relates to the presence in such feed of substances which impede detection of activity. Assay of enzymes such as β–glucanase, pentosanase and cellulase are normally based upon detection of reducing sugars produced as reaction products. Reducing sugars produced may be conveniently estimated by reaction with dinitrosalicylic acid. All feedstuffs naturally contain very high levels of such reducing sugars which are extracted along with the enzyme in the extraction buffer. Detection of reducing sugars produced in a subsequent enzyme assay is rendered impractical due to very large background sugar values.

APPROACHES TO IN-FEED ASSAY DEVELOPMENT

Various approaches have been adopted in attempts to overcome such difficulties. These include prolonged assay incubation times to allow detection of even low levels of activity, the use of artificial substrates which yield reaction products not naturally present in feed, removal of interfering substance prior to assay commencement, and detection of enzyme activity by monitoring disappearance of reaction substrate rather than production of product.

Various immunological approaches have also been attempted with very limited success. Such immunological approaches are based upon the use of antibodies in immunoassay systems designed to detect and quantify specific enzymes. The main obstacle seems to relate to the complex nature of the enzyme-containing feed which contains a multitude of constituents. Enzymes added will represent a small fraction of such substances. Some of the additional feed ingredients can interfere with the specificity of the immunological assay which makes detection and accurate assessment of enzyme levels present difficult if not impossible.

An additional drawback of this technique is its inability to assess the biological activity of the enzyme molecules. Immunoassays will only detect the presence of enzyme protein. This does not automatically correlate to biological activity. In many cases, enzyme molecules which have been inactivated may be detected by immunoassay though they may be catalytically inactive. Hence quantification by immunological means may not always reflect product potency.

Direct detection of catalytic activity
The use of synthetic substrates yielding easily distinguishable products to detect enzyme activity in feed is worthy of serious consideration. Synthetic substrates have been developed which may be used in the assay of a variety of enzymes (Table 1). Most such substrates contain a coloured label (chromogenic group) which is released by the action of the enzyme (Biely et al., 1985; Biely et al, 1988; McCleary and Shameer, 1987). The released colour may be readily detected by spectrophotometric methods subsequent to termination of the assay.

Several laboratories have developed or are in the process of developing feed enzyme assays based upon the utilization of a chromogenic substrate. While this facilitates detection of enzyme activity, there are a number of potential drawbacks. The synthetic substrate by definition does not represent the native substrate of the enzyme. Enzyme activity as

Table 1. Some synthetic chromogenic substrates and the corresponding enzymes which may be assayed using such substrates.

Enzyme	Chromogenic substrate
Protease	Azoalbumin
Cellulase	Cellulose coupled to remazol brilliant blue (RBB)
Hemicellulase	RBB–xylan
β–glucanase	Azo β–glucan
Amylase	Starch coupled to ostazin brilliant red

expressed towards the synthetic substrate may differ from that detected with the native substrate. In addition, enzyme extracts from feed would be expected to contain a quantity of native substrate. The possibility of competition between native and artificial substrates for catalytic conversion could generate complications if extraction of feeds of different composition yielded varying concentrations of the native substrate.

An additional approach relates to the removal of the native enzyme products present naturally in the enzyme-containing feed extract prior to commencement of enzyme assay. Most such substances are of low molecular weight and studies in our laboratory have shown that such substances may be easily and quickly removed by application of a gel filtration chromatographic step using Sephadex G-25. Application of a complex mixture to such a column results in the separation of the constituent molecules on the basis of their molecular weights. Molecules of high molecular weight such as enzymes are eluted from the column quickly while low molecular weight substances such as glucose and low molecular weight oligosaccharides are retained on the column for a longer period of time. Using this method, it is possible to recover virtually 100% of enzyme activity in the void volume while separating the enzyme from substances, generally endogenous products, which interfere with the detection of enzyme activity (Figure 3). The enzyme extract, now devoid of reducing sugars and allied substances which interfere with conventional assay methods may then be assayed by such methods.

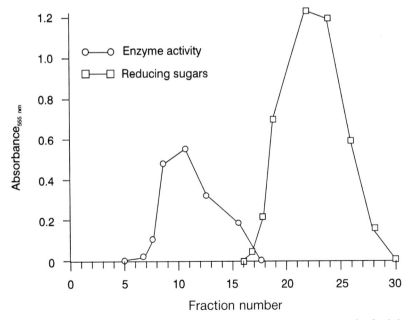

Figure 3. Separation of enzyme activity from reducing sugars present in feed by chromatography on Sephadex G-25

Radial diffusion of enzyme in an agar matrix

Perhaps the approach which has met with most success to date relates to assay of enzyme activity by monitoring disappearance of reaction substrates by methods of radial enzyme diffusion (RED). This approach entails the incorporation of the enzyme substrate into an agar gel, with subsequent introduction of the enzyme-containing feed extract into wells which have been generated in the gel. The enzyme then diffuses outward through the gel, and when incubated under the appropriate conditions will degrade the substrate contained therein. This method of detecting enzyme activity has been in existence since the 1950s (Dingle et al., 1953) and may be used to quantitate enzyme activity as the diameter of the zone of substrate degraded is proportional to the log of the enzyme activity present.

A prerequisite to the utilization of such an assay system is obviously that a clear visual distinction between degraded and undegraded substrate is evident. The finding of Wood (1981) that the dye Congo red interacts strongly with polysaccharides exhibiting contiguous $\beta(1\rightarrow 4)$ linked D–glucopyranosyl residues provides the basis for the development of sensitive assays for detection of enzyme activities such as cellulase and β–glucanase.

In the case of cellulase and β–glucanase, the agar plate employed contains carboxymethyl cellulose (CMC) and Lichenan respectively. Diffusion of cellulase or β–glucanase activity from wells in these gels results in enzyme-mediated substrate degradation. Subsequent staining with Congo red produces unstained circular zones around the wells where the substrate has been utilized. This contrasts strongly with the purple background of the undigested substrate. Other activities such as amylase and protease may be detected by similar methods and employing suitable stains in order to visualize the zones of substrate hydrolysis.

Quantitation of enzyme activity in finished feed basically involves spiking samples of feed devoid of added enzyme (i.e. control feed) with increasing amounts of the enzyme in question – normally within the range equivalent to from 0 to 2 kg enzyme/tonne. The enzyme is then extracted from the spiked feed and assayed by the plate method. A standard curve may be prepared by plotting the diameter of the zone of hydrolysis of the various samples versus the log of the enzyme activity present. A straight line results. Feed containing an unknown quantity of enzyme may be extracted and assayed in an identical fashion and the level of enzyme added may be calculated by reference to the standard curve. A photographic representation of an assay plate in which the extract samples for both a standard curve and an unknown sample were run, is illustrated in Figure 4. A typical standard curve as obtained for an in-feed β–glucanase enzyme in our own laboratory is illustrated in Figure 5.

While such an assay system is feasible and has been found to yield satisfactory results, it is characterized by at least one drawback; the requirement of a feed sample devoid of enzyme activity which can be used in the construction of a standard curve. Except for the lack of added enzyme, this feed sample should be identical to the sample of feed which is to be tested as this allows determination of any background enzyme activity which may be present naturally in that particular feed.

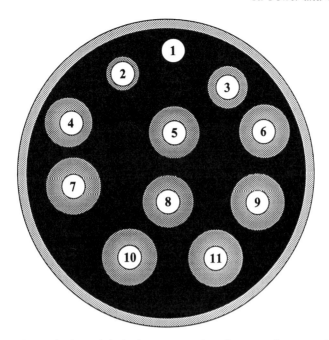

Figure 4. Determination of in-feed enzymes using the zone diameter of substrate hydrolysed. Well No.1: Buffer, No.2: Feed extract devoid of enzyme, Wells No.3–7: Spiked with various levels of enzyme (0.25, 0.5, 1.0, 1.5, 2.0 kg/tonne). Wells No.8–11 were loaded with an extract obtained from a commercial feed sample to which enzyme had been added by the manufacturer at a rate of 1 kg/tonne (see also frontispiece)

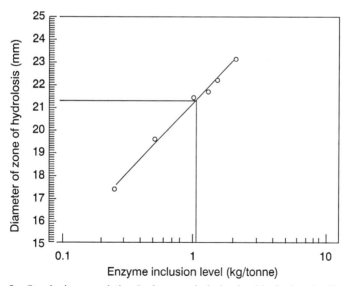

Figure 5. Standard curve relating β–glucanase inclusion level in feed to the diameter of the zone of substrate hydrolysis. Using the standard curve constructed, the inclusion level of β–glucanase in a commercially produced enzyme-containing feed may be determined. In this case the inclusion level was found to be 1.03 kg/tonne, i.e. 103% of the specified inclusion level

Many feeds will have low levels of such activity. However, the technique is sensitive, technically straightforward, inexpensive to perform, requires no sophisticated laboratory equipment, and it works.

Conclusions

Most countries are currently enacting specific legislation with regard to the inclusion of enzymes in animal feed. Enzyme dossiers which must be submitted to the appropriate regulatory authority must illustrate the safety and efficacy of the enzyme in question. Demonstration of the presence of the active substance, in the quantity specified in the finished feed is also required. Central to this is the development of a suitable assay system which can detect and quantify enzyme activity in finished feed.

Estimating enzyme activity in finished feed formulations has proven to be technically challenging. A number of different strategies have been pursued in this regard. Perhaps the approach which has yielded the most encouraging results to date entails assay by radial enzyme diffusion. This method is based upon the catalytic conversion of substrate by enzyme as it diffuses through a substrate containing gel. The method is quantitative in nature and results from our own laboratory confirm that it is suited to assay of enzymes in finished feed preparations.

References

Biely, P., Markovic, O., and Mislovicova, D., (1985). Anal. Biochem., 144, 147–151.
Biely, P., Mislovicova, D., Markovic, O., and Kalac, V., (1988), Anal. Biochem., 172, 176–179.
McCleary, B. and Shameer, I., (1987). J. Inst. Brew. 93, 87–90.
Dingle, J., Reid, W. and Solomon, E., (1953). The enzymatic degradation of pectin and other polysaccharides. Application of the cup plate assay to the estimation of enzymes. J. Sci. Food Agric. 4: 149–155.
Wood, P.J., (1981). The use of dye-polysaccharide interactions in β–glucanase assay. Carbohydr. Res. 94: C19.

USE OF NOVEL BIOMOLECULES TO ACTIVATE AND MAXIMIZE HEALTH AND PRODUCTION

DE NOVO DESIGNED SYNTHETIC PLANT STORAGE PROTEINS: ENHANCING PROTEIN QUALITY OF PLANTS FOR IMPROVED HUMAN AND ANIMAL NUTRITION

JESSE MICHAEL JAYNES
Demeter Biotechnologies, Ltd., Research Triangle Park, North Carolina USA

Background

The composition of storage proteins, a major food reservoir for the developing seed, determines the nutritional value of plants and grains when they are used as foods for man and domestic animals. The amount of protein varies with genotype or cultivar, but in general, cereals contain 10% of the dry weight of the seed as protein, while in legumes, the protein content varies between 20% and 30% of the dry weight. In many seeds, the storage proteins account for 50% or more of the total protein and thus determine the protein quality of seeds. Each year the total world cereal harvest amounts to some 1,700 million tons of grain (Keris et al., 1985). This yields about 85 million tons of cereal storage proteins harvested each year and contributes a majority of the total protein intake of humans and animals.

With respect to human and animal nutrition, most seeds do not provide a balanced source of protein because of deficiencies in one or more of the essential amino acids in the storage proteins. For example, humans require from foods, eight amino acids: isoleucine, leucine, lysine, methionine, phenylalanine, threonine, tryptophan and valine, to maintain a balanced diet. Consumption of proteins of unbalanced amino acid composition can lead to a malnourished state which is most often found in children in developing countries where plants are the major source of protein intake. Therefore, the development of nutritionally-balanced proteins for introduction into plants is of extreme importance.

Recently, many laboratories have attempted to improve the nutritional quality of plant storage proteins by transferring heterologous storage protein genes from other plants (Pederson et al., 1986). The development of recombinant DNA technology and the Agrobacterium-based vector system has made this approach possible. However, genes encoding storage proteins containing a more favorable amino acid balance do not exist in the genomes of major crop plants. Furthermore, modification of native storage proteins has met with difficulty because of their instability, low level of expression, and limited host range. One possible alternative would be the *de novo* design of a more nutritionally-balanced protein which retains certain characteristics of the natural storage proteins of plants.

Our initial work described the use of small fragments of DNA which encoded spans of protein high in essential amino acids (Jaynes et al., 1985; Yang et al., 1989). Subsequently, the genes encoding these protein domains were cloned into an existing protein and the expression level of this modified protein determined in transgenic potato plants. However, because of some of the problems mentioned above, the results were somewhat less than desirable (Yang et al., 1989).

There are at least two fundamental difficulties in achieving efficient expression of designed proteins. First, it is not yet known what stabilizes a protein against proteolytic breakdown; and second, the mechanisms for folding of an amino acid sequence into a biologically-stable tertiary structure have not yet been fully delineated. For the construction of DNP 1 (Designed Nutritional Protein), we focused on the design of a physiologically-stable as well as a highly nutritious, storage-protein-like, artificial protein. Based on what we have learned in the design and expression of lytic peptides for enhanced disease resistance in plants (Jaynes et al., 1992), it now seemed possible for us to design entirely artificial, stable, nutritionally-significant proteins to improve the nutritive quality of plants.

Amino acid requirements

The biosynthesis of amino acids from simpler precursors is a process vital to all forms of life as these amino acids are the building blocks of proteins. Organisms differ markedly with respect to their ability to synthesize amino acids. In fact, virtually all members of the animal kingdom are incapable of manufacturing some amino acids. There are twenty common amino acids which are utilized in the fabrication of proteins; and essential amino acids are those protein building blocks which cannot be synthesized by the animal. It is generally agreed that humans require eight of the twenty common amino acids in their diet. Protein deficiencies can usually be ascribed to a diet which is deficient in one or more of the essential amino acids. A nutritionally adequate diet must include a minimum daily consumption of these amino acids (Figure 1).

When diets are high in carbohydrates and low in protein over a protracted period, essential amino acid deficiencies result. The name given to this undernourished condition in humans is "Kwashiorcor" which is an African word meaning "deposed child" (deposed from the mother's breast by a newborn sibling). This debilitating and malnourished state, characterized by a bloated stomach and reddish-orange discolored hair, is more often found in children than adults because of their great need for essential amino acids during growth and development. In order for normal physical and mental maturation to occur, the above mentioned daily source of essential amino acids is a requisite. Essential amino acid content, or protein quality, is as important a feature of the diet as total protein quantity or total calorie intake.

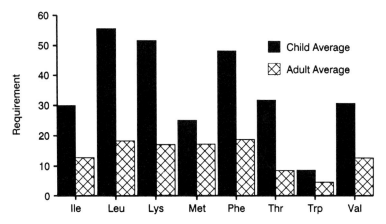

Figure 1. The average essential amino acid requirement for both children and adults in mg per kg body weight. Note that children, on a per weight basis, require more of the essential amino acids than do adults. This indicates the importance of diet to normal physical and mental maturation.

Some foods, such as milk, eggs, and meat, have very high nutritional values because they contain a disproportionately high level of essential amino acids. On the other hand, most foodstuffs obtained from plants possess a poor nutritional value because of their relatively low content of some or, in a few cases, all of the essential amino acids. Generally, the essential amino acids which are found to be most limiting in plants are isoleucine, lysine, methionine, threonine, and tryptophan (MLEAA) (Figure 2).

It has been difficult to produce significant increases in the essential amino acid content of crop plants utilizing classical plant breeding approaches. This is primarily due to the fact that the genetics of plant breeding is complex and that an increase in essential amino acid content may be offset by a loss in other agronomically important

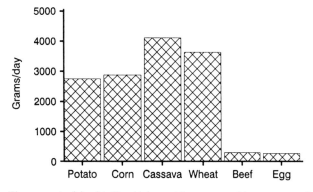

Figure 2. The amount of foodstuffs which must be consumed in grams per day in order to meet the minimum daily requirement of all essential amino acids.

characteristics. Also, it is probable that the storage proteins are very conserved in their structure and their essential amino acid composition would be little modified by these conventional techniques.

Structure and classification of natural storage proteins

Seed storage proteins can be characterized by several main features (Pernollet and Mosse, 1983): 1) their main function is to provide amino acids or nitrogen to the young seedling; 2) the general absence of any other known function; c) their peculiar amino acid composition in cereal and legume seeds; and d) their localization within storage organelles called protein bodies, at least during seed development. Several classes of storage proteins are generally recognized based on their solubilities in different solvents. Proteins soluble in water are called "albumins"; proteins soluble in 5% saline, "globulins"; and proteins soluble in 70% ethanol, "prolamins". The proteins that remain following these extractions are treated further with dilute acid or alkali, and are named "glutelins". Most cereals contain primarily prolamin-type proteins and can be classified into different groups on the basis of the relative proportions of prolamins, glutelins, and globulins, and the subcellular location of these proteins in the mature seed. The first group corresponds to the Panicoidae sub-family, the second group the Triticeae tribe, and the last one to oat and rice storage proteins.

The principal members of the Panicoideae sub-family are maize, sorghum, and millet. Their major storage proteins are prolamins (50 to 60% of seed protein) and glutelins (35 to 40% of seed protein; Pernollet and Mosse, 1983). Prolamins are stored within protein bodies, but glutelins are located both inside and outside these organelles. The Triticeae tribe which includes wheat, barley, and rye, differ from the Panicoideae mainly in storage protein localization and structure. In the starchy endosperm of the seeds belonging to this tribe, no protein bodies are left at maturity. Clusters of proteins are then deposited between starch granules, but are no longer surrounded by a membrane.

In legumes and most other dicotyledons, the major storage proteins are salt-soluble globulins (80%) and prolamins (10–15%). Globulins can be divided into vicillins and legumins (Agros, 1985), based on their sedimentation coefficient (7S/11S), oligomeric organization (trimeric/hexameric), and polypeptide chain structure (single chain/disulphide-linked pair of chains). In the legume seed cotyledon, protein bodies are embedded between starch granules (Pernollet and Mosse, 1983). They are membrane-bound organelles, a few microns in diameter, mainly filled with storage proteins and phytates. Besides storage proteins, protein bodies also contain other proteins, such as enzymes or lectins, although in lesser amounts.

The structure of soluble globulins was studied more than that of the insoluble prolamins and glutelins. Vicillin appears as a homo- or heterotrimer, sometimes able to associate into hexameric form. Soybean β–conglycin and french bean phaseolin (Bollini and Chrispeels, 1978) are the structurally best known vicillins. Recently, the three-dimensional structure of phaseolin was determined by X-ray crystallographic analysis (Lawrence et al., 1990). However, unlike other vicilins, the phaseolin trimer can associate into a dodecamer (tetramer or trimer) below pH 4.5. Each polypeptide of the trimeric form comprises two structurally similiar units each made up of a β–barrel and an α–helical domain.

Glycinin, the soybean legumin, has a quaternary structure that was suggested by Badley et al. (1975) to be twelve subunits packed in two identical hexagons. In general, the legumin molecule is a polymer formed by the association of six monomers. Each monomer consists of two subunits, acidic and basic. Sometimes these subunits are associated by disulfide linkages. On the other hand, arachin, the peanut legumin, was found to consist of different kinds of subunits. The arachin hexamer association does not need different kinds of subunits, which suggests that the subunits have a very similar structure.

The most studied storage proteins, in terms of structure, are the corn prolamines called zeins. These proteins perform no known enzymatic function. Three types of zeins (α, β and γ) (Esen, 1986) are synthesized on rough endoplasmic reticulum and aggregate within this membrane as protein bodies. The zein protein readily self-associates to form protein bodies and is insoluble in water even in low concentrations of salt. The presence of all types of zeins is not necessary for the formation of a protein body as a single type of zein can aggregate into a dense structure and is generally found at the surface of protein bodies (Lending et al., 1988; Wallace et al., 1988). The mechanism responsible for protein body formation is thought to involve hydrophobic and weak polar interactions between individual zein molecules (Wallace et al., 1988; Agros et al., 1982), while they require a high amount of ethanol in aqueous systems to maintain their strict molecular conformation (Agros et al., 1982).

Circular dichroic measurements, amino acid sequence analysis, and electron microscopy of a zein protein suggests that zein secondary structure to be primarily helical with nine adjacent, topologically anti-parallel helices clustered within a distorted cylinder (Agros et al., 1982; Larkins, 1983; Larkins et al., 1984). Polar and hydrophobic residues appropriately distributed along the helical surfaces allowing intra- and intermolecular hydrogen bonds and van der Waals interactions among neighboring helices, such that rod-shaped zein molecules could aggregate and then stack through glutamate interactions at the cylindrical caps. Because of this structure, zein is much less soluble under physiological conditions than the globulin phaseolin, and precipitation of insoluble zein in the tightly packed protein body may make them less available for proteolytic degradation (Greenwood and Chrispeels, 1985). The storage protein structures are adapted to a maximal packing within protein bodies (Pernollet and Mosse, 1983). Maximal packing is achieved in at least either of two ways. The folding of the polypeptide chain may favor the maximal packing of amino acids within the protein molecule, or the compacting of proteins is increased by the formation of closely packed quaternary structure. High degrees of polymerization can be observed in pearl millet pennisetin (Pernollet and Mosse, 1983) or zein (Lending et al., 1988; Wallace et al., 1988). Also, wheat prolamins and glutelin associate into aggregates arising in the formation of insoluble gluten.

These insoluble forms of protein deposits are osmotically inactive and stable during the long period of storage between the time of seed maturation and germination.

Regulation of storage protein genes

All storage proteins which have been investigated are encoded by multigene families (Bartels and Tompson, 1983; Crouch et al., 1983; Forde et al., 1985; Kasarda et al., 1984; Lycett et al., 1985; Rafalski et al., 1984; Slightom et al., 1983). The structure of these families varies, and in some cases, such as in wheat or barley, two major subgroups can be noted: the α– and γ–gliadins and the B- and C-hordeins, respectively (Forde et al., 1985; Kasarda et al., 1984; Rafalski et al., 1984). Within each subgroup, several subfamilies can be distinguished. Often short repeats account for at least part of the structure of the polypeptides. These repeats constitute links through which different subfamilies within the same species are related.

Storage protein genes, like most other plant genes characterized to date, are transcribed in a regulated rather than a constitutive fashion. Expression is frequently tissue-specific and(or) temporally regulated. Cis-acting DNA sequences involved in developmental and(or) tissue-specific regulation of gene expression can be defined by introducing plant storage protein gene regulatory regions coupled to bacterial reporter genes (Twell and Ooms, 1987; Wenzler et al., 1989, Marries et al., 1988; Chen et al., 1988), or by introducing entire or dissected genes (Colot et al., 1987; Chen et al., 1986) into a transgenic environment. Unfortunately, a transformation system for the nutritionally important cereal species has not yet been well established. Therefore, most regulation mechanisms have been studied with transgenic dicotyledonous plants. However, there is increasing evidence that gene expression is controlled, at least partly, by the interaction between regulatory molecules and short sequences that are present in the 5' flanking region of the gene.

The regulatory sequences of potato storage protein were investigated using transgenic potato plants. A 2.5 kb 5' flanking DNA fragment containing the promoter and the patatin gene was used to construct a transcriptional fusion gene with chloramphenicol acetyl transferase (CAT) or the β–glucuronidase (GUS) gene (Twell and Ooms, 1987; Wenzler et al., 1989). When reintroduced into potato, these chimeric genes were expressed in tubers, but not in leaves, stems or roots.

The expression pattern of storage protein genes of cereals is retained in tobacco, not only with respect to tissue, but also to temporal expression. The 5' upstream regions of wheat glutenin genes possess regulatory sequences that determine endosperm-specific expression in transgenic tobacco (Colot et al., 1987). Deletion analysis of the low molecular weight (LMW) glutenin sequence indicated that sequences present between 326bp and 160bp upstream of the transcriptional start point are necessary to confer endosperm-specific expression. Furthermore, cis-acting elements determining the regulation of each gene in the cluster are recognized by the tobacco trans-acting factor. Also cis-acting elements directing expression of one gene do not affect expression of neighboring genes. This was demonstrated by the transfer of a 17.1 kb soybean DNA containing a seed lectin gene with at least four non-seed protein genes to transgenic tobacco plants (Okamuro, 1986). The genes in this cluster were expressed in a manner similar to that in soybean; i.e., the lectin gene products accumulated in seeds, and the other genes were expressed in tobacco leaves, stems, and roots.

The expression of several DNA deletion mutants with a 257 bp 5′ flanking sequence of the α'-conglycin gene indicates that this region contained enhancer like elements (Chen et al., 1986). Only a low level of expression of the α' gene occurred in developing seeds of transgenic plants that contain the α' gene flanked by 159 nucleotides 5′ of the transcriptional start site. However, a 20 fold increase in expression occurred when an additional 98 nucleotides of upstream sequence were included. The DNA sequence between 143 and 257 contained five repeats of the sequence AA(G)CCCA, and played a role in conferring tissue-specific and developmental regulation. The 35S promoter containing this sequence in different positions and different orientations could enhance the expression of the CAT gene by 25 to 40 fold (Chen et al., 1988).

Trans-acting factors directly involved in storage protein gene regulation have not yet been reported. However, in some cases, the level of amino acids can control the expression of storage protein. Vegetative storage protein (VSP) gene expression in leaves, stems and seed pods is closely related to whether these organs are currently a sink for nitrogen or a source for mobilized nitrogen for other organs (Staswick, 1989). The leaves have a sensitive mechanism for detecting changes in sink demand of mobilizing reserves, and VSP gene expression can be rapidly adjusted accordingly. Sequestering excess amino acids in this way may prevent their accumulation to toxic levels.

Genetic engineering using *Agrobacterium tumefaciens*

One of the most significant recent advances in the area of plant molecular biology has been the development of the *Agrobacterium tumefaciens* Ti plasmid as a vector system for the transformation of plants. In nature, *A. tumefaciens* infects most dicotyledonous and some monocotyledonous plants by entry through wound sites. The bacteria bind to cells in the wound and are stimulated by phenolic compounds released from these cells to transfer a portion of their endogenous, 200 kb Ti plasmid into the plant cell (Weiler and Schroder, 1987). The transferred portion of the Ti plasmid, T-DNA, becomes covalently integrated into the plant genome where it directs the biosynthesis of phytohormones using enzymes which it encodes. The vir gene in the bacterial genome is known to be responsible for this process. In addition to vir gene products, directly repeating sequences of 25 bases called "border" sequences are essential, but only the right terminus has been shown to be used for T-DNA transfer and integration.

Expression of the T-DNA gene inside the plant results in the uncontrolled growth of these and surrounding cells, leading to formation of a gall (Weiler and Schroder, 1987). Ti plasmids, from which these disease-producing genes have been removed or replaced, are referred to as "disarmed" and can be used for the introduction of foreign genes into plants. The great size of the disarmed Ti plasmid and lack of unique restriction endonuclease sites prohibit direct cloning into the T-DNA. Instead, intermediate vectors such as pMON237 or pBI121 can be used to introduce genes into the Ti plasmid. Currently, two kinds of vector systems are available as intermediate vectors: cointegrating vectors and binary vectors. A cointegrating transformation vector must include a region of homology between the vector plasmid and the Ti plasmid. Once recombination occurs, the cointegrated plasmid is replicated by the Ti plasmid origin of replication. The cointegrate system, while more difficult to use, does offer advantages. Once the cointegrate has been formed, the plasmid is stable in Agrobacterium.

A binary vector contains an origin of replication from a broad host-range plasmid instead of a region of homology with the Ti plasmid. Since the plasmid does not need to form a cointegrate, these plasmids are considerably easier to introduce into Agrobacterium. The other advantage to binary vectors is that this vector can be introduced into any Agrobacterium host containing any Ti or Ri plasmid, as long as the vir helper function is provided. Using these systems, the gene regulation mechanism of storage proteins has been elucidated.

Improvement of nutritional qualities of plants

The amino acid composition of the cereal endosperm protein is characterized by a high content of proline and glutamine while the amount of essential amino acids, lysine and tryptophan in particular, is a limiting factor (Pernollet and Mosse 1983). In legumes, sulfur containing amino acids such as methionine and cystine are the major limiting essential amino acids for the efficient utilization of plant protein as animal or human food while roots and tubers are deficient in almost all of the essential amino acids.

There has been a great deal of effort to overcome these amino acid limitations by breeding and selecting for more nutritionally balanced varieties. Plants have been mutated in hopes of recovering individuals with more nutritious storage proteins. Neither of these approaches has been very successful, although some naturally occurring and artificially produced mutants of cereals were shown to contain a more nutritionally balanced amino acid composition. These mutations cause a significant reduction in the amount of storage protein synthesized and thereby result in a higher percentage of lysine in the seed; however, the softer kernels and low yield of such strains have limited their usefulness (Pernollet and Mosse, 1983). The reduction in storage protein also causes the seeds to become more brittle; as a result, these seeds shatter more easily during storage. The lower levels of prolamin also result in flours with unfavorable functional properties which cause brittleness in the baked products (Pernollet et al., 1983). Thus, no satisfactory solution has yet been found for improving the amino acid composition of storage proteins.

One direct approach to this problem would be to modify the nucleotide sequence of genes encoding storage proteins so that they contain high levels of essential amino acids. To achieve this aim, several laboratories have tried to modify and express storage proteins in the host plants. Modified storage proteins have been created and expressed by changing their codon sequences. In vitro mutagenesis was used to supplement the sulfur amino acid codon content of a gene encoding β–phaseolin, a *Phaseolus vulgaris* storage protein (Hoffman et al., 1988). The nutritional quality of β–phaseolin was increased by the insertion of 15 amino acids, six of which were methionine. The inserted peptide was essentially a duplication of a naturally occurring sequence found in the maize 15kd zein storage protein (Pederson et al., 1986). However, this modified phaseolin achieved less than 1% of the expression level of normal phaseolin in transformed seeds. Recently it has been found that this insertion was made in part of a major structural element of the phaesolin trimer (Lawrence et al., 1990). Therefore, an inclusion of 15 residues at this site could distort the structure at the tertiary and(or) quaternary level.

Lysine and tryptophan-encoding oligonucleotides were introduced at several positions into a 19kD α–type zein complementary DNA by oligonucleotide-mediated mutagenesis (Wallace et al., 1988). Messenger RNA for the modified zein was synthesized *in vitro* and injected into *Xenopus laevis* oocytes. The modified zein aggregated into structures similar to membrane-bound protein bodies. This experiment suggested the possibility of creating high-lysine corn by genetic engineering.

There are alternative approaches that might be more practical. One of these is to transfer heterologous storage protein genes that encode storage proteins with higher levels of the desired amino acids. For this purpose, a chimeric gene encoding a Brazil nut methionine-rich protein which contains 18% methionine has been transferred to tobacco and expressed in the developing seeds (Altenbach et al., 1989). The remarkably high level of accumulation of the methionine-rich protein in the seed of tobacco results in a significant increase in methionine levels of ~30%.

The maize 15 kd zein structural gene was placed under the regulation of French bean β–phaseolin gene flanking regions and expressed in tobacco (Hoffmann et al., 1987). Zein accumulation was obtained as high as 1.6% of the total seed protein. Zein was found in roots, hypocotyls, and cotyledons of the germinating transgenic tobacco seeds. Zein was deposited and accumulated in the vacuolar protein bodies of the tobacco embryo and endosperm. The storage proteins of legume seeds such as the common bean (*Phaseolus vulgaris*) and soybean (*Glycine max*) are deficient in sulfur-containing amino acids. The nutritional quality of soybean could be improved by introducing and expressing the gene encoding methionine-rich 15 kd zein (Pederson et al., 1986).

A 292 bp synthetic gene (HEAAE I = High Essential Amino Acid Encoding) which encoded a protein domain high in essential amino acid was expressed as a CAT-HEAAE I fusion protein in potato (Jaynes et al., 1986; Yang et al., 1989). However, structural instability limited the high level expression of this fusion protein in the potato system. Also, the content of essential amino acids was diluted to less than 40% of the original encoded protein by constructing this fusion.

There are several precautions that should be considered in engineering storage proteins (Larkins, 1983). First, *in vitro* mutational change must not be in regions of the protein that perturb the normal protein structure; otherwise, the proteins might be unstable. Second, when attempting to increase nutritional quality by introducing a gene encoding a heterologous protein in crop plants, it is important that the protein encoded by an introduced gene does not produce any adverse effects in humans or livestock, the ultimate consumers of the engineered seed proteins (Altenbach et al., 1989). Finally, it is critical that the amino acids present in the introduced protein are able to be utilized by the animal for growth and development.

De novo design of proteins

Recently, a new field in protein research, *de novo* design of proteins, has made remarkable progress due to a better understanding of the rules which govern protein folding and topology. Protein design has two components: the design of activity and the design of structure. This review will concentrate on the design of structurally stable storage protein-like proteins.

The usual approach for the design of helical bundle proteins consists of linking sequences with a propensity for forming an α-helix via short loop sequences to get linear polypeptide chains. This chain can fold into the predetermined 'globular type' tertiary structure in aqueous solution (Mutter 1988; DeGrado et al. 1989). α–Helical secondary structures are stabilized by interatomic interactions that can be classified according to the distance between interacting atoms in the sequence of the protein (DeGrado et al., 1989).

Short range interactions account for different amino acids having different conformational preferences. Both statistical (Chou and Fasman, 1978) and experimental (Sueki et al., 1984) methods show that residues such as Glu, Ala, and Met tend to stabilize helices, whereas residues such as Gly and Pro are destabilizing. However, these intrinsic preferences are not sufficient to determine the stability of helices in globular proteins.

Analysis of the free-energy requirements for helix nucleation and propagation indicates that peptides of 10 to 20 residues should show little helix formation in water (Bierzynski et al., 1982) when the Zimm-Bragg equation (Zimm and Bragg, 1959) is used, with parameters (s and S) determined by host-guest experiments where s is the helix nucleation constant, n is the number of H-bonded residues in the helix and S is an average stability constant for one residue:

$$sS^{n-1}/(S-1)$$

Nevertheless, the 13 amino acid C-peptide obtained from RNase A does show measurable helicity (~25%) at low temperature (Bierzynski et al., 1982; Brown and Klee, 1981). The stability of this peptide is 1000–fold greater than the value calculated from the Zimm-Bragg equation. Specific side-chain interactions, factors that are not considered in the Zimm-Bragg model, are responsible at least in part for the fact that the C-peptide is much more helical than predicted (Scherega, 1985).

Medium-range interactions are responsible for the additional stabilization of secondary structures (DeGrado et al., 1989). Interactions between the side-chains are regarded as important medium range interactions (Shoemaker et al., 1987; Marqusee and Baldwin, 1987). These include electrostatic interactions, hydrogen bonding, and the perpendicular stacking of aromatic residues (Blundell et al., 1986). An α–helix possesses a dipole moment as a result of the alignment of its peptide bonds. The positive and negative ends of the amide group dipole, point toward the helix NH_2-terminus and COOH-terminus, respectively, giving rise to a significant macrodipole. Appropriately charged residues near the ends of the helix can favorably interact with the helical dipole and stabilize helix formation. It was estimated that the electrostatic interaction between a pair of antiparallel α–helices is about 20kcal/mol less than a parallel α–helices pair (Hol and Sanders, 1981). Hydrogen bonds between side chains and terminal helical N-H and C=O groups also participate in the stabilization of helical structure (Richardson and Richardson, 1988; Presta and Rose, 1988; Richardson and Richardson, 1989).

Protein structures contain several long-range stabilizing interactions which include hydrophobic and packing interactions, and hydrogen bonds. Among these, the hydrophobic effect is a prime contributor to the folding and stabilizing of protein structures. The driving force for helix formation in RNase A arises from long-range interactions between C-peptide and S-protein, a large fragment of the protein from which C-peptide was excised (Komoriya and Chaiken, 1985).

The role of hydrophobic interactions in determining secondary structures was studied for a series of peptides containing only Glu and Lys in their sequence (DeGrado and Lear, 1985). Glu and Lys residues were chosen as charged residues for the solvent-accessible exterior of the protein to help stabilize helix formation by electrostatic interaction.

Stability of designed proteins

Hydrophobic residues often repeat every three to four residues in an α–helix and form an amphiphilic structure (DeGrado et al., 1989). Amphiphilicity is important for the stabilization of the secondary structures of peptides and proteins which bind in aqueous solution to extrinsic apolar surfaces, including phospholipid membranes, air, and the hydrophobic binding sites of regulatory proteins (Degrado and Lear, 1985). This amphiphilic secondary structure can be stabilized relative to other conformations by self-association. Therefore, short peptides often form the α–helix in water only because the helix is amphiphilic and is stabilized by peptide aggregation along the hydrophobic surface. Natural globular proteins are folded by a similar mechanism, involving hydrophobic interaction between neighboring segments of secondary structure (Presnell and Cohen, 1989). Using the concept of an amphiphilic helix, DeGrado and coworkers have successfully built peptide-hormone analogs with minimal homology to the native sequences. These peptides, like the native ones, are not helical in solution but do form helices at the hydrophobic surfaces of membranes.

Designed synthetic peptides have been used to show how hydrophobic periodicity in a protein sequence stabilizes the formation of simple secondary structures such as an amphiphilic α–helix (Ho and DeGrado, 1987). The strategies used in the design of the helices in the four-helix bundles are: 1) the helices should be composed of strong helix forming amino acids, and 2) the helices should be amphiphilic; i.e., they should have an apolar face to interact with neighboring helices and a polar face to maintain water solubility of the ensuing aggregates. The results show that hydrophobic periodicity can determine the structure of a peptide. Therefore, the peptides tend to have random conformations in very dilute solution, but form secondary structures when they self-associate (at high concentration) or bind to the air-water surface.

The free energy associated with dimerization or tetramerization of the designed peptides could be experimentally determined from the concentration dependence of the CD spectra for the peptides (DeGrado et al., 1989; Lear et al., 1988; DeGrado and Lear, 1985). At low concentrations, the peptides were found to be monomeric and have low helical contents, whereas at high concentration they could self-associate and stabilize the secondary structure. Therefore, possible hairpin loops between helices can affect the stability of the secondary structure by enhancing the self-association between the helical monomers. A strong helix breaker (Chou and Fasman, 1978; Kabsch and Sander, 1983; Sueki et al., 1984, Scheraga, 1978) was included as the first and last residue to set the stage for adding a hairpin loop between the helices. A single proline residue appeared capable of serving as a suitable link if the C and N terminal glycine residues are slightly unwound. Glycine lacks a β–carbon, which is essential for the reverse turn where positive dihedral angles are required. The pyrrolidone ring of proline constrains its f dihedral angle $-60°$. Thus, proline should be destabilizing at positions where significantly different backbone torsion angles are required. This amino acid, as well as glycine, has a high tendency to break helices and occurs frequently at turns (Creighton, 1987).

The direct evidence for stabilization of protein structure by adding the linking sequence was observed by comparing the guanidine denaturation curve for a monomer, dimer and tetramer (Degrado et al., 1989). The gene encoding tetrameric protein was expressed in *E. coli* and purified to homogeneity. In the series of mono-, di-, and tetramer, the stability toward guanidine denaturation increases concomitantly with the increase in covalent cross-links between helical monomer. At equivalent peptide concentrations, the midpoints of the denaturation curves occurred at 0.55, 4.5 and 6.5M guanidine for the mono-, di, and tetramer. Furthermore, as the number of covalent cross-links was increased, the curves became increasingly cooperative. Thus, the linker sequence

stabilized the formation of the four helix structures at low concentration of the peptides (<1mg/ml).
Structural stability of proteins is directly related to *in vivo* proteolysis (Parasell and Sauer, 1989). Proteolysis depends on the accessibility of the scissile peptide bonds to the attacking protease. The sites of proteolytic processing are generally in relatively flexible interdomain segments or on the surface of the loops, in contrast to the less accessible interdomain peptide bonds (Neurath, 1989). This suggests that the stability of the folded state of the protein is the most important determinant of its proteolytic degradation rate. The effect of a folded structure on the proteolytic degradation has been proven by several experiments. First, proteins that contain amino acid analogs or are prematurely terminated are often degraded rapidly in the cells (Goldberg and St. John, 1976). Second, there are good correlations between the thermal stabilities of specific mutant proteins and their rates of degradation in *E. coli* (Pakula and Sauer, 1986, Parasell and Sauer, 1989). Finally, second-site suppressor mutations that increase the thermodynamic stability of unstable mutant proteins have also been shown to increase resistance to intracellular proteolysis (Pakula and Sauer, 1989). The solubility of proteins could also affect their proteolytic resistance as some proteins aggregate to form inclusion bodies that escape proteolytic attack (Kane and Hartley, 1988).

Metabolic stability is another factor influencing the *in vivo* stability of proteins. Usually, damaged and abnormal proteins are metabolically unstable *in vivo* (Finley and Varshavsky, 1985; Pontremoli and Melloni, 1986). In eukaryotes, covalent conjugation of ubiquitin with proteins is essential for the selective degradation of short-lived proteins (Finley and Varshavsky, 1985). It was found that the amino acid at the amino-terminus of the protein determined the rate of ubiquitination (Bachmair et al., 1986). Both prokaryotic and eukaryotic long-lived proteins have stabilizing amino acids such as methionine, serine, alanine, glycine, threonine, and valine at the amino terminus end. On the other hand, amino acids such as leucine, phenylalanine, aspartic acid, lysine, and arginine destabilize the target proteins.

Designed nutritional proteins

We designed the synthetic protein DNP 1 to contain a high content of those amino acids which are essential to the diet of animals. The optimized content of essential amino acids for this new protein was obtained empirically by determining the amounts of essential amino acids necessary for normal metabolism of the animal. We also determined the "deficiency values" or the ratios of deficient essential amino acids for the 10 primary crops animals consume throughout the world (Figure 3). From these data, we then found the ratio of essential amino acids needed to totally complement each particular plant foodstuff. We merely averaged these values and came up with a set of numbers we call the 'Average Ratio for All Crops Idealized to the DNP 1 Monomer' (Figure 4). This set of numbers represents the ratio of essential amino acids necessary

De novo designed synthetic plant storage proteins

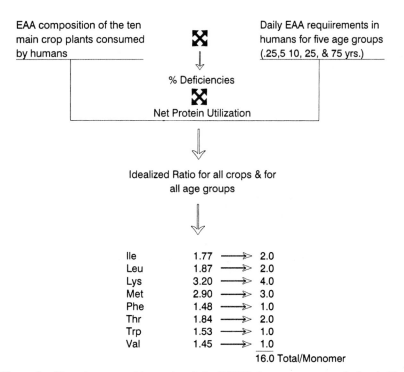

Figure 3. How the composition ratio of the HDNP 1 monomer was calculated. The ratio is an overall average which takes into consideration all daily requirements for five age groups; the deficiencies of the ten major crop plants people consume: rice, wheat, barley, sorghum, maize, potato, sweet potato, plantain, cassava, and taro; and how well these plant foodstuffs are utilized as a protein source.

to complement the deficiencies found in all 10 crops for all human age groups.

From the above set of numbers, we designed the DNP 1 protein for humans (HDNP 1). It has 1.8 times more of the essential amino acids compared with zein or phaseolin. The difference in MLEAA is much higher, containing three times more than phaseolin and 6.5 times more than zein. The helical region of HDNP 1 is amphipathic (hydrophobic residues clustered on one face of the helix while hydrophilic residues are found on the other face) and is stabilized by several GLU – LYS salt bridges (Figure 5). The helix breaker Gly-Pro-Gly-Arg has been used as a turn sequence. The design results in an antiparallel tetramer which achieves an extraordinarily stable secondary and tertiary structure even at low concentration.

The structural stability of a protein is important in determining its susceptibility to proteolysis. Most native proteins are relatively resistant to cleavage by proteolytic enzymes, whereas denatured proteins are much more sensitive (Pace and Barret, 1984). Several findings suggest that the stability of a folded protein is an important determinant of its rate of degradation. Therefore, in addition to improved nutritional quality, HDNP 1 has been designed to have a stable storage protein-like

J.M. Jaynes

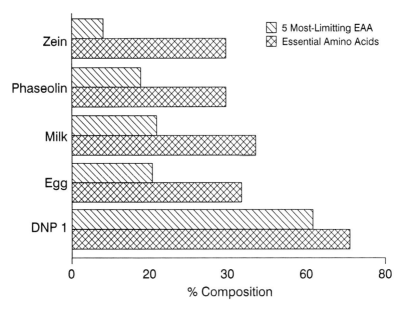

Figure 4. Percentage of essential amino acids (EAA) and percentage of most limiting essential amino acids (MLEAA) in DNP 1 tetramer compared with natural proteins.

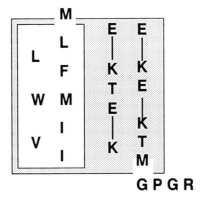

Figure 5. A depiction of the amphiphilicity of the HDNP 1 monomer where hydrophobic amino acids are in the white rectangle and hydrophilic in the shaded rectangle. Note the interaction between the Glu (E) and Lys (K) residues (dark lines depict salt-bridges).

structure in plants. Its design is based on the structurally well-studied corn storage zein proteins (Z19 and Z22), which comprise of nine repeated helical units (Agros et al., 1982). Each helical unit of zein, 16 to 26 amino acids long, is flanked by turn regions and forms an antiparallel helical bundle. Most of the amino acids in the helices are hydrophobic residues. On the other hand, HDNP 1 comprises four helical repeating units, each 20 amino acids long (Figure 6). Increased gene copy number

143

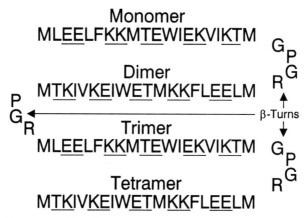

Figure 6. The amino acid sequence of the HDNP 1 tetramer. Hydrophilic amino acids are underlined and β–turns so indicated.

by concatenation can increase the protein yields. At the same time, gene concatenation gives the increased molecular mass of the encoded protein. Such an increase in size and concatenation can significantly stabilize an otherwise unstable product (Shen, 1984).

The 284bp gene encoding this novel peptide was chemically synthesized and cloned into an *E. coli* expression vector. This gene contains plant consensus sequences at the 5′ end of the translation initiation site to optimize the expression of proteins in vivo. It was placed under the control of the 35S cauliflower mosaic virus (CaMV) promoter in order to permit the constitutive expression of this gene in tobacco.

Prediction of the structure of HDNP 1

The secondary structures of the HDNP 1 monomer and tetramer were predicted by PREDICT-SECONDARY in β–SYBYL. The percentage of α–helix content predicted by information-theory showed a higher α–helix content compared with the other two prediction methods (Bayes-statistic and neural-net) in PREDICT-SECONDARY. The predicted secondary structures by information-theory gave 100% helical content for the monomer and 74% for the tetramer.

However, the accuracy of the three widely used prediction methods ranged from 49% to 56% for prediction of three states; helix, sheet, and coil (Kabsch and Sander, 1983). This inaccuracy might be due to the small size of the database and(or) the fact that secondary structure is determined by tertiary interactions which are not included in the local sequences. For further predictions of structure, the structures predicted by information-theory were energy minimized using SYBYL MAXIMIN2.

A perfect amphiphilic α–helical conformation was predicted for the HDNP 1–monomer after minimization. The tertiary structure of the HDNP 1–tetramer after minimization showed the antiparallel conformation as was designed. These minimization results suggested the high probability of stable secondary structure (α-helix and β-turn) formation of the HDNP 1-monomer and -tetramer. However, it might be too early to conclude the tertiary structure of the HDNP 1-tetramer with minimization only, without considering longer range interactions which are the most important determinant of protein folding.

The predominant driving force for folding the HDNP 1-tetramer might be the longer range hydrophobic interactions between the α-helical monomers, because HDNP 1 was designed to be amphiphilic. However, no currently available force field for the minimization of tertiary structure contains these parameters and could not give the perfect predicted tertiary structure of the protein. Actually, minimization schemes alone have failed to predict chain folding accurately (Fasman, 1989). Therefore, we might be able to obtain indications of a much more stable secondary and tertiary structures if we considered this factor for the minimization of HDNP 1-tetramer.

Structural analysis of HDNP 1 protein

The structural stability of HDNP 1-monomer and tetramer could not be determined by minimization only. Therefore, the stability of the α-helical secondary structure of HDNP 1-monomer was investigated. HPLC analysis of the gel filtered synthetic HDNP 1-monomer showed that purity was more than 90% and amino acid analysis of the purified fraction gave the expected molar ratios. This fraction was also analyzed by mass spectrometry, and the molecular weight peak corresponding to the HDNP 1-monomer (2896.5) was present. Since the structural stability of HDNP 1-monomer and tetramer could not be determined by minimization only, the stability of the α-helical secondary structure of HDNP 1-monomer was investigated by circular dichroism (CD) analysis. CD spectra of HDNP 1-monomer showed the typical pattern of α-helical proteins with double minima at 208 and 222 nm in aqueous solution (data not shown). The stability of the secondary structure can be induced by the inter-molecular interaction between the helical chains (DeGrado et al., 1989). Therefore, stable aggregation between monomers, presumably through hydrophobic interactions, could stabilize the helical structure. Besides, proper packing of the apolar side chains and proper electrostatic interaction might play important roles in stabilizing the secondary structure of HDNP 1. The stable interaction among the monomeric HDNP 1 molecules is an important determinant for the proper folding into the tertiary structure of the HDNP 1-tetramer. Therefore, the self-association capability of the HDNP 1-monomers was investigated by using size exclusion chromatography. The hydrodynamic behavior of this peptide showed that it was aggregated into a hexamer form with an apparent molecular weight of about 17 kD. This hexameric aggregate could be maintained in either low or high ionic strength solutions. This result provides proof of the stable globular type tertiary structure formation of tetrameric HDNP 1.

Three potential β–turn (Gly-Pro-Gly-Arg) sequences were inserted between four monomers for the HDNP 1–tetramer construction. The β–turn could play an important role for structural stability of the HDNP 1–tetramer when it is expressed in vivo. It can also help stabilize tertiary structure formation. The interactions between the helical monomers might be much faster due to the proximate effect when they are connected. This proximate effect might be critical for folding at the low concentrations of HDNP 1–tetramer that are possible when they are expressed *in vivo*. At the same time, the stability of the secondary structure is increased by the hydrophobic interactions between helical monomers. In addition, this β–turn sequence has a tryptic digestion site (Gly-Arg) which could increase the digestibility of this protein when it is consumed by animals.

The stability of the folded structure of a protein has a close relation to its proteolytic degradation rate (Pace and Barret, 1984; Pakula and Sauer, 1986; Parasell and Sauer, 1989; Pakula and Sauer, 1989). In this respect, we expected high stability of folded HDNP 1–tetramer against proteolytic degradation when it is expressed *in vivo*. Stable quaternary structure is essential for the formation of protein bodies of storage proteins in zein or phaseolin (Lawrence et al., 1990). These higher order structures can be achieved through the interaction and close packing of the stable tertiary structures. The major driving force for this quaternary structure formation is also hydrophobic interaction between the tertiary structures. At this moment, designing and predicting the quaternary structure is not easy but our data suggest that precisely this might be occurring with HDNP 1–tetramer.

Introduction of HDNP 1 gene into tobacco

The correct insertion and orientation of the pBI derivative containing the HDNP 1 tetramer was screened by EcoRI and HindIII digestion (it was found in *E. coli* that the most stable form of the gene was the tetramer form). The EcoRI digestion gave a fragment of the expected size, 3.2 kb, which consisted of 3'NOS of HDNP 1 and the GUS gene (data not shown). Also, the HDNP 1 gene with its 35S promoter and 3'NOS sequences was detected as a 1.4 kb band by HindIII digestion. Stable transformation of the HDNP 1 gene into *A. tumefaciens* LBA4404 was confirmed by HindIII digestion of isolated plasmid DNA. It could be isolated from Agrobacterium and detected by enzyme digestion because pBI121 is a binary vector. Leaf discs, transformed with LBA4404 carrying the HDNP 1 gene, gave about five to seven shoots two to three weeks after infection. A total of 565 kanamycin-resistant shoots were regenerated from 120 leaf discs. These shoots were excised from the leaf discs and transferred to new media to grow several more weeks and then transferred to rooting media. After three weeks in rooting medium, 126 rooted shoots were analyzed for β–glucuronidase (GUS). Root tips of 56 out of 126 plants showed various levels of GUS activity. Not all the kanamycin-resistant shoots showed the GUS positive result. Although kanamycin resistance was due to the expression of neomycin phosphotransferase (NTP II gene), regeneration of nontransgenic shoots in the presence of kanamycin has been reported. Therefore, escapes from the screening based on kanamycin sensitivity might have occurred in the nontransformed plants, making them kanamycin resistant.

Thirty six plantlets which showed high levels of β–glucuronidase activity were transplanted into "Jiffy" pots. After establishment of the plants, a more accurate fluorogenic assay for GUS activity was done to quantify the expression level of this gene (data not shown). Some of these transformed tobacco plants showed higher levels of β–glucuronidase activity compared with other plants. The level of expression might be primarily affected by whether the gene is incorporated into an active or inactive site of chromatin. Activity of chromatin, methylation of DNA and nuclease hypersensitivity are closely related to each other. It has been found that the nuclease hypersensitive sites correlate to active transcription (Gross and Garrard, 1987). The degree of methylation of DNA is inversely related to gene expression. Furthermore, if the gene is located near the plant's endogenous promoter or enhancer sites, the level of expression of this gene will be increased by these near-by enhancing factors. Therefore, the difference in the levels of GUS activity between the transformed plants might be due to this positional effect, which was determined by the sites of incorporation of this gene into the tobacco genome.

Expression of HDNP 1 in transformed plants

Polyclonal antibody raised against synthetic HDNP 1 monomer was used to detect the production of stable HDNP 1 protein in tobacco. High levels of the tetrameric form (11.2 kd) of the HDNP 1 protein were detected from plant #17 by western blot analysis (data not shown). Therefore, direct correlation was found between gene copy number, number of genetic NPTII loci, GUS expression, accumulation of HDNP 1 transcript and protein expression level in the case of plant #17. Some heterologous seed proteins undergo specific degradation when expressed in transgenic plants. A significant amount of the immunoreactive protein accumulated in tobacco seed expressing the phaseolin gene is smaller than the final processed protein (Sengupta et al., 1985). A similar result was found when β–conglycinin was expressed in transgenic petunia (Beachy et al., 1985). In contrast to these results, the HDNP 1 protein appears to be quite stable in transgenic tobacco plants.

Amino acid and total protein analyses were conducted on leaf tissue from several of the transgenic plants which produced detectable levels of HDNP 1. Surprisingly, we found that the overall levels of all amino acids were increased with some of the plants being remarkably high. (Figure 7). This rather disconcerting result has been repeated numerous times and the overall levels of all amino acids in the transgenic plants remain significantly elevated. Other methods of determining overall protein content have been used with similar trends observed. For example, comparison of total protein densitometric values derived from SDS-PAGE of equivalent samples (on a weight basis) yield the same results

De novo designed synthetic plant storage proteins

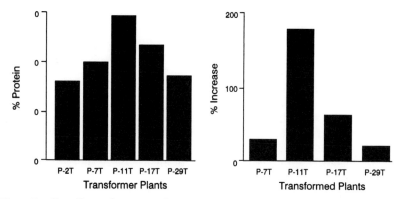

Figure 7. Overall protein content determined by amino acid analysis from control (P-2 TC) and various HDNP 1 tetramer plants. These data were derived from seedlings obtained from transformed mother plants. A minimum of four separate assays were used and the variation was no more than 30%. Percentage increase of HDNP 1 plants over the control is also shown.

(data not shown). At this time, we can offer no definitive explanation. Therefore, in addition to being a very stable protein in a plant cell, HDNP 1 must function as a general 'protein-stabilizer' and reduces overall protein turnover without apparent deleterious effects to the plants, since there is no observable difference in growth characteristics in the plants producing high amounts of HDNP 1 over control plants.

Conclusions

HDNP 1 is the first reported *de novo* designed protein expressed in plant systems. We have found that it is possible to construct a protein of high nutritional value mimicking the well known physical characteristics of plant storage proteins. Plans for future studies of tetrameric HDNP 1 begin with further analysis of its structural properties such as: 1) stability against proteolytic attack, and 2) solubility and aggregation pattern. Additionally, we are studying intensively the import of its apparent ability to reduce overall protein turnover. Subsequent plans are to pursue high level tissue-specific expression of this gene in the more economically important plants such as potato, soybean, and selected cereals.

It should be clear that the application of the above-described techniques for protein design could have a profound impact on how humans feed themselves and their livestock. It is possible that within the very near future (less than five years) traditional food and feed plants could be produced that contain the full-complement of essential amino acids to meet the specific needs of each species considered. Thus, foundation seed-stock of so-called genetically-engineered "human-corn", "swine-corn", "poultry-corn", etc. could be provided to farmers for high-level production, distribution, and sale to the food and feed industry.

References

Agros, P., Pederson, K., Marks, D. and Larkins, B.A. 1982. A structural model for maize zein proteins. J. Biol. Chem. 257: 9984–9990.

Agros, P., Naravana, S.V.L., and Nielsen, N.C. 1985. Structural similarity between legumin and vicillin storage proteins from legumes. The EMBO J. 4: 1111–1117.

Altenbach, S.B., Pederson, K.W., Meeker, G. Staraci, L.C., and Sun, S.S.M. 1989. Enhancement of the methionine content of seed proteins by the expression of seed proteins by the expression of a chimeric gene encoding a methionine-rich protein in transgenic plants. Plant Mol. Biol. 13: 513–522.

Bachmair, A., Finley, D., and Varshavsky, A. 1986. In vivo halflife of a protein is a function of its amino-terminal residue. Science 234: 179–186.

Badley, R.A., Atkinson, D., Hauser, H., Oldani, D., Green, J.P., and Stubbs, J.M. 1975. The structure, physical and chemical properties of the soybean protein glycinin. Biochim. Biophys. Acta. 412: 214–228.

Bartels, D., and Tompson, R.D. 1983. The characterization of cDNA clones coding for wheat storage proteins. Nucleic Acid Res. 11: 2961–2977

Beachy, R.N., Chen, Z.L., Horsch, R.B., Rogers, S.G., Hoffman, N.J., and Fraley, R.T. 1985. Accumulation and assembly of soybean β-conglycinin in seeds of transformed petunia plants. EMBO J. 4: 3047–3053.

Bierzynski, A., Kim, P.S., and Baldwin, R.L. 1982. A salt bridge stabilizes the helix formed by isolated c-peptide of RNAse A. Proc. Natl. Acad. Sci. U. S. A. 79: 2470–2474.

Blundell, T.L., Thornton, S.J., Burley, S.K., and Petsco, G.A. 1986. Atomic interactions. Science 234: 1005–1009.

Bollini, R. and Chrispeels, M.J. 1978. Characterization and subcellular localization of vicillin and phyto-hemaglutinin, the two major reserve proteins of *Phaseolus vulgaris*. Planta 142: 291–298.

Brown, J.E. and Klee, W.A. 1971. Helix-coil transition of the isolated amino terminus of ribonuclease. Biochemistry 10: 470–476.

Chen, Z.L., Pan, N.S., and Beachy, R.N. 1988. A DNA sequence element that confers seed-specific enhancement of a constitutive promoter. The EMBO J. 7: 297–302.

Chen, Z.L., Schuler, M.A. and Beachy, R.N. 1986. Functional analysis of regulatory elements in a plant embryo-specific gene. Proc. Natl. Acad. Sci. U. S. A. 83: 8560–8564.

Chou, P.Y. and Fasman, G.D. 1978. Prediction of the secondary structure of proteins from amino acid sequence. Adv. Enzymol. 47: 45–148.

Colot, V., Robert, L.S., Kavanagh, T.A., Beavan, M.W. and Tompson, R.D. 1987. Localization of sequences in wheat endosperm protein genes which confer tissue-specific expression in tobacco. The EMBO J. 6:3559–3564.

Creighton, T.E. 1984. Proteins. New York: Freeman.

Crouch, M., Tenberge, K., Simone, N.E., and Ferl, R. 1983. Sequence of the 1.7K storage protein of *Brassica napus*. Mol. Appl. Genet. 2:273–283.

Degrado, W.F., and Lear, J.D. 1985. Induction of peptide conformation at apolar/water interfaces. J. Am. Chem. Soc. 107: 7684–7689.

Degrado, W.F., Wasserman, Z.R., and Lear, J.D. 1989. Protein design, a minimalist approach. Science 241: 622–628.

Esen, E. 1986. Separation of alcohol-soluble proteins (zeins) from maize into three fractions by differential solubility. Plant Physiol. 80: 623–627.

Fasman, G. 1989. Protein conformational prediction. Trends in Biochem. 14: 295–299.

Finley, D. and Varshavsky, A. 1985. The ubiquitin system:functions and mechanisms. Trends Biochem. Sci. 10: 343–346.

Forde, B.G., Kreis, M., Williamson, M.S., Fry, R.P. and Pywell, J. 1985. Short tandem repeats shared by B- and C-hordein cDNAs suggest a common evolutionary origin for two groups of cereal storage protein genes. The EMBO J. 4: 9–15.

Goldberg, A.L., and St John, A.C. 1976. Intracellular protein degradation in mammalian and bacterial cells: part 2. Annu. Rev. of Biochem. 45:747–803.

Greenwood, J. S., and Chrispeels, M.J. 1985. Correct targeting of the bean storage protein phaseolin in the seeds of transformed tobacco. Plant Physiol. 79: 65–71.

Gross, D.S., and Garrard, W.T. 1987. Poising chromatin for transcription. Trends in Biochem. 12: 293–296.

Ho, S.P., and Degrado, W.F. 1987. Design of a 4-helix bundle protein: synthesis of peptides which self-associate into helical protein. J. Am. Chem. Soc. 109: 67516758.

Hoffmann, L.E., Donaldson, D.D., and Herman, E.M. 1988. A modified storage protein is synthesized. processed, and degraded in the seeds of transgenic plants. Plant Mol. Biol. 11: 717–729.

Hoffmann, L.E. Donaldsopn, D.D. Bookland, R. Rashka, K and Herman, E.M. 1987. Synthesis and protein body deposition of maize 15-kd zein in transgenic tobacco seeds. The EMBO J. 6: 3213–3221.

Hol, W, G., and Sander, H.C. 1981. Dipole of the α–helix and β–sheet: their role in protein folding. Nature. 294: 532–536.

Horsch, R.B., Fry, J., Hoffmann, N., Neidermeyer, J., Rogers, S.G. and Fraley, R.T. 1988. *In:* Plant Mol. Biol. Manual ed. S.B. Gelvin and R.A. Schilperoort, Dordrecht: Kluwer Academic.

Jaynes, J.M., Nagpala, P., Destefano, L., Denny, T., Clark, C., and Kim, J-H. 1990. Expression of a de novo designed peptide in transgenic tobacco plants confers enhanced resistance to *Pseudomonas solanacearum* infection. Plant Science: 89, 43–53.

Jaynes, J.M., Yang, M.S., Espinoza, N.O., and Dodds, J.H. 1986. Plant protein improvement by genetic engineering: use of synthetic genes. Trends in Biotechnol. 4: 314–320.

Jones, J.D.G., and Gilbert, D.E. 1987. T-DNA structure and gene expression in petunia plants transformed by *Agrobacterium tumefaciens* C58 derivatives. Mol. Gen. Genet. 207: 478–485.

Kabsch, W., and Sander, C. 1983. How good are predictions of protein structure? FEBS lett. 155: 179–182.
Kane, J.F. and Hartley, D.L. 1988. Formation of recombinant protein inclusion bodies in *Escherichia coli*. Trends in Biotechnol. 6: 95–101.
Kasarda, D.D., Okita, T.W., Bernardin, J.E., Baecker, P.A., and Nimmo, C.C. 1984. DNA and amino acid sequences of alpha and gamma gliadins. Proc. Natl. Acad. Sci. U.S.A. 81: 4712–4716.
Keris, M., Shewry, P.R., Forde, B.G., Forde, G. and Miflin, J. 1985. Structure and evolution of seed storage proteins and their genes with particular reference to those of wheat, barley and rye. Oxford Survey of Plant Mol. and Cell Biol. 2: 253–317.
Komoriya, A., and Chaiken, J.M. 1982. Sequence modeling using semisynthetic ribonuclease S.J. Biol. Chem. 257: 2599–2604.
Larkins, B.A. 1983. Genetic engineering of seed storage protein. *In:* Genetic Engineering of Plants B. A. Larkins (ed.), pp. 93–120. New York: Prenum.
Larkins, B.A., Pederson, K, Mark, M.D., and Wilson, D.R. 1984. The zein protein of maize endosperm. Trends in Biochem. 9: 306–308.
Lawrence, M.C., Suzuki, E., Varghes, J.N., Davis, P.C., Van Donkelaar, A. Tulloch, P.A. and Collman, P.M. 1990. The three-dimensional structure of the seed storage protein phaseolin at 3 Å resolution. The EMBO J. 9: 9–15.
Lear, J.D., Wasserman, Z.R. and Degrado, W.F. 1988. Synthetic amphiphilic peptide model for protein ion channels. Science. 240: 1177–1181.
Lending, C.R., Kriz, A., Larkins, B.A. and Bracker, C.E. 1988. Structure of maize protein bodies and immunocytochemical localization of zeins. Protoplasma 143: 51–62.
Lycett, G.W., Cory, R.D., Shirsat, A.H., Richards, D.M., and Boulter, D. 1985. The 5′-flanking regions of three pea legumin genes: comparison of DNA sequences. Nucleic Acids Res. 13: 6733–6743.
Marqusee, S. and Baldwin, R. 1987. Helix stabilization by GLU-LYS salt bridges in short peptides of de novo design. Proc. Natl. Acad. Sci. U.S.A. 84: 8898–8902.
Marries, C., Gallois, P., Copley, J. and Keris, M. 1988. The 5′ flanking region of a barley B hordein gene controls tissue and developmental specific CAT expression in tobacco plants. Plant Mol. Biol. 10: 359–366.
Mutter, M. 1988. Nature's rules and chemist's tools: a way for creating novel proteins. Trends in Biochem. 13: 260–264.
Pace, C.N. and Barret, A.J. 1984. Kinetics of tryptic hydrolysis of the arginine-valine bond in folded and unfolded ribonuclease T1. Biochem. J. 219: 411–417.
Pakula, A.A. and Sauer, R.T. 1986. Bacteriophage 1 Cro mutation: effect on activity and intracellular degradation. Proc. Natl. Acad. Sci. U.S.A. 82: 8829–8833.
Pakula, A.A. and Sauer, R.T. 1989. Amino acid substitutions that increase the thermal stability of the l Cro protein. Proteins. 5: 202–210.
Parasell, D.A. and Sauer, R.T. 1989. The structural stability of a

protein is an important determinant of its proteolytic susceptibility in *Escherichia coli.* J. of Biol. Chem. 264: 7590–7595.

Pederson, K., Agros, P., Naravana, S.V.L., and Larkins, B.A. 1986. Sequence analysis and characterization of a maize gene encoding a high-sulfur zein protein of Mw 15,000. J. Biol. Chem. 201: 6279–6284.

Pernollet, J.C. and Mosse, J. 1983. Structure and location of legume and cereal seed storage protein. Seed Proteins. Phytochemical Soc. of Eur. Sym. Series 20: 155–187.

Pontremoli, S. and Melloni, E. 1986. Extralysosomal protein degradation. Annu. Rev. Biochem. 55: 455–481.

Presnell, S.R., and Cohen, F.E. 1989. Topological distribution of a four–β–helix bundle. Proc. Natl. Acad. Sci. U.S.A. 86: 6592–6596.

Presta, L.G. and Rose, G.D. 1988. Helix signals in proteins. Science 240: 1632–1641.

Rafalski, J.A. Scheets, K., Metzler, M., and Peterson, D.M. 1984. Developmentally regulated plant genes: the nucleotide sequence of a wheat gliadin genomic clone. The EMBO J. 3: 1409–1415.

Richardson, J.S. and Richardson, D.C. 1988. Amino acid preferences for specific locations at the ends of α–helices. Science. 240:1648–1652.

Richardson, J.S. and Richarson, D.C. 1989. The de novo design of protein structures. Trends in Biochem. 14: 304–309.

Sanders, P.R., Winter, J.A., Barnason, A.R. and Rogers, S.G. 1987. Comparison of cauliflower mosaic virus 35S and nopaline synthetase promoters in transgenic plants. Nucleic Acids Res 15: 1543–1558.

Scheraga, H. 1978. Use of random copolymers to determine helix-coil stability constants of the naturally occuring amino acids. Pure. Appl. Chem. 50: 315–324.

Scheraga, H.A. 1985. Effect of side chain-backbone electrostatic interaction on the stability of α–helices. Proc. Natl. Acad. Sci. U.S.A. 82:5585–5587.

Scott, R.J., and Draper, J. 1987. Transformation of carrot tissue derived from proembryogenic suspension cells: a useful model system for gene expression studies in plants. Plant Mol. Biol. 8: 265–274.

Sengupta, G.C., Reichert, N.A., Baker, R.F., Hall, T.C. and Kemp, J.D. 1985. Developmentally regulated expression of the bean β–phaseolin gene in tobacco seed. Proc. Natl. Acad. Sci. U.S.A. 82: 3320–3324.

Shen, S-H. 1984. Multiple joined genes prevent product degradation in *E. coli.* Proc. Natl Acad. Sci. U.S.A. 81: 4627–4631.

Slightom, J.L., Sun, S.M., and Hall, T.C. 1983. Complete sequence of french bean storage protein gene: phaseolin. Proc. Natl. Acad. Sci. U.S.A. 80: 1897–1901.

Staswick, P.E. 1989. Preferential loss of an abundant storage protein from soybean pods during seed development. Plant Physiol. 90: 1251–1255.

Stockhaus, J., Eckes, P., Blau, A., Schell, J., and Willmitzer, L. 1987. Organ-specific and dosage-dependent expression of a leaf/stem specific gene from potato after tagging and transfer into potato and tobacco plants. Nucleic Acids Res. 15: 3479–3491.

Sueki, M., Lee, S., Power, S.P., Denton, J.B., Konishi, Y., and Scheraga, H. 1984. Helix-coil stability constants for the naturally occuring amino acids in water. Macromolecules. 17: 148–155.

Twell, D. and Ooms, G. 1987. The 5' flanking DNA of a patatin gene directs tuber specific expression of a chimeric gene in potato. Plant Mol. Biol. 9: 365–375.
Wallace, J.C., Galili, G., Kawata, E.E., Cuellar, R.E., Shotwell, M.A., and Larkins, B.A. 1988. Aggregation of lysine containing zeins into protein bodies in *Xenopus oocytes*. Science. 240: 662–664.
Wenzler, H.C., Mignery, G.A. Fisher, L.M., and Park, W.D. 1989. Analysis of a chimeric class I potatin-GUS gene in transgenic potato plants: high level expression of tubers and sucrose-inducible expression in cultured leaf and stem explants. Plant Mol. Biol. 12: 41–50.
Yang, M.S., Espinoza, N.O., Dodds, J.H., and Jaynes, J.M. 1989. Expression of a synthetic gene for improved protein quality in transformed potato plants. Plant Science. 64: 99–111.
Zimm, B.H. and Bragg, J.R. 1959. Theory of the phase transition between helix and random coil in polypeptide chains. J. Chem. Phys. 31:526–535.

POTENTIAL FOR MANIPULATING THE GASTROINTESTINAL MICROFLORA: A REVIEW OF RECENT PROGRESS

SCOTT A. MARTIN

Departments of Animal and Dairy Science and Microbiology
University of Georgia, Athens, Georgia, USA

Introduction

Much interest has been generated over the past few years aimed at evaluating alternative means to manipulate the gastrointestinal microflora in production livestock. Motivation for examining these alternatives comes from increasing public scrutiny about the use of antibiotics in the animal feed industry as well as the need for a safe food supply. The recent deaths of several individuals due to the consumption of undercooked meat products in some western sections of the United States attracted a lot of media attention. As a consequence, there is a sense of urgency by the United States government to do more to address the problem of food contamination by pathogenic microorganisms.

The fact that meat as well as other animal products is susceptible to infection by pathogenic microorganisms is not a new revelation. It has been previously estimated that the annual cost of medical expenses and lost productivity in the United States associated with *Salmonella* infections resulting from the consumption of contaminated meat and poultry is $1.0 billion (Roberts, 1988). Annual losses due to human foodborne salmonellosis in Canada and Europe are also in the millions of dollars (Stavric and D'Aoust, 1993). Because there is a high correlation between the increased consumption of poultry products and the incidence of human salmonellosis worldwide (Stavric and D'Aoust, 1993), several laboratories around the world are interested in manipulating the poultry intestinal microflora to alleviate problems associated with *Salmonella* contamination. Research is also being conducted to address the problems associated with other pathogenic microorganisms such as *Campylobacter, Escherichia coli*, and *Listeria*. Due to a limited amount of available information in cattle and swine, this review will primarily focus on strategies being investigated to combat the pathogen contamination problem in poultry.

Poultry gastrointestinal microflora

Compared with what is known about the different microorganisms that inhabit the rumen, relatively little is known regarding the physiology of specific microorganisms found in the poultry gastrointestinal tract. One reason for the lack of information may be due to the perception that, unlike the ruminant, these microorganisms do not play an important role in the digestion of feedstuffs. The cecum appears to contain the largest number of bacteria in poultry and most of these are, not surprisingly, strict anaerobes (Barnes et al., 1972). Several groups of anaerobic streptococci as well as Gram-negative and Gram-positive nonsporeforming anaerobes including species of *Bacteroides*, *Fusobacterium*, *Eubacterium*, *Propionibacterium*, *Bifidobacterium*, and *Peptostreptococcus* have been identified (Barnes and Impey, 1972; Salanitro et al., 1974a,b). In addition, sporeforming rods (*Clostridium*) and *Escherichia coli* have been isolated (Salanitro et al., 1974a). It appears that many of these cecal isolates are predominantly saccharolytic, with some strains having proteolytic activity (Barnes and Impey, 1972). Unlike the active fiber-degrading population found in the rumen, the ability to hydrolyze cellulose and(or) xylan (hemicellulose) does not seem to be prevalent in any of these poultry cecal isolates (Barnes and Impey, 1972).

Even though information is limited detailing physiological mechanisms in poultry cecal bacteria, much research has been conducted using these microorganisms to prevent *Salmonella* colonization in chicks. It is well accepted that newly hatched chicks are more susceptible to *Salmonella* colonization than adult birds due to the lack of a developed normal intestinal flora. Therefore, several approaches have been studied with varying degrees of success to address this problem: (1) use of undefined and defined bacterial cultures (competitive exclusion) obtained from pathogen-free adult chicken fecal or cecal matter, (2) incorporation of sugars into the feed and water, and (3) using a combination of undefined or defined bacterial cultures plus sugars.

Competitive exclusion

UNDEFINED CULTURES

The concept of competitively excluding *Salmonella* from the gut of chickens was introduced over 20 years ago (Nurmi and Rantala, 1973). Competitive exclusion (CE) is a complex interaction of microbes, nutrients, and host factors that selectively excludes specific groups or genera/species/strains of microorganisms from colonizing the intestinal tract (Blankenship et al., 1990). More simply, CE involves creating a set of environmental conditions that favors the "good" bacteria (normal intestinal microflora) and selects against the "bad" bacteria (*Salmonella*). In general, CE involves treating chicks with aqueous slurries of *Salmonella*-free fecal or cecal matter from healthy adult chickens or anaerobic bacteriological cultures of the fecal/cecal material (Blankenship et al., 1990). Many studies have shown that this method is effective in reducing *Salmonella* colonization in challenged birds (Nurmi and Rantala, 1973; Mulder and Bolder, 1991; Stavric and D'Aoust, 1993).

Application of these cultures in laboratory experiments is by gavage into the crop. However, this is inconvenient with large numbers of birds so CE treatment can be applied via the drinking water or by spraying hatching eggs or hatchlings with CE preparations (Stavric and D'Aoust, 1993).

The mode of action of these cultures is complex and at this time poorly understood. It has been hypothesized that the bacteria isolated from pathogen-free birds occupy available attachment sites on the intestinal mucosa and prevent attachment and subsequent colonization by *Salmonella* (Impey and Mead, 1989). In addition, these bacteria produce short chain volatile fatty acids (VFA) and lower the oxidation–reduction potential of their environment (Barnes et al., 1979; Barnes et al., 1980). It is believed that both of these factors serve to inhibit the growth of *Salmonella* because low oxidation–reduction potentials have been reported to increase the antibacterial activity of short chain VFA (Meynell, 1963). It should be noted that bacteria-free filtrates have proven to be unprotective (Snoeyenbos et al., 1978). Most likely a combination of many factors, rather than one or two specific factors, are involved in preventing growth of *Salmonella* in birds treated with CE cultures.

Several field studies involving large numbers of birds have been conducted to evaluate the efficacy of CE treatment. Wierup et al. (1988) observed that the incidence of *Salmonella* colonization in broilers was dramatically reduced with CE treatment involving 2.86 million birds (144 flocks) in Sweden. Only one of the 144 flocks was found to be salmonellae contaminated. Similarly, Mulder and Bolder (1991) reported that both the number of flocks that were colonized as well as the incidence of contaminated carcasses by naturally occurring *Salmonella* were significantly reduced by CE treatment in the Netherlands. These results indicate that CE treatment seems to be successful in reducing *Salmonella* contamination in a large number of birds.

Because commercial poultry feeds can contain a variety of different antimicrobial, anticoccidial, and(or) growth-promoting feed additives, research has been conducted to evaluate the effects of some of these feed additives on CE cultures. In a recent review, Stavric and D'Aoust (1993) outlined which feed additives were detrimental to CE cultures as well as which additives had no effect. In some cases, treatment with antibiotics improved CE-dependent protection.

Due to the ability of bacteria, including *Salmonella*, to penetrate egg shells as well as the high incidence of *Salmonella* contamination in commercial broiler and breeder hatcheries, alternative applications of CE into the egg (in ovo) are being explored (Cox et al., 1990; 1991a; 1991b). One method inoculated two dilutions of an undefined CE culture onto the air cell membrane or beneath the air cell membrane of an 18-day-old incubating hatching egg (Cox et al., 1991b). The results from this approach have been promising in that birds receiving CE treatment into the air cell were resistant to *Salmonella typhimurium* challenge (Table 1). Injection of diluted CE cultures onto the air cell resulted in hatchability rates that were similar to commercial hatchability rates (Cox et al., 1991b), but injection of this culture beneath the air cell membrane strongly reduced hatchability (Table 2). These results show promise in

Table 1. Resistance of day-of-hatch chicks to a 10^6 oral challenge of *Salmonella typhimurium* following in ovo injection of CE into the air cell and beneath the air cell membrane on the 18th day of incubation[a].

Treatment	Number of chicks colonized/challenged
Untreated	11/12
AC(T)[1]	0/4
AC(M)[2]	4/5
AM(M)[3]	2/2

[a]Cox et al. (1991b).
[1]AC(T) = injection site was on the air cell membrane and CE culture was diluted 1:1,000.
[2]AC(M) = injection site was on the air cell membrane and CE culture was diluted 1:1,000,000.
[3]AM(M) = injection site was beneath the air cell membrane and CE culture was diluted 1:1,000,000.

Table 2. Effect of in ovo CE culture administration on hatchability[a].

In ovo Treatment	Hatchability*
Noninjected	96
AC[1] – Undiluted	56
AC(T)	81
AC(M)	78
AM[2] – Undiluted	0
AM(T)	0
AM(M)	48

[a]Cox et al. (1991b).
[1]AC-Injection site was onto the air cell membrane. AC(T) = CE culture diluted 1:1,000; AC(M) = CE culture diluted 1:1,000,000.
[2]AM-Injection site was beneath the air cell membrane. AM(T) = CE culture diluted 1:1,000; AM(M) = CE culture diluted 1:1,000,000.
*Out of 100 fertile eggs.

using CE cultures in ovo to protect the chick from environmental sources of *Salmonella* without adversely affecting hatchability.

DEFINED CULTURES

Even though the use of undefined cultures has shown promise, regulatory agencies in many countries are hesitant to approve the use of these cultures (Stavric and D'Aoust, 1993). Much of the concern is due to the fact that undefined mixtures of microorganisms are used and this leads to questions regarding the possible transmission of human and(or) avian pathogens in these cultures. Efforts are currently underway in several laboratories to develop defined CE cultures. This is important in ensuring that a commercial product is consistent from one batch to another and would make licensing much easier (Blankenship et al., 1990;

Mead, 1991). However, there have been problems with defined cultures providing protection similar to the undefined cultures and protective activity being lost during subculturing and cold storage (Mead, 1991; Stavric and D'Aoust, 1993). In addition, the most effective mixtures of defined cultures contain too many different microorganisms to be economically feasible for product development (Mead, 1991).

Several research groups have isolated pure bacterial cultures from the fecal and cecal contents of adult birds. Once these bacteria are isolated, they are used individually or mixed together and used to evaluate protective activity against *Salmonella* in chicks. In vitro experiments have shown that as few as one (Gram-positive coccus) or two (*Veillonella, Enterococcus durans*) bacterial isolates were effective in inhibiting the growth of *S. typhimurium* and pathogenic *Escherichia coli* in the presence of 2.5% lactose (Hinton et al., 1991; Hinton et al.,1992a). The reason for using lactose is to stimulate the production of both lactic acid and short chain VFA. It is believed that lactate leads to a decrease in cecal pH and increases the concentration of undissociated VFA, which are more bacteriostatic and bactericidal than dissociated VFA (Corrier et al., 1990). However, it is likely that the accumulation of lactic acid may also inhibit the growth of enteropathogens, particularly in the upper intestinal tract of chickens (Hinton et al., 1992b). These in vitro studies help to identify what factors may be involved in altering the growth of pathogens in vivo. This type of information is needed in order to fully understand the mechanism of protection and allow for the development of a consistent defined CE culture.

Another approach has been to combine the various isolates in order to mimic the known bacterial composition of the cecal microflora (Stavric, 1992). Unfortunately, only when these defined mixtures contain large numbers of bacteria (i.e., 50) is protection against *Salmonella* equal to that observed with undefined CE cultures (Stavric et al., 1985; Stavric, 1992; Stavric and D'Aoust, 1993). When fewer numbers of bacterial strains are used in these mixtures, they are generally less protective (Stavric and D'Aoust, 1993). According to Stavric and D'Aoust (1993) the major problems in developing defined culture treatments include: (1) a lack of a sound scientific basis for the selection of potentially protective strains, (2) inadequate selective isolation media for the recovery of minor bacterial components in cecal materials, and (3) the need for reliable diagnostic schemes for the identification of strains as reflected in the general inability to speciate more than 25% of the bacterial components isolated from the ceca of adult birds. Collectively, these problems emphasize the need for more research in order to fully understand the fundamental aspects of how these CE cultures work.

To address the problem of inadequate in vitro culture systems for selecting and maintaining cultures of defined microorganisms, continuous-flow culture was recently used to select for a defined bacterial culture from the chicken cecum (Nisbet et al., 1993). This method has the advantage of being able to simulate environmental conditions (i.e., dilution rate) found in the chicken cecum. Based on this method using lactose as the primary carbon source, it was possible to maintain a defined mixture of 11 indigenous Gram-negative and Gram-positive cecal

Table 3. Composition of a defined bacterial culture obtained by continuous-flow culture of cecal contents from an adult broiler chicken[a].

Gram-positive cocci (Facultative anaerobes)
Enterococcus avium
Enterococcus faecalis A
Enterococcus faecalis B
Lactococcus lactis

Gram-positive bacilli (Facultative anaerobes)
Lactobacillus animalis
Lactobacillus strain CMS

Gram-negative bacilli (Facultative anaerobes)
Citrobacter freundii
Escherichia coli
Escherichia fergusonii

Gram-positive bacilli (Obligate anaerobes)
Bifidobacterium animalis
Propionibacterium acidipropionici

[a]Nisbet et al. (1993).

bacteria (Table 3; Nisbet et al., 1993). Nine of the isolates are facultative anaerobes, while the other two isolates are obligate anaerobes. Furthermore, a combination of these bacteria and dietary lactose effectively controlled *S. typhimurium* colonization in chicks (Nisbet et al., 1993).

Soluble carbohydrates

In addition to lactose, several reports have shown promising results regarding the use of other soluble sugars, particularly mannose, to reduce colonization of the chicken intestinal tract by *S. typhimurium*. Initial studies using one-day-old chicks investigated the effects of various sugars on the in vitro colonization of the small intestine by *S. typhimurium* (Oyofo et al., 1989a). The addition of D-mannose, arabinose, methyl-α–D-mannoside, and galactose significantly reduced adherence of *S. typhimurium* to chick small intestine epithelial cells (Table 4). Methyl-α–D-mannoside and D-mannose were much more inhibitory than galactose. These researchers hypothesized that the decreased adherence was due to D-mannose interfering with the mechanism(s) used by *S. typhimurium* to attach to the intestinal epithelial cells. Adherence to the intestinal mucosa is a prerequisite for colonization and invasion of hosts by bacteria (Gibbons and Van Houte, 1975). In several pathogenic bacteria it appears that adherence to the intestinal epithelial cells is mediated by a mannose-specific substance present on the bacterial surface (Freter and Jones, 1976). Therefore, if mannose is present in the intestinal tract in sufficient concentrations, it may be possible to block the mannose-specific receptors on the surface of pathogenic bacteria and reduce intestinal colonization.

Table 4. Effect of various sugars on adherence of *Salmonella typhimurium* (ST-10) to 1-day-old chicken epithelial cells[a].

Carbohydrate tested[1]	Adherence[2]	Inhibition (%)
Control (no sugar)	52.0	...
Methyl-α-D-mannoside	3.3	94
D-Mannose	5.5	90
Arabinose	9.4	82
Galactose	20.3	62

[a]Oyofo et al. (1989a).
[1]Carbohydrate (2.5% wt/vol) was added with *S. typhimurium* (ST-10) to chicken small intestinal cells. Binding assay was carried out at 37°C for 30 min.
[2]Adherence is expressed as $\bar{x} \pm SD$ (n=6) of *S. typhimurium* bound per segment of chicken small intestinal cells ($\times 10^3$).

Based on these in vitro results, experiments were performed to evaluate the effect of incorporating 2.5% mannose into the drinking water of broiler chickens (Oyofo et al., 1989b). When these chickens were orally challenged with *S. typhimurium*, they were significantly ($P < .001$) less colonized than the control animals (Table 5). Further studies examined the effects of several other sugars (dextrose, lactose, maltose, sucrose) on *S. typhimurium* colonization of the ceca of broiler chickens and only lactose gave results similar to those observed for mannose (Oyofo et al., 1989c). No changes in weight gain were noted with incorporation of different sugars into the drinking water. These authors postulated that the mechanism of action associated with the lactose-mediated inhibition of *S. typhimurium* colonization was different than that for mannose. It was suggested that lactose promotes the growth of lactose-fermenting bacteria that compete with *S. typhimurium* for colonization sites in the intestine or, as described above, produce VFA and(or) lactate that are toxic to *S. typhimurium* (Oyofo et al., 1989c). Corrier et al. (1991) found in broiler chicks that a combination of anaerobic cecal microflora (undefined culture) and lactose decreased cecal colonization by *S. typhimurium* better than either treatment alone (Table 6) and this observation is fairly consistent in the scientific literature.

Table 5. Effect of D-mannose treatment on *Salmonella typhimurium* colonization of the cecum of broiler chickens[a].

		Number of chickens colonized/group				
		1st Replication		2nd Replication		3rd Replication	
Treatment Group		n	% Colonized	n	% Colonized	n	% Colonized
1	Control (water)	0		0		0	...
2	Control (mannose)	0		0		0	..
3	*Salmonella* (water)	22/28	78	23/28	82	26/28	8
4	*Salmonella* (mannose)	8/29	28***	6/28	21***	12/28	43***

[a]Oyofo et al. (1989b).
***$P<.001$ between Groups 3 and 4.

Table 6. Effect of providing lactose in the drinking water and inoculation of anaerobic cultures of cecal flora on *Salmonella typhimurium* colonization of broiler chicks[a,b].

Treatment Group	Log$_{10}$*Salmonella*/g Cecal Content	*Salmonella* Positive Cultures/Total (%)
1. Controls	5.74	10/10 (100)
2. Anaerobic cultures	4.12	13/14 (93)
3. Lactose (2.5%) in water	3.96	9/11 (82)
4. Lactose (2.5%) in water and anaerobic cultures	0.36	2/14 (14)

[a]Lactose was provided from 1 to 10 days of age. Chicks that received anaerobic cultures were inoculated orally on day of hatch. All groups were challenged with 10^6 *S. typhimurium* at three days of age.
[b]Corrier et al. (1991).

Collectively, these results suggest that the addition of mannose or lactose to the drinking water may be a simple means of significantly reducing *S. typhimurium* colonization in chickens. The economic question then becomes which sugar is the least expensive and readily available? Oyofo et al. (1989c) favors lactose over mannose because of its availability and very low cost. For example, 100 gm of D-mannose costs $46.00 versus 1000 gm of lactose at $9.50 (Sigma Chemical Co., St. Louis, MO). Another source of lactose may be the incorporation of dried whey or nonfat milk into the feed or drinking water of chickens. Both byproducts contain lactose and were effective in reducing *S. typhimurium* numbers in young broiler chickens (DeLoach et al., 1990).

Oligosaccharides

Recently, the effect of fructooligosaccharide (FOS) on the ability of *S. typhimurium* to grow and colonize the chicken gut was studied. Fructooligosaccharide consists primarily of one to three fructose residues attached to a sucrose molecule and is found naturally in relatively high concentrations in onions, wheat, rye, and barley (Bailey et al., 1991). Because FOS can enhance the growth of some intestinal bacteria (*Lactobacillus, Bifidobacterium*), there has been some interest in evaluating the effects of FOS on *Salmonella* colonization in chickens (Hidaka et al., 1986; Mitsuoka et al., 1987; Bailey et al., 1991). When 0.375% FOS was fed to chicks, little influence on *Salmonella* colonization was observed (Bailey et al., 1991). However, with 0.75% FOS 12% fewer FOS-fed birds were colonized with *Salmonella* compared with control birds. Furthermore, a combination of 0.75% FOS plus an undefined CE culture resulted in only 19% colonization of chickens challenged with *Salmonella* versus 61% in chicks given only the CE culture. Incorporation of FOS (0.375%) in the absence of CE culture into broiler diets had little consistent effect on growth rate, feed utilization, mortality, carcass dressing percentage, abdominal fat content, or incidence or severity of salmonellae contamination of processed broiler carcasses (Waldroup et al., 1993).

Based on the dramatic effect D-mannose has on the attachment and subsequent intestinal colonization by *Salmonella* (see above), there has been commercial interest in evaluating the effects of mannan-based oligosaccharides (MOS) on gastrointestinal microbiology. Since mannan is more resistant to microbial degradation than FOS, it is believed that MOS treatment would be more specific for beneficial lactic acid producing gastrointestinal bacteria such as *Lactobacillus* and *Bifidobacterium* (Alltech, Inc.). Furthermore, growth of some pathogenic bacteria appears to be depressed in the presence of MOS (Alltech, Inc.). A good source of this compound is yeast (i.e., *Saccharomyces cerevisiae*) due to the presence of mannan in the cell wall of these microorganisms (Spencer and Gorin, 1973). Little published information is available describing the effect of MOS on gastrointestinal microflora. However, preliminary studies have shown that addition of MOS to milk replacer tended to improve the rate of gain in Holstein calves and decrease the incidence of respiratory disease (Newman et al., 1993). Preliminary results have also been positive with the addition of MOS to poultry and swine diets (Alltech, Inc.).

Final thoughts

There is great potential for manipulating gastrointestinal microflora in production livestock to improve feedstuff utilization, production efficiency, and reduce intestinal colonization by pathogenic microorganisms. However, in order to unlock this potential researchers need to fully understand the fundamental aspects associated with these microbes if progress is to be made in the future. Scientists frequently get criticized for spending too much time studying "trivial" details or having "their nose in the test tube", but this focused approach is needed if we are to understand the "mechanism of action" involved in biological processes. Because the mechanism of protection of CE cultures is not known, the development of defined CE cultures will be limited as will the application of this technique. In my opinion, if progress is to be made in the area of defined CE cultures, research needs to focus on understanding fundamental mechanisms associated with predominant poultry cecal bacteria.

References

Bailey, J.S., L.C. Blankenship, and N.A. Cox. 1991. Effect of fructo-oligosaccharide on *Salmonella* colonization of the chicken intestine. Poultry Sci. 70:2433.

Barnes, E.M., and C.S. Impey. 1972. Some properties of the nonsporing anaerobes from poultry caeca. J. Appl. Bacteriol. 35:241.

Barnes, E.M., G.C. Mead, D.A. Barnum, and E.G. Harry. 1972. The intestinal flora of the chicken in the period 2 to 6 weeks of age, with particular reference to the anaerobic bacteria. Brit. Poultry Sci. 13:311.

Barnes, E.M., C.S. Impey, and B.J.H. Stevens. 1979. Factors affecting the incidence and anti-salmonella activity of the anaerobic caecal flora of the young chick. J. Hyg. (Cambridge) 82:263.

Barnes, E.M., C.S. Impey, and D.M. Cooper. 1980. Manipulation of the crop and intestinal flora of the newly hatched chick. Am. J. Clin. Nutr. 33:2426.

Blankenship, L.C., N.A. Cox, J.S. Bailey, and N.J. Stern. 1990. Competitive exclusion cultures in chickens. *In:* BioZyme Service Through Science Technical Symposium, St. Joseph, MO. p. 37.

Corrier, D.E., A. Hinton, Jr., R.L. Ziprin, R.C. Beier, and J.R. DeLoach. 1990. Effect of dietary lactose on cecal pH, bacteriostatic volatile fatty acids, and *Salmonella typhimurium* colonization of broiler chicks. Avian Dis. 34:617.

Corrier, D.E., A. Hinton, Jr., R.L. Ziprin, R.C. Beier, and J.L. DeLoach. 1991. Effect of dietary lactose and anaerobic cultures of cecal flora on *Salmonella* colonization of broiler chicks. In: L.C. Blankenship (Ed.) Colonization Control of Human Bacterial Enteropathogens in Poultry. Academic Press, Inc., New York. p. 299.

Cox, N.A., J.S. Bailey, J.M. Mauldin, and L.C. Blankenship. 1990. Research note: presence and impact of *Salmonella* contamination in commercial broiler hatcheries. Poultry Sci. 69:1606.

Cox, N.A., J.S. Bailey, J.M. Mauldin, L.C. Blankenship, and J.L. Wilson. 1991a. Research note: extent of salmonellae contamination in breeder hatcheries. Poultry Sci. 70:416.

Cox, N.A., J.S. Bailey, and L.C. Blankenship. 1991b. Alternative administration of competitive exclusion treatment. *In:* L.C. Blankenship (Ed.) Colonization Control of Human Enteropathogens in Poultry. Academic Press, Inc., New York. p. 105.

DeLoach, J.R., B.A. Oyofo, D.E. Corrier, L.F. Kubena, R.L. Ziprin, and J.O. Norman. 1990. Reduction of *Salmonella typhimurium* concentration in broiler chickens by milk or whey. Avian Dis. 34:389.

Freter, R., and G.W. Jones. 1976. Adhesive properties of *Vibrio cholera*: nature of the interaction with mucosal surfaces. Infect. Immun. 14:246.

Gibbons, R.J., and J. Van Houte. 1975. Bacterial adherence in oral microbial ecology. Annu. Rev. Microbiol. 29:19.

Hidaka, H., T. Edida, T. Takazawa, T. Tokunga, and Y. Tashiro. 1986. Effects of fructooligosaccharides on intestinal flora and human health. Bifidobacteria Microflora 5:37.

Hinton, A., Jr., G.E. Spates, D.E. Corrier, M.E. Hume, J.R. DeLoach, and C.M. Scanlan. 1991. In vitro inhibition of the growth of *Salmonella typhimurium* and *Escherichia coli* 0157:H7 by bacteria isolated from the cecal contents of adult chickens. J. Food Prot. 54:496.

Hinton, A., Jr., D.E. Corrier, and J.R. DeLoach. 1992a. In vitro inhibition of *Salmonella typhimurium* and *Escherichia coli* 0157:H7 by an anaerobic Gram-positive coccus isolated from the cecal contents of adult chickens. J. Food Prot. 55:162.

Hinton, A., Jr., D.E. Corrier, and J.R. DeLoach. 1992b. Inhibition of the growth of *Salmonella typhimurium* and *Escherichia coli* 0157:H7 on chicken feed media by bacteria isolated from the intestinal microflora of chickens. J. Food Prot. 55:419.

Impey, C.S., and G.C. Mead. 1989. Fate of salmonellas in the alimentary tract of chicks pre-treated with a mature caecal microflora to increase colonization resistance. J. Appl. Bacteriol. 66:469.

Mead, G.C. 1991. Developments in competitive exclusion to control *Salmonella* carriage in poultry. *In:* L.C. Blankenship (Ed.) Colonization Control of Human Bacterial Enteropathogens in Poultry. Academic Press, Inc., New York. p. 91.

Meynell, G.G. 1963. Antibacterial mechanisms of the mouse gut. II: the role of Eh and volatile fatty acids in the normal gut. Brit. J. Exp. Pathol. 44:209.

Mitsuoka, T., H. Hidaka, and T. Eida. 1987. Effect of fructooligosaccharides on intestinal microflora. Die Nahrung 31:5–6, 427.

Mulder, R.W.A.W., and N.M. Bolder. 1991. Experience with competitive exclusion in the Netherlands. *In*: L.C. Blankenship (Ed.) Colonization Control of Human Bacterial Enteropathogens in Poultry. Academic Press, Inc., New York. p.77.

Newman, K., K. Jacques, and R. Buede. 1993. Effect of mannanoligosaccharide on performance of calves fed acidified and non-acidified milk replacers. J. Anim. Sci. 71(Suppl. 1):271.

Nisbet, D.J., D.E. Corrier, C.M. Scanlan, A.G. Hollister, R.C. Beier, and J.R. DeLoach. 1993. Effect of a defined continuous-flow derived bacterial culture and dietary lactose on *Salmonella typhimurium* colonization in broiler chickens. Avian Dis. 37:1017.

Nurmi, E., and M. Rantala. 1973. New aspects of *Salmonella* infection in broiler production. Nature 241:210.

Oyofo, B.A., R.E. Droleskey, J.O. Norman, H.H. Mollenhauer, R.L. Ziprin, D.E. Corrier, and J.R. DeLoach. 1989a. Inhibition by mannose of in vitro colonization of chicken small intestine by *Salmonella typhimurium*. Poultry Sci. 68:1351.

Oyofo, B.A., J.R. DeLoach, D.E. Corrier, J.O. Norman, R.L. Ziprin, and H.H. Mollenhauer. 1989b. Prevention of *Salmonella typhimurium* colonization of broilers with D-mannose. Poultry Sci. 68:1357.

Oyofo, B.A., J.R. DeLoach, D.E. Corrier, J.O. Norman, R.L. Ziprin, and H.H. Mollenhauer. 1989c. Effect of carbohydrates on *Salmonella typhimurium* colonization in broiler chickens. Avian Dis. 33:531.

Roberts, T. 1988. Salmonellosis control: estimated economic costs. Poultry Sci. 67:936.

Salanitro, J.P., I.G. Fairchilds, and Y.D. Zgornicki. 1974a. Isolation, culture characteristics, and identification of anaerobic bacteria from the chicken cecum. Appl. Microbiol. 27:678.

Salanitro, J.P., I.G. Blake, and P.A. Muirhead. 1974b. Studies on the cecal microflora of commercial broiler chickens. Appl. Microbiol. 28:439.

Snoeyenbos, G.H., O.M. Weinack, and C.F. Smyser. 1978. Protecting chicks and poults from salmonellae by oral administration of "normal gut microflora." Avian Dis. 22:273.

Spencer, J.F.T., and P.A.J. Gorin. 1973. Mannose-containing polysaccharides of yeast. Biotech. Bioeng. 15:1.

Stavric, S., T.M. Gleeson, B. Blanchfield, and H. Pivnick. 1985. Competitive exclusion of *Salmonella* from newly hatched chicks by mixtures of pure bacterial cultures isolated from fecal and cecal contents of adult birds. J. Food Prot. 48:778.

Stavric, S. 1992. Defined cultures and prospects. Int. J. Food Microbiol. 55:245.

Stavric, S., and J.-Y. D'Aoust. 1993. Undefined and defined bacterial preparations for the competitive exclusion of *Salmonella* in poultry – a review. J. Food Prot. 56:173.

Waldroup, A.L., J.T. Skinner, R.E. Hierholzer, and P.W. Waldroup. 1993. An evaluation of fructooligosaccharide in diets for broiler chickens and effects on salmonellae contamination of carcasses. Poultry Sci. 72:643.

Wierup, M., M. Wold-Troell, E. Nurmi, and M. Hakkinen. 1988. Epidemiological evaluation of the *Salmonella*-controlling effect of a nationwide use of a competitive exclusion culture in poultry. Poultry Sci. 67:1026.

MANNAN-OLIGOSACCHARIDES: NATURAL POLYMERS WITH SIGNIFICANT IMPACT ON THE GASTROINTESTINAL MICROFLORA AND THE IMMUNE SYSTEM

K. NEWMAN

North American Biosciences Center, Nicholasville, Kentucky USA

Introduction: functions of oligosaccharides

The interaction between nutrition and disease has long been known, but we are just beginning to understand that this relationship is far more complex than originally thought. Not only is the food we eat important to health by providing the building blocks of the immune system, but certain compounds that we eat may help immunity directly. It is common knowledge that certain allergies are caused by the food we eat. Recently, there has been incredible interest in the role of complex carbohydrates in disease prevention and treatment. For several years it was assumed that carbohydrates could not account for anything other than energy sources because of the apparent simplicity of the structure of the monosaccharide backbone. However, consider that two identical monosaccharides can bond to form 11 distinct disaccharides whereas two identical amino acids can make only one dipeptide. Structural complexity of the oligosaccharides explains the AB and O blood types in humans as well as why bacterial infections caused by K99 *Escherichia coli* occur in calves, piglets and lambs but not humans.

MANNAN-OLIGOSACCHARIDES: PATHOGEN ADSORPTION, IMMUNOMODULATION

One of the most fascinating of these oligosaccharides is mannan-oligosaccharide (MOS). MOS is a glucomannoprotein complex derived from the cell wall of yeast. The activity of this compound as we understand it today is two-fold: adsorption of enteric pathogens and immunomodulation. Adsorption of pathogenic bacteria is not a novel concept. Considerable evidence has accumulated over the past 15 years to show that certain components of bacterial surfaces called lectins are involved in the onset of enteric and urinary disease by mediating adherence of the bacteria to epithelial cells in these sites (Abraham and Beachey, 1985; Beachey, 1981; Firon et al., 1987; Roberts, 1984; and Sharon and Ofek, 1986). Although there are a number of lectins that

Table 1. Attachment of bacterial isolates to mannans[a].

Strain	No. of positive strains	No. of strains tested
Escherichia coli	54	118
Salmonella enteritidis	4	4
Salmonella typhi	4	6
Salmonella typhimurium	4	13
Proteus morganii	11	11
Klebsiella pneumoniae	15	16
Citrobacter diversus	36	36
Serratia marcescens	12	12
Aeromonas hydrophila	5	7

[a]Adapted from Mirelman et al., 1980.

are specific for sugars such as galactose and fucose, lectins specific for mannose predominate in intestinal pathogens (Table 1). These lectins are located on the exterior of the cell and are associated with the pili or fimbriae of bacteria. Bacterial cells with pili that are specific for mannose attach to mannose-containing cells in the intestinal tract. Once these cells are attached, they can then colonize the tract and cause disease. Mannan-oligosaccharide provides a mannose-rich source for attachment which will adsorb bacteria that would otherwise attach to the gut wall. Since MOS is not degraded by digestive enzymes, it passes through the tract with the pathogen(s) attached, preventing colonization.

Several studies have been conducted examining the role of mannans and their derivatives on binding of pathogens to epithelial cells in the GI tract. *E. coli* with mannose-specific lectins did not attach to mammalian

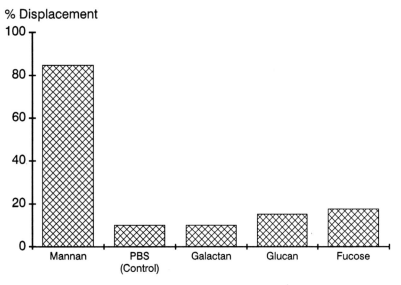

Figure 1. Ability of certain sugars to displace *E. coli* from epithelial cells. All sugars except fucose are derivatives of the sugar listed (Adapted from Ofek and Beacheyk, 1987).

cells when mannose was present (Salit and Gotschlich, 1977). In vitro studies examining *E. coli* bound to epithelial cells could be displaced from these epithelial cells within 30 minutes when exposed to a mannan derivative (Figure 1). Glucose and galactose derivatives had no effect on attached bacteria. This is important because it shows that not only can mannans prevent the attachment of pathogens to the gut, but they may also "clean-up" already attached pathogens. In poultry infected with *Salmonella typhimurium*, mannose in the feed significantly decreased colonization in the cecum (Figure 2).

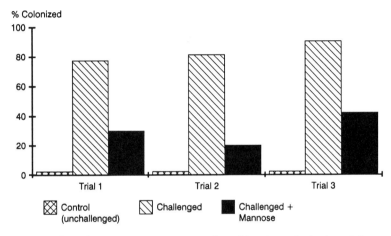

Figure 2. Effect of D-mannose treatment on *S. typhimurium* colonization of the cecum of broiler chicks (Adapted from Oyofo et al., 1989).

An overview of the immune system

Carbohydrate specificity for attachment is not limited to bacteria. Certain toxins, viruses and eukaryotic cells also attach to surfaces by recognition of certain sugars on the surface of other cells (Sharon and Lis, 1993). For example, leukocytes (white blood cells) exhibit carbohydrate attachment. Leukocytes are part of the immune system that defend the body from infection by leaving the general circulation and migrating into the tissues. When a tissue becomes infected, it secretes cytokines such as interleukin-1 and tumor necrosis factor. The cytokines cause the endothelial cells in the venules to express selectins which protrude from the venule surface. Passing white blood cells attach to these selectins by the lectin-carbohydrate mechanism described earlier. The leukocyte can then leave the bloodstream by pressing between adjacent endothelial cells. This mechanism is indispensable to leukocyte infection-fighting. Cellular immunity is mediated by leukocytes in the blood. The specificity of the cellular immune response depends on T cells which are a subset of the lymphocytes and originate in the thymus.

Antibody production is another form of host defense to infection. Antibodies are produced by B cells that form in the bone marrow. Antibodies are carried through the blood and lymph systems and represent what has historically been called the humoral immune system. Most immune responses involve the interaction between the humoral and cellular branches of the immune system. Because of this interaction, differentiation between immune responses may now be inappropriate. B cells recognize antigens due to antibodies that they carry on their cell surface. T cells identify protein fragments of antigens of eight to 15 amino acids brought to the surface of a cell by Major Histocompatibility Complex (MHC). Once the foreign peptide is identified, the T cell divides and secretes chemical signals called lymphokines, that in turn, activate other components of the immune system such as B cells.

Oligosaccharides and the immune system

In order to completely understand the role that oligosaccharides play in the immune system, much more research is necessary. However, certain generalized facts are recognized, e.g. adjuvant and immunomodulatory effects can be considered as two components of immunostimulation. An adjuvant is a substance that increases the immunological response to a vaccine, drug or antigen. Adjuvants can be added to vaccines to slow the absorption and increase their effectiveness (Figure 3). The ability of lipopolysaccharides of microbial origin to increase the resistance to infection by bacteria, fungi, viruses and parasites was first demonstrated by Rowley (1956) using the cell wall of Gram negative bacteria. An adjuvant effect to lipopolysaccharides has been shown to enhance both humoral and cell-mediated immunity (Le Garrec, 1986).

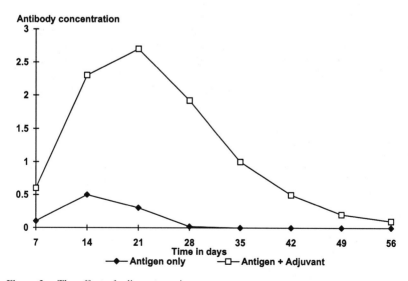

Figure 3. The effect of adjuvants on immune response.

Adjuvant activity is not the only manner in which lipopolysaccharides affect the immune system. In certain types of pneumonia, it has long been known that the lipopolysaccharide capsule of *Streptococcus pneumoniae* is the major antigen that the immune system recognizes and to which it elicits a response. The type 3 pneumococcal polysaccharide is an unbranched polymer of alternating units of glucose and glucuronic acid. Individuals that survive an infection caused by that *S. pneumoniae* become immune to future infections caused by that specific organism. This is a specific immunity response and is due to the formation of antibodies to the specific surface polysaccharides of that strain of *S. pnuemoniae*.

In addition to adjuvant and antigen properties, oligosaccharides containing mannose may also affect the immune system by stimulation of the liver to secrete mannose binding protein. This protein binds to the capsule of bacteria and triggers the complement cascade (Janeway, 1993).

Glucomannoprotein also has a role in the immune system. Early work injecting the material into mice, rats and rabbits resulted in an immediate decrease in properdin levels followed by an increase to concentrations up to three times greater than the initial properdin levels (Pillemer and Ross, 1955). Properdin is an important component of the alternate pathway of complement fixation. It is therefore not surprising that glucomannoprotein increases the bactericidal activity of animal sera (Blattberg, 1956).

Trial results: the practical response

Knowledge of the roles played by oligosaccharides has been put to practical use in livestock, poultry and aquaculture diets. Initial in vitro work with Bio-Mos (mannan-oligosaccharide) confirmed the ability of this material to adsorb specific pathogenic bacteria (Table 2). As was seen with the mannose table shown earlier, Bio-Mos has the ability to adsorb a high percentage of the pathogens examined including certain salmonella and clostridial species, and *E. coli* K88. Field trials with Bio-Mos have shown a decrease in condemnations in broilers and turkeys (Table 3). Bio-Mos has also proved beneficial when added

Table 2. Adsorption of certain enteric bacteria to Bio-Mos.

Bacterial species	Adsorption
Clostridium paraputrificum	Weakly positive
C. butyricum	Positive
C. sporogenes	Positive
E. coli K88	Positive
Salmonella enteritidis	Positive
Salmonella pullorum	Positive
Salmonella choleraesuis	Negative

Newman, 1994, Unpublished data.

Table 3. Effect of Bio-Mos on percent condemnations in poultry.

			% Condemned	
Species	Location	# of birds on trial	Control	Bio-Mos
Broilers	Canada	22,000	1.499	1.345
Broilers	Canada	73,000	1.86	1.71
Turkeys	Minnesota	51,000	2.49	1.67

Table 4. Effect of Bio-Mos on performance of starter pigs receiving a medicated diet[1,2].

	Control	Bio-Mos
Number of animals	36	36
Days on feed	14	14
Start wt (kg)	8.04	7.87
End wt (kg)	12.8	12.76
Gain (kg)	4.76	4.89
Feed consumed (kg)	7.42	7.18
Feed:Gain	1.56	1.47

[1]Banminth/Meccadox.
[2]King, 1993 (Personal communication.)

along with Banminth/Meccadox to pig starter diets. In a trial done in Iowa, starter pigs had improved gain and feed conversion when Bio-Mos was incorporated into the ration (Table 4). In calves fed a commercial milk replacer, Bio-Mos supplementation resulted in lower fecal coliform concentrations and decreased respiratory disease incidence than unsupplemented calves (Figures 4 and 5).

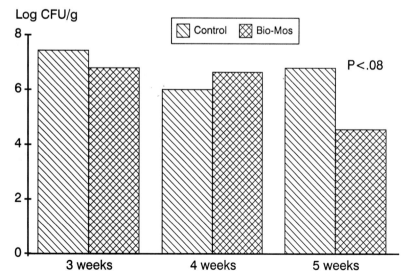

Figure 4. Fecal coliform concentrations in calves receiving a commercial milk replacer with or without Bio-Mos (2g per day) (Newman et al., 1993).

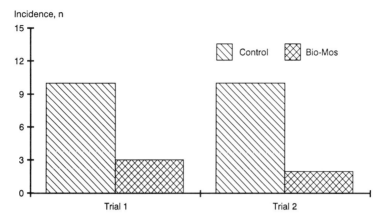

Figure 5. Incidence of respiratory disease in calves receiving a commercial milk replacer with or without Bio-Mos (Adapted from Newman et al., 1993).

Enhanced resistance to disease has also been demonstrated when glucans have been added to aquaculture feeds. Briefly, glucans increased the survival rate of salmon and trout fry exposed to the bacteria *Aeromonas salmonicida*; the organism that causes furunculosis in fish. Furunculosis in salmonids is a general bacteremia characterized by lesions throughout the internal viscera and represents a major problem to the producer when present.

Summary

The interaction between oligosaccharides and the immune system is just beginning to be understood. It is clear from previous work that mannans demonstrate the ability to bind certain bacteria and prevent these organisms from attaching and colonizing the GI tract. In addition, mannans and glucans elicit an immunomodulating effect as is shown in work with calves and in trout. This immunological benefit is clearly not species specific due to the diversity of the species examined that demonstrate some benefit from supplementation with mannan-oligosaccharide. The exact nature of this activity is just beginning to be understood and represents pioneering science in the interrelationship between nutrition and immunity.

References

Abraham, S.N. and E.H. Beachey. 1985. In: J.I. Gallin and A.S. Fauci (ed.), Host Defense Mechanisms, vol. 4. Raven Press, New York.
Beachey, E.H. 1981. J. Infect. Dis. 143:325–345.
Blattberg, B. 1956. Proc. Soc. Exp. Biol. Med. 92:745.
Firon, N., S. Ashkenazi, D. Mirelman, I. Ofek and N. Sharon. 1987. Infect. and Immun. 55:472–476.

Grant, G., L.J. More, N.H. McKenzie and A. Pusztai. 1983. Br. J. Nutr. 50:207.
Janeway, C.A. 1993. Scientific American. September. 73–79.
Le Garrec. 1986. Comp. Immun. Microbiol. Infect. Dis. 9:137–141.
Mirelman, D., G. Altmann and Y. Eshdat. 1980. J. Clinical Microbiol. 11:328–331.
Ofek, I. and E.H. Beachey. 1978. Infect. and Immun. 22:247–254.
Oyofo, B.A., J.R. DeLoach, D.E. Corrier, J.O. Norman, R.L. Ziprin and H.H. Mollenhauer. 1989. Poultry Sci. 68:1357–1360.
Pillemer, L. and O.A. Ross. 1955. Science. 121:732.
Roberts, J.A. 1984. Am. J. Kidney Dis. 4:103–117.
Rowley, D. 1956. Br. J. Exp. Path. 37:223.
Salit, I.E. and E.C. Gotschlich. 1977. J. Exp. Med. 146:1182–1194.
Sharon, N. and H. Lis. 1993. Scientific American. Jan. 82–89.
Sharon, N. and I. Ofek. 1986. In: D. Mirelman (ed.), Microbial lectins and agglutinins. John Wiley and Sons, Inc., New York.

STUDY OF THE USE OF YUCCA EXTRACT (DE-ODORASE) IN THE FATTENING OF BOARS

KLAUS ENDER, GERDA KUHN, KARIN NÜRNBERG
*Research Institute for the Biology of Agricultural Animals,
Dummerstorf, Germany*

Presented at Alltech's 9th Annual Symposium on Biotechnology in the Feed Industry, April 26–28, 1993

Introduction

Boars have better growth performance than barrows, utilize feed more efficiently and yield a carcass lower in fat. Also, due to less excretion of nitrogen and phosphorous, feeding boars reduces environmental concerns. From the animal protection point of view, avoidance of castration is welcomed. A problem, however, with fattening boars instead of barrows is the emission of sex-specific odours during preparation of the meat. Boar odour is a complex phenomenon formed by several chemical reactions. The two compounds 5-androst-16-en-3-one (androstenone) and 3-methyl indole (skatole) are regarded as the main elements in boar odour (Bonneau, 1982; Lundström *et al.*, 1988). Hannson *et al.* (1980) showed that 36% of the smell can be attributed to androstenone, 33% to skatole and 7% to indole. Hannson *et al.* (1980) also found that most people have a more sensitive reaction to the odour of skatole than to androstenone, so skatole has a more important influence on the smell of the meat than the sexual odour caused by androstenone and other compounds.

The subject of the fattening boar has been much discussed recently. The Fresh Meat Guidelines (64/433/EEC) released by the EC have influenced this discussion. These stipulate that from 1 January 1993 meat from non-castrated male pigs up to a slaughter weight of 80 kg (100 kg liveweight) can generally be classified as suitable. Additionally, heavy boar carcasses can be classified as suitable for consumption if a recognized test method rules out distinct differences in odour. In Danish abattoirs carcasses are selected according to skatole content. A practical method for regulation of androstenone is being developed in Germany. Extensive studies have been undertaken to find ways of preventing or reducing boar odour, however nothing conclusive has been found to date. The meat processing industry is concerned that the market position of pork could be threatened by odour-tainted meat reaching the shops. A reduction in the quantity of compounds that create boar odour would significantly reduce this risk.

Study of the use of yucca extract (De-Odorase) in the fattening of boars

Objectives

Skatole is found in the large intestine as the result of the breakdown of tryptophan by microorganisms. Skatole presence in the intestine is not confined to boars; but due to a higher amino acid metabolism boars have a higher concentration of skatole than barrows. As in the case of androstenone, skatole is found in adipose tissue. The NH_2 or NH groups of molecules are characteristic of the accumulation of tryptophan and skatole. Headon (1992) demonstrated that *Yucca schidigera* extract in the product De-Odorase has ammonia-binding characteristics. The hypothesis for the trial reported here was that skatole or NH groups of intermediary metabolism can be bound by the yucca extract at the site of origin (i.e., the large intestine) before they can be deposited in muscle and fat. The trial examined whether the boar odour due to skatole could be reduced through feeding De-Odorase. Because De-Odorase can be used as a feed additive, it was a further matter of interest to examine the effects of its inclusion on carcass quality.

Materials and methods

The trial involved 104 fattening boars and 35 castrated pigs of the genetic construction DE × DL with the castrated pigs serving as a comparison. Weaned pigs (average weight of 28 kg) were assigned to pens of eight to 10 animals. Two adjacent pens sharing a common feed trough constituted a trial group. Feeding was semi-*ad libitum* and utilized liquid feed. (Table 1).

Table 1. Composition of the basal diet.

Nutrient	g/kg dry matter
Dry matter	880.0
Metabolizable energy, MJ	12.5
Crude protein	193.4
Lysine	10.5

The De-Odorase preparation was given immediately before each feeding at a rate of 120 g/t of feed added to the mix container. The effect of De-Odorase was evaluated in boars finished at 95, 105 (both boars and barrows) and 120 kg (Table 2).

Animals were slaughtered when target liveweights were reached. Carcasses were weighed and the percentage of muscle determined using a regression based on apperative methods. Intramuscular fat content of the longissimus dorsi muscle was measured using the Soxhlet method (without acid decomposition). Skatole content of backfat was established

Table 2. Treatment groups.

Group	n	Sex	End weight	Treatment
1	18	boar	95	Control
2	16	boar	95	De-Odorase
3	17	boar	105	Control
4	17	boar	105	De-Odorase
5	18	boar	120	Control
6	18	boar	120	De-Odorase
7	17	barrow	105	Control
8	18	barrow	105	De-Odorase

spectroscopically in keeping with the modification of Mortensen and Sörenson's method (1984). A sensory test for boar odour was conducted using backfat and internal fat from the test animals. Samples were cut in cubes and heated in Erlenmeyer flasks on a hot plate. Odour emitted was assessed using a 4-point scale by a panel of six people:

Scale	Response
0	No boar odour
1	Weak boar odour
2	Medium boar odour
3	Strong boar odour

Statistical evaluation used the GLM procedures of SAS (Statistical Analysis System). Significant differences in the LS means were determined using a t-test.

Results

EFFECTS OF DE-ODORASE IN BOARS AT DIFFERENT FINAL WEIGHTS

The principal result of the trial was a marked reduction in skatole content of the adipose tissue of the boar with the addition of De-Odorase to the feed (Table 3). Average reduction in skatole was 25% across all weight groups. In the 95 kg liveweight group, the reduction was 33% ($P<0.05$) with a decrease from 0.24 µg/g to 0.16 µg/g of adipose tissue. For the 105 and 120 kg weight groups the relative differences were 27 and 22%, respectively. Furthermore, under condition of higher increases, animals at the same weight given De-Odorase had a significantly reduced muscle meat content. The difference in the middle weight group was 2.6% ($P<0.05$). The higher adipose tissue content of these animals is reflected in slightly higher values for the intramuscular fat.

The animals were graded according to the limit of 0.24 µg/g of skatole in adipose tissue used in Denmark (Table 4). According to this, the proportion of nonobjectionable carcasses increased with use of De-Odorase. Less than 0.24 µg skatole/g adipose tissue in boars at 95 and 105 kg final liveweight resulted in an improvement of about 20% in the number of acceptable carcasses based on skatole content. Reductions in skatole in the 120 kg boars increased the proportion of acceptable carcasses by 5%.

Table 3. The effect of De-Odorase on carcass quality and boar odour at different end weights.

		Carcass weight (kg)	Muscle (%)	Intramuscular fat (%)	Skatole (μg/g)	Sensory internal fat	Sensory fat
95 kg							
Control	Mean	75.58	57.50	0.74	0.24[a]	0.9	0.8
(n=18)	SE[1]	1.15	0.72	0.06	0.03	–	–
De-Odorase	Mean	77.49	57.28	0.68	0.16[b]	0.6	0.5
(n=16)	SE[1]	1.22	0.76	0.06	0.03	–	–
105 kg							
Control	Mean	82.87	58.20	0.79	0.22	1.1	0.8
(n=17)	SE[1]	1.18	0.74	0.06	0.03	–	–
De-Odorase	Mean	83.67	55.56[b]	0.88	0.16	1.0	0.7
(n=17)	SE[1]	1.18	0.74	0.06	0.02	–	–
120 kg							
Control	Mean	94.80	58.06	0.66	0.27	1.0	1.0
(n=18)	SE[1]	1.15	0.72	0.06	0.02	–	–
De-Odorase	Mean	94.84	56.10	0.75	0.21	1.1	1.2
(n=18)	SE[1]	1.15	0.72	0.06	0.02	–	–
Total							
Control	Mean	84.42	57.92[a]	0.73	0.24[a]	1.0	0.9
(n=53)	SE[1]	0.67	0.42	0.03	0.02	–	–
De-Odorase	Mean	85.34	56.32[b]	0.77	0.18[b]	0.9	0.8
(n=51)	SE[1]	0.68	0.43	0.03	0.01	–	–

[1]standard error
*n=16
[ab] Means differ, P<0.05

Table 4. Animal grading (%) according to skatole limit (24 µg/g adipose tissue).

		≤ 0.24 µg/g	>0.24 µg/g	% change
95 kg	Control	62.50	37.50	
	De-Odorase	81.25	18.75	18.75
120 kg	Control	56.25	43.75	
	De-Odorase	82.35	17.65	26.10
120 kg	Control	55.56	44.0	
	De-Odorase	61.1	38.89	40.44 5.55

These results indicated that the effect of De-Odorase decreased as the weight of the animal increased. This could be caused by a less intensive protein metabolism and more intestinal flora in the case of the heavier animals. The sensory examination confirmed the results of the chemical analysis on principle. At a final liveweight of approximately 95 kg, the fat samples of the animals given DeOdorase had on average a considerably higher value than the untreated animals, whereas after 120 kg there was no effect. The unchanging high average grade of the heavy animals could be an indicator that the components of the boar odour due to androstenone are increasingly important. The most promising result as regards the reduction of boar odour is with the addition of De-Odorase to lighter weight fattening boars.

COMPARATIVE EFFECTS OF DE-ODORASE IN BOARS AND BARROWS

The relationship between the decrease in the amount of muscle meat in the carcass and the inclusion of De-Odorase can be observed with both males and females (Table 5). The effect is much less pronounced with fatter barrows than with boars. Also for this reason there was no increase in the amount of intramuscular fat in the longissimus dorsi muscle of barrows.

Table 5. The effect of De-Odorase on carcass quality of boars and barrows with the same weight at the end of the fattening period (carcass weight 82.5 kg).

		Carcass weight (kg)	Intramuscular fat (%)	Skatole (μg/g)	Sensory internal fat	Sensory backfat
Boars						
Control	Mean	58.26[a]	0.78	0.22	1.1	0.8
(n=17)	SE[1]	0.66	0.10	0.02	–	–
De-Odorase	Mean	55.71[b]	0.87	0.16	1.0	0.7
(n=17)	SE[1]	0.66	0.10	0.02	–	–
Barrows						
Control	Mean	50.87	1.24	0.16	0.2	0.1
(n=17)	SE[1]	0.66	0.10	0.02	–	–
De-Odorase	Mean	50.57	1.10	0.14	0.3	0.1
(n=18)	SE[1]	0.64	0.10	0.02	–	–

[1]Standard error

De-Odorase reduced skatole content of boars by 0.06 μg/g adipose tissue (27%). The difference was close to significance at $P < 0.05$. At 0.16 μg/g adipose tissue the level of the untreated barrow is reached. Further decrease in skatole content through the use of De-Odorase is limited in the case of barrows because treated animals differ from the untreated animals by only 0.02 μg/g adipose tissue (12%). Boars given De-Odorase had skatole levels equal to barrows (0.16 μg/g).

It is notable that the analysed skatole contents of boars and barrows do not correspond to the differences in the sensory values. At the same skatole levels, fat samples from barrows are significantly better. This is proof that apart from skatole other substances, eg. androstenone or indole, are responsible for boar odour.

Conclusions

The effects of *Yucca schidigera* extract addition to fattening diets at 120 g/kg on carcass quality and the sex-specific boar odour of 104 fattening boars and 35 castrated pigs of different end liveweights (95, 105, 120 kg) was examined. The following effects were observed:

1. Skatole content in adipose tissue was reduced by 0.08 µg/g (i.e., 33%) in boars at 95 kg final weight.
2. Sensory values were improved in boars at 95 kg final weight.
3. The effect of De-Odorase decreased with increasing carcass weight (overall average of 27%).
4. Boars given De-Odorase reached the same skatole levels as barrows.
5. De-Odorase improved weight with less muscle meat content.

References

Bonneau, M. 1982. Compounds responsible for boar taint, with special emphasis on androstenone: a review. Livest. Prod. Sci. 9:687.

Hansson, K.E., Lundström, K., Fjelkner-Modig, S., Persson, J. 1980. The importance of androstenone and skatole for boar taint. Swedish J. of Agricultur. Res. 10:167.

Headon, D.R. 1992. Eine Fallstudie zur Ermittlung der Glycokomponenten und Enzyme, die zur Herabsetzung der Umweltbelastung beltragen. 6, Europ. Alltech-Symposium, Würzburg 3:3.

Lundström, K., Malmfors, B., Malmfors, G., Stern, S., Petersson, H., Mortensen, A.B., Sörensen, S.E. Skatole. 1988. Androstenone and taint in boars fed two different diets. Livest. Prod. Sci. 18:55.

Mortensen, A.B., Sörensen, S.E. 1984. Relationship between boar taint and skatole determined with a new analysis method. Proc. of the 30th European Meeting of Meat Research Workers, Bristol, 394.

ENVIRONMENTAL ASPECTS OF ANIMAL PRODUCTION: THE DEPENDENCE OF LIVESTOCK PRODUCTION ON, AND IMPACT ON, ENVIRONMENTAL QUALITY

WATER QUALITY AND NUTRITION FOR DAIRY CATTLE

DAVID K. BEEDE
Dairy Science Department, University of Florida
Gainesville, Florida, USA

Introduction

Doubtless, water is the most important dietary essential nutrient. Loss of about 1/5 of body water is fatal. Lactating dairy cows need larger proportions of water relative to body weight (BW) than most livestock species because water comprises 87% of milk. Factors influencing daily water intake and requirements include physiological state, milk yield (MY), dry matter intake (DMI), body size, rate and extent of activity, diet composition including types of feedstuffs (e.g., concentrate, hay, silage or fresh forage), ambient temperature, and other environmental factors (e.g., humidity and wind velocity). Other factors affect how much of a particular water supply is consumed; included are salinity, and sulfate and chloride contents, dietary sodium (Na) content, temperature of water, frequency and periodicity of watering, social or behavioral interactions of animals and other quality factors such as pH and toxic substances. This chapter addresses factors affecting water quality, practical considerations of water nutrition, needs during heat stress, and effects of chilling drinking water during hot weather for dairy cattle.

Water quality and factors affecting performance

The most extensive review of water quality factors was by the NRC (1974) subcommittee on nutrients and toxic substances in water for livestock and poultry. Water quality is an extremely important issue both in terms of quality of drinking water provided to livestock and in terms of quality of water leaving production units as a potentially renewable resource. The latter is not considered in this paper.

Five criteria can be considered when evaluating drinking water quality: organoleptic, physio-chemical, substances present in excess, toxic compounds, and microorganisms (primarily bacteria). Organoleptic factors (e.g., odor and taste) may be readily detectable by the animal, but are of little direct consequence to health or productivity unless water consumption is affected dramatically. Physio-chemical properties, i.e., pH, total dissolved solids, hardness, and total dissolved oxygen are used to classify broadly water sources and generally do not present direct health risks but may indicate underlying problems. Excess of some chemicals normally found in water (i.e., nitrates, iron, Na, sulfates, and flourine) may be health risks or depress water consumption. Toxic substances may be common in water but generally are below dangerous concentrations; examples are arsenic, cyanide, lead, mercury, hydrocarbons, organochlorides, and organophosphates. Maximum bacterial concentrations of potable water for humans are regulated; however, less control is exercised for drinking water for livestock. Therefore, potential hazards might exist in some circumstances.

Salinity and total dissolved solids

Salinity refers to the amount of dissolved salts present in water. First consideration generally is sodium chloride but other dissolved inorganic constituents such as carbonates, sulfates, nitrates, potassium (K), calcium (Ca), and magnesium (Mg) are in the same category. These constituents may affect osmotic balance of animals and generally are measured as total dissolved solids (TDS). Table 1 adapted from NRC (1974) provides a guide to use of saline waters.

Table 1. Guide to use of saline waters for dairy cattle.

TDS[1] (mg/l, or ppm)	Comment
Less than 1,000 [fresh water]	Presents no serious burden to livestock.
1,000 – 2,999 [slightly saline]	Should not affect health or performance, but may cause temporary mild diarrhea.
3,000 – 4,999 [moderately saline]	Generally satisfactory, but may cause diarrhea, especially upon initial consumption.
5,000 – 6,999 [saline]	Can be used with reasonable safety for adult ruminants. Should be avoided for pregnant animals and baby calves.
7,000 – 10,000 [very saline]	Should be avoided if possible. Pregnant, lactating, stressed or young animals can be affected negatively.
>10,000 [approaching brine]	Unsafe, should not be used under any conditions.

[1]TDS = total dissolved solids

Jaster et al. (1978) studied effects of water salinity on lactation. Tap water (196 ppm) was compared with drinking water containing 2,500 ppm dissolved sodium chloride offered to 12 lactating cows averaging 82 lb MY/day at the beginning of the experiment. Water intake was 7% greater, but feed intake and MY tended to be less for cows offered high saline water. Minerals in blood and milk, and diet digestibilities were similar between treatments, but urine and fecal Na, and urine chloride (Cl) were higher for cows offered drinking water containing high amounts of salt. Other studies suggested no effect of drinking water containing 15,000 ppm (Heller, 1933) and 10,000 ppm (Frens, 1946) sodium chloride, or 5,895 ppm sodium sulfate. However, Frens (1946) reported reduction in MY with 15,000 ppm sodium chloride. Also in Arizona, Holstein heifers showed an increasing preference for drinking water as salinity approached 2,000 ppm, but water consumption dropped dramatically when salinity was greater than 2,500 ppm (Wegner and Schuh, 1974). In growing beef heifers, consumption of drinking water with 12,500 to 20,000 ppm sodium chloride caused symptoms of salt toxicity (Weeth and Haverland, 1961).

In a recent study in Saudi Arabia during warm weather, Challis *et al.* (1987) found that desalination of drinking water originally containing about 4,400 ppm TDS, of which 2,400 ppm were sulfates, improved MY by 28% (77 vs 60 lb/day), increased water intake 20 % and grain intake 32% compared with high saline water. Desalinated water contained 441 ppm TDS. When high saline water was re-introduced, MY dropped 13.2 lb/day during the first week. This study suggests that a combination of high TDS in drinking water and hot weather can be particularly deleterious for lactating cows.

Maximum tolerable concentrations of sulfates in drinking water were investigated in Nevada. Growing cattle were affected adversely by 3,493 ppm sulfate in their drinking water. Feed and water intakes were reduced and methemoglobin concentration was increased (Weeth and Hunter, 1971). In a subsequent study, 1,462 ppm or 2,814 ppm sulfate-water made by adding sodium sulfate to tap water, reduced hay intake, but not water consumption of Hereford heifers compared with controls (110 ppm sulfate in drinking water) (Weeth and Capps, 1972). Rate of BW gain was reduced by water containing the highest sulfate concentrations. Drinking water with 2,814 ppm sulfate increased methemoglobin concentrations and significantly altered renal function. These researchers concluded that the tolerable concentration of sulfate in drinking water for growing beef cattle in summer in Nevada was near 1,450 ppm.

In a follow-up study designed to define more accurately maximum tolerable concentrations of inorganic sulfate in drinking water, Digesti and Weeth (1976) offered 110, 1,250 or 2,500 ppm sulfate in drinking water, with higher concentrations added as sodium sulfate. Feed consumption, water intake and growth rate of beef heifers were not affected by higher sulfate drinking water during the 90-day experiment. No overt toxicity was detected. However, heifers given water with 1,250 or 2,500 ppm sulfate tended to accumulate more methemoglobin and sulfhemoglobin without a decrease in concentration of total blood hemoglobin. Sulfate loading did not cause diuresis, although renal filtration of sulfate was increased by the highest sulfate treatment. It appeared that 2,500 ppm sulfate in drinking water was tolerated by these heifers without adverse effects, and it was suggested to represent a safe tolerance concentration. In a companion study, sulfate (3,317 ppm estimated rejection threshold) in drinking water was found to be more unpalatable than Cl (5,524 ppm estimated rejection threshold). Recent evidence suggests that high dietary intakes of the anions, sulfate and Cl, can perturb acid–base balance of cattle (Wang and Beede, 1992). Abnormally high intakes of these anions in drinking water likely are responsible for detrimental effects on animal health and productivity. Their negative influence likely is more marked than that of high Na intake.

If anti-quality factors (e.g., high TDS, Cl or sulfates) are suspected of affecting animal performance, concentrations in drinking water should be determined. Water intake can be estimated from equation of Murphy *et al.* (1983) or by using water meters. It may be feasible to adjust amounts of minerals supplemented in the diet so that total intake (from water plus feed) is more nearly those recommended. Alternatively, processes to reduce concentrations of minerals in water (e.g., dilution, ion exchange or distillation) may be possible, though possibly costly.

NITRATES (NITRITES)

Nitrate poisoning (nitrite poisoning) results from conversion of nitrate to nitrite by ruminal microorganisms and subsequent conversion of hemoglobin to methemoglobin by nitrite in blood. This reduces dramatically the oxygen carrying capacity of blood and can result in asphyxiation in severe cases.

One 35-month study in Wisconsin compared reproductive efficiency and lactational performance of a 54-cow herd in which drinking water contained 19 ppm or 374 ppm nitrate (added as potassium nitrate; Kahler *et al.*, 1975). During the first 20 months of study there were no effects on reproductive function as assessed by incidences of abortions, retained fetal membranes, cystic ovaries, observed heats, services per conception, and first-service conception rates. During the latter 15 months of study cows drinking higher nitrate-containing water had higher services per conception (1.7 vs 1.2) and lower first-service conception rates. Average MY was not different between groups but, total MY during the 36-month study was somewhat lower for cows drinking higher nitrate-containing water, probably due to increased dry period length resulting from lower conception. No effects on blood hemoglobin, methemoglobin, vitamins A or E, or liver vitamin A concentrations were detected.

Though nitrates have not been a major concern in drinking water of dairy cattle in the past, it may be an important future consideration. This poisoning has been reported when cattle drink from ponds or ditches contaminated by run-off from heavily fertilized crop land or pastures. Drinking water with above normal nitrate concentrations in combination with feeds containing high nitrate concentrations may pose an important practical concern in specific situations. There is a lack of information upon which to base definitive standards. Table 2 gives general guidelines (NRC, 1974).

WATER HARDNESS AND pH

Hardness often is confused with salinity or TDS, but the two are not necessarily related meaningfully. For example, high saline waters may contain an abundance of Na salts of Cl and sulfate and yet be quite soft if relatively low concentrations of Mg and Ca are present. Concentrations of these two cations primarily are responsible for degree of hardness of water. Hardness (Ca plus Mg) classifications include: soft (0 to 60 ppm), moderate (61 to 120 ppm), hard (121 to 180 ppm) and very hard (181 ppm

Table 2. Concentrations of nitrates and nitrate-nitrogen in drinking water and expected response.

Ion in water, ppm		
NO_3	NO_3-N	Comment
0–44	10	No harmful effects.
45–132	10–20	Safe if diet is low in nitrates and nutritionally balanced.
133–220	20–40	Could be harmful if consumed over a long period of time.
221–660	40–100	Dairy cattle at risk; possible death losses.
661–800	100–200	High probability of death losses; unsafe.
Over 800	Over 200	Do not use; unsafe.

and greater; NRC, 1974). Some laboratory analyses may list hardness in terms of grains; 1.0 grain per gallon is equal to 5.8418×10^{-3} ppm.

Apparently degree of hardness does not affect animal health or productivity. Over 30 years ago researchers in Washington compared influence of hard (116.4 ppm of calcium carbonate equivalents) and soft (8.4 ppm) water on MY of dairy cows (Blosser and Soni, 1957). Calcium plus Mg concentrations were 33 ppm for hard water and about 1.2 ppm for soft water. No differences were detected due to degree of hardness in 4% fat corrected milk (FCM) yield, water intake or water consumed/lb MY. Similarly, Graf and Holdaway (1952) in Virginia found no effects of hard water (290 ppm of Ca plus Mg) on MY, BW changes, water intake or ratio of water intake to MY compared with soft (0 ppm) water offered for 57 days.

Drinking water with pH between 6 to 9 is thought to be acceptable to livestock (NRC, 1974). Information on potential deleterious effects outside this range was not found.

OTHER POTENTIALLY TOXIC COMPOUNDS AND ORGANISMS

NRC (1974) provided guidelines of upper limit concentrations of potentially toxic substances in drinking water. Studies are limited on effects of these compounds on lactation and secretion into milk. Clinical cases have been reported where Pb and Hg were determined to cause toxicity. Table 3 gives safe upper limits of several toxic substances.

Chlorinating water can increase intake because bacteria present in pipes and waterers are reduced. Commercial dry pellet chlorinators are available which can be connected at the well to service the whole dairy. Reid (1992) recommended in certain situations that all water receptacles be chlorinated weekly. Household bleach at 2 to 3 ounces per 50 gallon water capacity was recommended.

Gorham (1964) reported that at least six species of blue-green algae poisoned cattle drinking water from a lake. However, the causative agents were not identified specifically. It was recommended that water with plentiful algae growth not be offered to cattle. Shading of water troughs and frequent sanitation also can help minimize algae growth.

Table 3. Safe upper limit concentrations of some potentially toxic substances in drinking water of livestock (NRC, 1974)[1].

Substance	Upper limit mg/liter (ppm)	Substance	Upper limit mg/liter (ppm)
As	0.2	Fe	Not defined[2]
Ba	Not defined	Pb	0.1
Cd	0.05	Mn	Not defined
Cr	1.0	Hg	0.01
Co	1.0	Mo	Not defined
Cu	0.5	Ni	1.0
Cyanide	Not defined	V	0.1
Fl	2.0	Zn	25.0

[1]These concentrations generally are far below that required to cause death of half the test subjects (LD_{50}) administered these substances.
[2]Experimental evidence not available to make definitive recommendations.

Water nutrition

FUNCTIONS AND METABOLISM

Water is ubiquitous within the body and is a great solvent. It is chemically neutral; thus, ionization of most substances occurs more freely in water than other media. Water serves as a medium for dispersion or suspension of colloids and ions within the body, and is necessary for maintaining osmotic balance. It functions as a medium for processes of digestion (hydrolysis), absorption, metabolism, milk and sweat secretion, and elimination of urine and feces. It provides a medium for transport of nutrients, metabolites, hormones, and gases and is a lubricant and support for various organ systems and the fetus. A special role is in heat exchange and maintenance of heat balance because of its high thermal conductivity, allowing rapid transfer of heat. High latent heat of vaporization allows cows to transfer significant heat from their bodies to the environment with only a small loss of water volume; high heat capacity provides a thermal buffer by conserving body heat in cold climates and conserving body water in warm environments.

Water balance is affected by total intake of water and losses arising from urine, feces, milk, saliva, sweating, and vaporization from respiratory tissues. Amounts lost via various routes are affected by amount of milk produced, ambient temperature, humidity, physical activity of the animal, respiratory rate, water consumption and dietary factors (e.g., Na or N contents).

WATER INTAKE AND REQUIREMENTS

Water requirements of dairy cattle are met from three sources: (1) that ingested as drinking water, (2) that contained on or in feed consumed, and (3) that resulting from metabolic oxidation of body tissues. Murphy *et al.* (1983) developed a prediction equation to estimate intake of drinking water. Data were from the first 16 weeks of lactation of

19 multiparous Holstein cows (average BW 1276 lb) averaging 73 lb MY/day. Diet was approximately 40% corn silage and 60% concentrate, dry basis. Sodium intake varied because sodium bicarbonate was fed to part of the cows. Factors included in the prediction equation were DMI (lb/day), MY (lb/day), Na intake (g/day), and weekly average minimum environmental temperature (°F). Ranges and averages for independent factors from the data set used to develop the equation were: 7.7 to 112.3, 72.9 lb MY/day; 11.4 to 59.9, 41.8 lb DMI/day; 12 to 153, 74 g Na intake/day; and, 9.0 to 68.0, 46.5°F, average minimum temperature.

The equation was: water intake (lb/day)

$= 0.90 \times$ (MY, lb/d) $+ 1.58 \times$ (DMI, lb/day)
$+ 0.11 \times$ (Na intake, g/day)
$+ 2.64 \times$ (°F/1.8 – 17.778, average minimum temperature) $+ 35.25$.

Table 4 depicts relative influence of these factors on drinking water intake. Values of factors used generally are within ranges of data utilized to develop the prediction equation. Actual average minimum temperature (°F) was characterized within cool (Dec, Feb, and Apr) and warm (Jun, Aug, and Oct) season categories from weather data at Gainesville, FL (Whitty *et al.*, 1991). Milk yield and DMI were estimated, with typical expected declines in DMI in warm season when MY was 60 lb/cow/day or more. Sodium intake (g/day) was calculated based on specified DMI and dietary Na concentrations of 0.18% (NRC, 1989) or 0.50% which would be typical of diets with supplemental Na-containing buffer. Water contained on or in feeds consumed was not considered in prediction; water content of experimental diets to develop the equation was about 38% (Murphy *et al.*, 1983). Water intake in gallons per day can be calculated by multiplying lb/day by 0.1198.

The prediction equation indicates that intake of drinking water changes 0.90 lb for each 1.0 lb change in MY, 1.58 lb for each 1.0 lb change in DMI, 0.11 lb for each 1 g change in Na intake, and 1.47 lb for each 1°F change. Thus, DMI has the most relative influence on water intake. However, absolute magnitude of change of various factors has direct bearing on how much water intake will be affected. For instance, because possible range in MY is greater than DMI it could affect water intake more (Table 4). Based on the prediction equation, cattle consume 0.19 gallon of water for each lb increase of DMI and 0.11 gallon for each lb increase in MY when the other three independent variables of the equation are held constant.

Sodium has a relatively small influence (3 to 4% increase) on water intake when Na content increased from 0.18 to 0.50% of diet DM (Table 4). Using 70-year average minimum temperatures of Feb and Aug, water intake increased about 25% during the warmer month when DMI, MY and Na intake were the same. Winchester and Morris (1956) found a relatively constant ratio of 3 lb of water consumed/lb of DMI within temperature range of 0 to 41°F. However, water intake per unit of DMI accelerated rapidly as ambient temperature rose above 41°F, reaching over 7 lb of water/lb DMI at 90°F.

Table 4. Predicted daily water intake of dairy cattle as influenced by MY, DMI, Na intake, and average minimum temperature (season and month)[1,2].

			Cool Season						Warm Season		
			Dec	Feb	Apr				Jun	Aug	Oct
Min. temperature			51.2	43.5	51.0				62.8	71.0	69.6
Max. temperature			75.6	68.6	76.7				87.9	91.6	89.1
Avg. temperature			63.4	56.1	63.8				75.2	81.3	79.3
MY (lb)	DMI[3] (lb)	Na[4] intake (g)	Water intake, gal/day			MY (lb)	DMI[3] (lb)	Na[4] intake (g)	Water intake, gal/day		
0	25	20	12.6	11.2	12.6	0	25	20	14.6	16.1	15.8
20	38	31	17.4	16.0	17.3	20	38	31	19.4	20.8	20.6
		86	18.1	16.7	18.0			86	20.1	21.6	21.4
40	40	33	19.9	18.6	19.9	40	40	33	22.0	23.4	23.1
		91	20.7	19.3	20.6			91	22.7	24.2	24.0
60	45	37	23.1	21.7	23.0	60	44	36	24.9	26.3	26.1
		102	23.9	22.6	23.9			100	25.8	27.2	27.0
80	50	41	26.2	24.9	26.2	80	46	38	27.5	28.9	28.7
		114	27.2	25.8	27.2			104	28.3	29.8	29.6
100	55	45	29.4	28.0	29.3	100	48	39	30.0	31.5	31.2
		125	30.4	29.1	30.4			109	30.9	32.4	32.2
120	60	49	32.5	31.2	32.5	120	50	41	32.6	34.0	33.8
		136	33.7	32.3	33.6			114	33.5	35.0	34.8

[1]Drinking water intake predicted from equation of Murphy et al. (1983); equation uses average minimum temperature (°F).
[2]Average minimum, maximum and average monthly temperatures are 70-yr averages for specified months at Gainesville, FL (Whitty et al., 1991).
[3]Estimated DMI.
[4]First row within each MY by DMI by average minimum temperature category is with dietary Na % = 0.18 (NRC, 1989); second row is with Na % = 0.50, typical of feeding a Na-containing buffer.

OTHER FACTORS AFFECTING WATER INTAKE

Dietary moisture content

Dairy cattle consuming typical air-dry (about 90 per cent DM) diets consume less than 1 gallon of water from feed daily, depending on feed intake. This quantity is small compared with drinking water (Table 4). By comparison, when cattle consume pastures, silages, and liquid feeds, a substantial portion of water needs is provided. A typical diet for lactating cows containing 50% water would result in intake of 50 lb (6 gallon) of water if feed intake was 100 lb, as-fed; this would be equal to about 17 to 23% of predicted drinking water intake depending on MY and average minimum temperature, based on equation of Murphy et al. (1983). Belgium workers found a negative relationship between total water intake and DM content of ration when evaluated at constant DMI (Paquay et al., 1970). In an equation developed from several pasture experiments, total water intake was affected negatively by DM content of the ration, and positively by

DMI and mean temperature (Stockdale and King, 1983). Davis *et al.* (1983), investigating feeding value of wet brewers grains, showed that total water consumed (drinking water intake plus that derived from the ration) decreased about 26% as total ration moisture content increased from 30.7 to 53.6%. Drinking water intake *per se* declined 37% over this range of ration moisture contents. However, this effect may have been more a function of actual DMI, because as total ration moisture content increased from 30.7 to 53.6%, actual DMI declined 24%. Substantial influence of DMI on drinking water intake was evident.

Metabolic water
When organic compounds are oxidized by animals, hydrogen molecules go toward formation of metabolic water. During metabolic oxidation, water yields (ml/g tissue) are 1.07 from fat, 0.40 from protein, and 0.50 from carbohydrate. This can account for as much as 15% of total water intake (Chew, 1965), which is substantially more than from consumption of an air-dry ration. Although oxidation (e.g., protein catabolism) contributes metabolic water, there also are increased demands for water for respiration, heat dissipation and urine excretion associated with oxidative processes. Thus, generation of metabolic water is not adequate to cover other demands associated with oxidation. Additional sources of water (e.g., drinking or feed-borne water) are required for metabolic oxidation.

Drinking behavior, waterer characteristics and stray voltage
Pattern of water consumption is associated with feeding pattern (Nocek and Braund, 1985). When four first lactation cows were fed one, two, four or eight times daily, peak hourly water intake was associated with peak times of DMI. Cows would alternate the intake of feed and water.

Given the opportunity, peaks of drinking can be associated with milking. Typically, greater consumption is observed immediately after milking. Therefore, it seems judicious to provide abundant water to cows immediately after milking such as in the return lanes. Some dairies provide water cups or troughs for cows in the milking parlor. However, field observations suggest intake at this location is not appreciable (Beede, personal observation). This may be because water was quite cool (about 55°F) and thus not as acceptable to cows (Wilks *et al.*, 1990). A field observation in southeastern Georgia suggested that cows preferred warmer water (about 80°F) coming from a heat-exchange unit in the milking parlor compared with well water (about 65°F) in summertime. Water temperatures between 60 and 80°F appear most acceptable to dairy cattle.

The type of water receptacle may affect drinking behavior. Compared on a herd basis in Europe, cows drank less frequently from water troughs than from water cups (bowls) (Castle and Thomas, 1975). Total daily drinking time ranged from 2.0 to 7.8 min, with longest time found for a herd which had only water cups. Drinking rate ranged from 1.2 to 6.5 gallon per min. The lowest rate was with water cups. Total daily intake was highest with drinking cups, but this likely was biased by herd and diet differences.

Filling rate of water cups can affect water intake and is largely a function of pipe size and water pressure. Reid (1992) noted that during renovation of a tie-stall barn, 2 inch PVC water pipe replaced 1 inch galvanized pipe. Larger diameter pipe facilitated more rapid filling of water cups and MY increased 3 lb during summer in Wisconsin when cows were housed 14 hr/day. Actual flow rates into water cups before and after the change were not given. In Sweden, water intake behavior from cups with flow rates of .5, 1.8 and 3.2 gallon per min was examined (Anderrson et al., 1984). Time spent drinking decreased from 37 to 11 and 7 min/day as flow rate increased. Frequency of drinking episodes was 40, 28 and 30 times/day. As flow rate increased, total water intake increased from 20.4 to 22.0 and 23.3 gallons per day. However, MY and composition and DMI were not affected by flow rate into water cups. In this experiment, each cow of a pair sharing the same water cup was classified as dominant or submissive based on videotaped behavior of frequent confrontations in drinking episodes. Submissive cows consumed 7% less water and ate 9% less hay than dominant cows. Milk fat percentage and FCM yield also were lower for submissive cows.

Use of water cups in most large herds is relatively infrequent for obvious reasons. However, watering troughs with adequate accessibility and flow rates are important, because cows tend to drink in groups associated with other events (e.g., feeding or after milking). Therefore, adequate linear dimension and filling rate of the water receptacle are required to accommodate the group's needs. Otherwise, more submissive cows may not have adequate opportunities to consume water and may not return to the water trough at a later time.

Potential for stray voltage at or around water tanks or other receptacles is worth considering. In a recent study, lactating Holstein cows adapted to being subjected to 3 volts AC or less between the water bowl and the rear feet (Gorewit et al., 1989). About 91% of cows subjected to 4 to 6 volts adapted within 2 days so that there was no change in water intake. However, some cows refused to drink within the first 36 hr of subjection to 4 volts or more and treatment was ended. There also was a direct relationship between amount of voltage applied and time required for cows to adapt and consume their first gallon of water. When 6 volts were applied, over 11 hr were required before the first gallon of water was consumed. Probably stray voltage is not a prevalent problem. However, it should be evaluated if lactational performance and water intake are less than expectations. Use of water heaters in cold climates may contribute to stray voltage problems.

On a practical basis, it seems obvious that a fresh, clean, abundant, easily accessible supply of drinking water must be available at all times to dairy cattle. However, based on numerous farm visits, this is not always the case (Beede, personal observation). If a herd or group is not performing to expectations, one of the first factors that should be evaluated and monitored is the drinking water.

David K. Beede

WATER NUTRITION OF YOUNG CALVES

During early life (0 to 3 weeks), suckling calves consume 0.20 to 0.40 gallon of water daily via milk or milk replacer. Young calves fed liquid diets consume more water per unit DMI than do older cattle fed dry diets (ARC, 1980). Water intake accelerates as calves begin to consume larger amounts of dry feeds. Providing free access to drinking water increased DMI and BW gain (37% increase) of young calves fed a liquid diet (89% water) compared with calves fed only liquid diet (Kertz *et al.*, 1984). There was no indication that offering free-choice drinking water produced scours. Sometimes producers do not offer free-choice drinking water to young calves. This practice appears detrimental and may be perilous, e.g., in heat-stressing environments when the physiologic demand for water to aid in thermoregulation may be higher. Additionally, the notion that restriction of free-choice drinking water enhances intake of liquid milk replacers seems equally dangerous when demands from stress are great. Performance and health of young calves is superior when fresh drinking water is offered free-choice at all times. Proper and timely sanitation of water receptacles obviously is extremely important.

DRINKING WATER IN HOT WEATHER AND EFFECTS OF CHILLED WATER

Water needs during heat stress
Many large dairy herds are located in warm climates. Water unequivocally is the most important nutrient for lactating dairy cows in heat-stressing environments. Water is the primary medium for dissipation of excess body heat through lungs and skin in addition to that required in milk. USDA research showed total water loss from the body increased by 58% in nonlactating cows maintained at 86 compared with 68°F. Much of the increase was due to increased (176%) secretion of water through skin as sweat (McDowell and Weldy, 1967). Concomitantly, loss of water in feces decreased 25%, but increased 54% and 26% via respiratory and urinary routes at 86 compared with 68°F. For lactating cows in climate chambers (64 vs 86°F), drinking water consumption increased 29% at the warmer temperature; fecal water loss dropped 33%, but loss of water via urine, skin surface and respiratory evaporation increased 15, 59 and 50% (McDowell, 1972). Marked increases in water intake were observed starting at 81 to 86°F with lactating cows (Winchester and Morris, 1956; NRC, 1981). Cows also consumed less water in high humidity than lower humidity environments, probably because of reduced DMI and dampened ability to employ evaporative heat loss mechanisms.

Surprisingly little is known about actual requirements for water during heat stress. Numerous factors, such as rate of feed intake and physical form of the diet, physiological state, breed of animal, and quality, accessibility and temperature of water, likely affect intake during heat stress (NRC, 1981). Studies in climate chambers suggested that water needs under heat stress are 1.2- to 2-fold higher than required of cows producing in the thermal comfort zone. Using the prediction equation of Murphy et al. (1983), intake of drinking water increased 1.25-fold in Aug compared with Feb for the same MY by DMI by Na intake category (Table 4). Under natural conditions, particularly with potential for plentiful natural ventilation and sweating, water expenditure may be even greater.

Inadequate provision of water decreases milk production faster and more dramatically than any other nutritional factor. If milk production drops dramatically, particularly during summer, water supply should be evaluated. All too often, dirty water tanks, or improper placement of waterers may be the culprit. A good guideline is, "Based on appearance of cleanliness, would you be willing to drink from the tank? If the answer is no, it is not clean enough for cows." A second frequent problem is that waterers or tanks are placed too far away from shade where cows spend their time during the hottest part of the day. If cows must choose between shade and walking to an unshaded watering station, they stay in the shade. During this time cows use much of their available body water to dissipate heat through evaporation, reducing that available for synthesis of milk. Waterers should be placed in shade in close proximity to cows; this also keeps water from getting hot, from solar radiation which can reduce water intake.

Drinking water temperature
Researchers at Texas A & M University compared cooling effects of chilled drinking water (50, 61, 72 and 82°F; Stermer et al., 1986). All water was withheld for 6 hr before offering to cows. Inner ear temperature was reduced more with 50 than 82°F water. However, 50°F water was only 32% effective in reducing body temperature and authors were doubtful if the effect was maintained long enough (about 2.2 hr) to keep body temperature from rising above the upper critical temperature. There were no differences in MY of cows offered drinking water at various temperatures. This, coupled with estimated cost to chill water from 82 to 72°F ($.049/cow/day) or to 50°F ($.125/cow/day) led to the conclusion that there probably was no advantage to chilling drinking water for lactating cows. In another study, cows offered chilled (50°F) water had greater DMI (15% increase) and produced more 3.5% FCM (11% increase) than those drinking 82°F water (Milam et al., 1986).

Cows were offered 51 or 81°F water on a 24 hr basis in a switchback design (Wilks et al., 1990). Cows offered cooler water consumed more feed (3%), drank more water (7.7%) and had reduced respiration rates and rectal temperatures. Milk yield was increased 4.8% for cows consuming chilled water. An alternative to chilling water may be to insulate water tanks to maintain a lower water temperature if it comes from the well (or other source) relatively cool. Measurements in Florida indicated a temperature range of well water of 73 to 79°F immediately after pumping, considerably cooler than the high temperature treatments used in most of the Texas A and M University experiments. The practical approach seems to be to prevent well water from warming after it is pumped and stored above ground.

Cows exhibited preference when offered 50 or 86°F drinking water cafeteria-style (Wilks et al., 1990). Respiration rates and rectal temperatures were reduced with cooler water. However, cows preferred to drink warmer water given the choice, with over 97% of total water consumed being warmer water. Over 70% of cows drank only warm water. It was concluded that if chilled drinking water was offered as way to cool cows, it must be the only source of drinking water available. Otherwise, cows may wait to drink until a time when warmer water is available.

Well water (77°F) or chilled (59°F) drinking water was offered on a large Florida dairy (Bray et al., 1990). Cows were kept in open lots with cooling ponds and two shade structures per lot. Cows did not have access to feed and water under shades. Four watering stations (unshaded) were in each lot. Over 1100 cow-period observations were collected in a reversal design from Jun through Sep. Mean daily minimum, maximum, and average ambient temperatures and mean daily minimum relative humidity were 68.4, 91.0 and 79.7°F, and 58%. Under these conditions there was no difference in MY (61.4 vs 61.8 lb/day) for cows offered well or chilled water.

A similar experiment was performed the following year on another commercial dairy where cows were kept in feeding barns with fans and sprinklers, and had access to feed and water continuously (Bray et al., 1991). Mean daily minimum, maximum, and average ambient temperatures and mean daily minimum relative humidity were 66.3, 91.3 and 78.8°F, and 50.5%. There were about 175 cows per treatment and drinking water temperatures were well water (75 to 80°F) and chilled water (52 to 57°F) in a two month reversal experiment. Water consumption from total group measurements averaged 21.7 and 23.2 gallon per cow per day for well water and chilled water. There was no difference in intravaginal temperatures as detected by thermal couples with radiotransmitters. Average MY was 63.1 and 64.2 lb/d, for well and chilled water treatments, but were not significantly different.

In a survey of over 200 drinking water tanks on 31 dairies in central Florida in summer (Giesy, 1990 cited by Bray et al., 1991), overall average water temperature was 86°F, and ranged from 77 to 97°F. Shading tanks lowered temperatures somewhat with average water temperatures of 87, 85, 81 and 81°F when tanks were unshaded, shaded during the morning, shaded during the afternoon or continuously shaded. Average temperature of fresh water at the tank inlet was 82°F, and was affected by the distance water traveled before entering the tank. Fresh water inlet temperature to the tank was higher (82°F) if pipe servicing the tank was above ground for 200 ft or more, compared with less than 100 ft (79°F). Volume capacity of tank relative to number of cows it serviced affected tank water temperature. When tank capacity was less than 1 gallon per cow, water temperature was 82°F. At tank capacities of 1 to 3, 4 to 9, 10 to 19, 20 to 39 and over 40 gallon per cow, temperatures of water in tanks were 85, 86, 87, 88 and 91°F. This information emphasizes benefit of relatively small volume, rapidly filling tanks for cows in warm climates. Access to sufficient linear space of an abundant water supply probably is more important than tank capacity.

Main consideration for water nutrition during hot weather is to provide an easily accessible source of clean drinking water in close proximity to cows. This should be in the shade so that water in the tank or waterer is not heated excessively above that temperature which it comes from the well. Additionally, chilling drinking water probably is not warranted except where water comes from the well at temperatures above 86°F or where water cannot be kept reasonably cool by shade, specifically designed drinking water receptacles, and (or) insulation.

Summary and conclusions

Water is indispensable for life and is the most important dietary essential nutrient for dairy cattle. Lactating cows require a larger portion of water relative to their BW because milk is 87% water. Water intake and requirements are influenced by physiological state, rate of MY and DMI, BW, composition of diet and environmental factors. The water intake prediction equation of Murphy *et al.* (1983) is useful to estimate water intake requirements. Dry matter intake and MY have large influences in the equation. Diet moisture content, cow behavior, physical characteristics of water receptacle, and ambient temperature also affect water intake.

Water nutrition of young calves fed liquid diets is important and there is no justifiable reason not to provide free-choice drinking water. Feed intake and growth rate have been increased by offering free-choice drinking water. Under heat-stressing conditions water needs are increased 1.2- to 2-fold. Chilled drinking water did not consistently improve lactational performance under commercial conditions and was not economically advantageous. Drinking water temperature above 86°F may reduce consumption.

Several factors should be evaluated if problems with the drinking water supply are a suspicion. Abnormally high concentrations of Cl and sulfate are often times of most concern. Nitrates and hardness of water have not been detrimental factors.

An abundant, continuous, clean source of drinking water for all classes of dairy animals is crucial!

A portion of this chapter was originally published in: "Water for Dairy Cattle," page 260 in *Large Dairy Herd Management*. (H.H. Van Horn and C.J. Wilcox, eds.), Management Services American Dairy Science Assoc., Champaign, IL, 1992.

References

Agriculture Research Council. 1980. The nutrient requirements of ruminant livestock. London, Commonwealth Agricultural Bureaux.

Anderrson, M., J. Schaar, and H. Wiktorsson. 1984. Effects of drinking water flow rates and social rank on performance and drinking behaviour of tied-up dairy cows. Livest. Prod. Sci. 11:599.

Blosser, T.H., and B.K. Soni. 1957. Comparative influence of hard and soft water on milk production of dairy cows. J. Dairy Sci. 40:1519.

Bray, D.R., D.K. Beede, M.A. DeLorenzo, D. Wolfenson, R.G. Giesy, and R.A. Bucklin. 1990. Environmental modification update. Proc. FL Dairy Prod. Conf., p 100.

Bray, D.R., D.K. Beede, M.A. DeLorenzo, D. Wolfenson, R.G. Giesy, R.A. Bucklin, R. Nordstedt, and S. Means. 1991. Environmental modification update. Proc. FL Dairy Prod. Conf., p. 134.

Castle, M.E., and T.P. Thomas. 1975. The water intake of British Friesian cows on rations containing various forages. Anim. Prod. 20:181.

Challis, D.J., J.S. Zeinstra, and M.J. Anderson. 1987. Some effects of water quality on the performance of high yielding cows in an arid climate. Vet. Rec. 120:12.

Chew, R.M. 1965. Water metabolism of mammals. In: Chap. 2. Physiological Mammalogy. Vol. II, W.V. Mayer and R.G. Van Gelder (Eds.), New York, Academic Press.

Davis, C.L., D.A. Grenwalt, and G.C. McCoy. 1983. Feeding value of pressed brewers grains for lactating dairy cows. J. Dairy Sci. 66:73.

Digesti, R.D., and H.J. Weeth. 1976. A defensible maximum for inorganic sulfate in drinking water of cattle. J. Anim. Sci. 42:1498.

Frens, A.M. 1946. Salt drinking water for cows. Tijdschr. Diergeneeskd. 71 (No. 1):6(C.A. 44:5972,1950).

Gorewit, R.C., D.J. Aneshansley, D.C. Ludington, R.A. Pellerin, and X. Zhao. 1989. AC voltages on water bowls: effects on lactating Holsteins. J. Dairy Sci. 72:2184.

Gorham, P.R. 1964. Toxic algae. In: Algae and Man, D.F. Jackson ed., Plenum Press, NY. p. 307.

Graf, G.H. and C.W. Holdaway. 1952. A comparison of "hard" and commercially softened water in the ration of lactating dairy cows. J. Dairy Sci. 35:998.

Heller, V.G. 1933. The effect of saline and alkaline waters on domestic animals. Okla. Agr. Exp. Sta. Bull. 217.

Jaster, E.H., J.D. Schuh, and T.N. Wegner. 1978. Physiological effects of saline drinking water on high producing dairy cows. J. Dairy Sci. 61:66.

Kahler, L.W., N.A. Jorgensen, L.D. Satter, W.J. Tyler, J.W. Crowely, and M.F. Finner. 1975. Effect of nitrate in drinking water on reproductive and productive efficiency of dairy cattle. J. Dairy Sci. 58:771.

Kertz, A.F., L.F. Reutzel, and J.H. Mahoney. 1984. Ad libitum water intake by neonatal calves and its relationship to calf starter intake, weight gain, feces score and season. J. Dairy Sci. 67:2964.

McDowell, R.E. 1972. Improvement of Livestock Production in Warm Climates. W. H. Freeman and Co., San Francisco, CA.

McDowell, R.E. and J.R. Weldy. 1967. Water exchange of cattle under heat stress. Biometeorol. 2:414.

Milam, K.Z., C.E. Coppock, J. W. West, J.K. Lanham, D.H. Nave J.M. Labore, R.A. Stermer, and C.F. Brasington. 1986. Effects of drinking water temperature on production responses in lactating Holstein cows in summer. J. Dairy Sci. 69:1013.

Murphy, M.R., C.L. Davis and G.C. McCoy. 1983. Factors affecting water consumption by Holstein cows in early lactation. J. Dairy Sci. 66:35.

National Research Council. 1974. Nutrients and toxic substances in water for livestock and poultry. Natl. Acad. Sci., Washington, DC.

National Research Council. 1981. Effect of environment on nutrient requirements of domestic animals. Natl. Acad. Press, Washington, DC.

National Research Council. 1989. Nutrient requirements of dairy cattle, 6th rev. ed. Natl. Acad. Sci., Washington, DC.

Nocek, J.E. and D.G. Braund. 1985. Effect of feeding frequency on diurnal dry matter and water consumption, liquid dilution rate, and milk yield in first lactation. J. Dairy Sci. 68:2238.

Paquay, R., R. De Baere, and A. Lousse. 1970. Statistical research of the fate of water in the adult cow. 2. The lactating cow. J. Agric. Sci. (Camb.) 75:251.

Reid, D.A. 1992. Water: What you see is not always what you get. In: Prod. 24th Ann. Conv. Am. Assoc. Bovine Practitioners. No. 24:145.

Stermer, R.A., C.F. Brasington, C.E. Coppock, J.K. Lanham, and K.Z. Milam. 1986. Effect of drinking water temperature on heat stress of dairy cows. J. Dairy Sci. 69:546.

Stockdale, C.R., and K.R. King. 1983. A note on some of the factors that affect the water consumption of lactating dairy cows at pasture. Anim. Prod. 36:303.

Wang, C., and D.K. Beede. 1992. Effects of ammonium chloride and sulfate on acid-base status and calcium metabolism of dry Jersey cows. J. Dairy Sci. 75:820.

Weeth, H.J., and L.H. Haverland. 1961. Tolerance of growing cattle for drinking water containing sodium chloride. J. Anim. Sci. 20:518.

Weeth, H.J., and J.E. Hunter. 1971. Drinking of sulfate-water by cattle. J. Anim. Sci. 32:277.

Weeth, H.J., and D.L. Capps. 1972. Tolerance of growing cattle for sulfate-water. J. Anim. Sci. 34:256.

Wegner, T.N., and J.D. Schuh. 1974. Effect of highly mineralized livestock water supply on water consumption and blood and urine electrolyte profiles in dairy cows. J. Dairy Sci. 57:608 (Abstr).

Whitty, B., R. Hill, and D. McCloud. 1991. Weather Office, Agronomy Dept. Institute of Food and Agricultural Sciences, Univ. FL, Gainesville.

Wilks, D.L., C.E. Coppock, J.K. Lanham, K.N. Brooks, C.C. Baker, W.L. Bryson, R.G. Elmore, and R.A. Stermer. 1990. Responses of lactating Holstein cows to chilled drinking water in high ambient temperatures. J. Dairy Sci. 73:1091.

Winchester, C.F. and M.J. Morris. 1956. Water intake rates of cattle. J. Anim. Sci. 15:722.

ASCITES – A METABOLIC CONDITION?

J.D. SUMMERS
Department of Animal and Poultry Science, University of Guelph, Guelph, Ontario, Canada

Ascites and sudden death syndrome

Ascites, which used to be seen mainly at high altitudes, has become a common problem with meat-type chickens around the world. Birds usually show right ventricle enlargement, heart congestion, and in many cases fluid accumulation in the abdominal cavity. While birds may die showing varying degrees of the above mentioned symptoms, others die showing little if any fluid accumulation. Why birds come down with ascites is not known, however there appears to be something limiting oxygen uptake from the lungs (eg., high altitude, poor ventilation, respiratory disease, etc.) which, in turn, triggers the condition.

If oxygen supply is insufficient to oxygenate blood in the lungs, then smooth muscle in blood vessel walls contracts thus restricting the blood vessel diameter. To aid the lungs in ensuring sufficient oxygen delivery to the various body tissues, under stress conditions the kidney produces a hormone that stimulates red blood cell and haemoglobin production. However, there is a small disadvantage in that the blood becomes more viscous and thus more resistant to flow.

The above adaptive responses cause the right ventricle to work harder in order to move a more viscous blood through a blood vessel with a smaller diameter thus resulting in higher resistance in the lung vascular system. The right ventricle of the heart adapts to this increased work load by increasing in size. This hypertrophy of the right ventricle, along with a congested condition, is typical with the ascites syndrome. Swelling with increased venous pressure eventually causes organ damage and exudate to leak from these organs. The liver is particularly vulnerable to such increased pressure and eventually fills the body cavity with fluid producing a typical ascitic condition.

Another major metabolic problem, which at times seems to have much in common with ascites, is "sudden death syndrome" (SDS), often referred to as "flip-overs". SDS was a metabolic problem noted in many places around the world for several decades, and it still constitutes a significant proportion of mortality in meat chickens. Today, though the incidence of SDS does not appear to have increased, ascites is definitely on the increase. Most pathologists say no relationship exists between SDS and ascites, however the syndromes are similar in that they both affect the cardiovascular system. In general, SDS characteristically affects well grown male broilers which die suddenly and are usually found dead with no evidence of any disease. Prior to death the birds lose balance, start violent wing flapping and have strong muscular contractions. Birds affected with SDS typically have a full digestive tract and often congestion and oedema in the lungs. Cardiac failure is generally assumed to be the cause of death.

In contrast to SDS, birds affected with ascites, (also usually males) are generally small and have a pale, shrunken comb. The abdomen is extended with fluid and an increased respiratory rate is often noted. However, it is not uncommon to find birds dying from right ventricular failure before clinical signs of ascites are seen. These can usually be distinguished from birds dying from SDS by the presence of heart lesions.

It is possible that both SDS and ascites are different degrees of a similar metabolic disorder and that the cardiovascular system is involved in both conditions. The right ventricle is affected with ascites, while SDS appears to result from sudden cardiac collapse.

BIOCHEMICAL EFFECTS OF ASCITES AND SDS

Odom et al. (1991) suggested that a primary deficiency in the growth of the vascular system of the lungs predisposes the chick to hypertension which, when exposed to various stress conditions, results in ascites. Other pathological and clinical findings have been well described for both ascites and SDS, on numerous occasions, but there has been a limited amount of data reported on the biochemical implications of these conditions.

A relationship between dietary fat and SDS and(or) ascites has been reported by several workers (Riddell and Orr, 1980, Rotter et al., 1985). Rotter et al. (1985) proposed that low levels of fat could lead to low levels of arachadonic acid resulting in decreased synthesis of prostaglandins which are involved in modulating cardiac rhythm, heart rate and contractile force. Since biotin is involved in the synthesis of arachadonic acid, reports suggesting the involvement of biotin in reducing SDS (Hulan et al., 1980; Kratzer et al., 1985) may have some validity.

Recently Chung et al. (1993) reported that calcium uptake was lower for cardiac muscle with birds dying from SDS than in normal birds (Figure 1). They suggested that their data support the hypothesis that SDS in broilers is a cardiac dysfunction and that dietary fat is implicated in the syndrome.

A condition similar to SDS had been reported in Australia in broiler breeders (Hopkinson et al. 1983, 1984). The authors suggest that the condition was probably metabolic in origin, affected acid–base balance, and appeared to respond to supplemental dietary potassium.

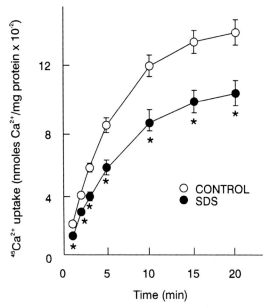

Figure 1. Calcium uptake of the cardiac sarcoplasmic reticulum from control and sudden death birds (From Chung et al., 1993)

LACTIC ACID METABOLISM, ACID–BASE BALANCE

It is of interest to review some of the metabolic problems reported for large animals that appear associated with "acidosis-like conditions". Ruminant animals changed abruptly to a high soluble carbohydrate diet can suffer from acute indigestion often resulting in death (Hungate et al., 1952; Ahrens, 1967). It has been shown that afflicted animals exhibit a rapid rise in lactic acid content in the rumen and a marked drop in rumen pH, while congestion of the lungs is often noted.

Lameness in race horses has been shown to be caused in part by high levels of lactic acid in blood (Asheim et al., 1974). Feeding corn starch to horses not only produced laminitis and severe lameness, but also resulted in the occurrence of severe cardiovascular alterations during the onset of the condition Garner et al. (1975). Garner et al. (1977) orally dosed 31 horses with corn starch. During the test 21 developed severe laminitis and lived, five did not develop laminitis, and five died. Of the five that died three showed no signs of laminitis prior to death. Plasma lactate levels were highest for the horses that died and lowest for those not exhibiting laminitis (Figure 2).

Circulatory collapse reported with acute death in horses and laboured breathing due to congestion are symptoms similar to those seen with broilers where SDS is a problem. A series of short experiments was undertaken where a 20% lactic acid solution was pipetted into the crop of two to three week old birds (Summers et al., 1987). Response was

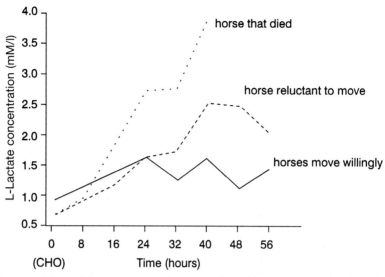

Figure 2. Plasma L-lactate concentration versus time of horses dosed with carbohydrate (From Garner et al., 1977)

variable, but a significant number of birds in some of the tests showed the typical squawking, wing flapping and flipping over typical of SDS. Some of the birds flipped within 30 seconds of dosing.

Various types of diets were fed to the chicks and while variable responses were noted, observations would suggest that birds fed high glucose diets began to flip earliest when dosed with lactic acid, followed by chicks fed high corn starch, followed by birds fed high fat diets. Birds that died from SDS had higher blood lactate levels than survivors.

Whitehead and Randall (1982) reported a significant number of birds dying from fatty liver and kidney syndrome (FLKS) also exhibited symptoms of SDS. While the authors offered no explanation as to the apparent association between FLKS and SDS; their work demonstrated quite clearly that biotin was not directly involved in SDS, but could decrease the apparent incidence by preventing FLKS (Table 1). A marked elevation in blood lactic acid was reported to take place in birds affected with FLKS (Table 2) (Balnave et al. 1977). They postulated that the clinical symptoms of FLKS may be related to a lactic acidosis in the bird. While further work (Balnave and Pearce, 1979) confirmed that the high lactate levels seen with FLKS were a consequence rather than a contributing factor of FLKS, their work demonstrated the marked increase in serum lactate levels that can take place in the chicken in a relatively short period of time.

The chicken, having a large proportion of white muscle, has the ability to produce relatively large quantities of lactic acid, especially during periods of rapid muscular activity. Lactic acid is also a fermentation product in the crop of the chicken, its concentration being dependent on the amount and type of feed present (Eyssen et al., 1962). Bolton (1965) showed that there was a marked increase in crop lactic acid content six hours of after feeding starved cockerels (Table 3).

Table 1. Number of deaths up to 7 weeks of age when fed a low biotin (96 μg/kg) diet with and without supplemental biotin[1].

Biotin supplement	Mortality (number from 270 chicks)					
	FLKS alone	FLKS and SDS	Total FLKS		SDS alone	
			Male	Female	Male	Female
−	10	5	3	12	1	0
−(+ other vitamins*)	30	6.7	12	24.7	0	.3
+	3	0	3	0	0	0
+ (+other vitamins*)	0	3.5	2.5	1	0	0

*Average of several treatments.
[1]From Whitehead and Randall (1982).

Table 2. Serum lactate levels (μmol/ml) for broiler pullets starved for 18 hours at 28 days of age when fed a FLKS diet.

Treatment	Serum lactate, μmol/ml
FLKS diet Birds with FLKS	17.3
" without FLKS	12.5
" + tallow	11.6
" + biotin	7.4

From Balnave et al. (1977)

Table 3. Composition of crop contents at intervals of time after feeding two types of diet[1].

	Time after feeding, h			
	1		6	
	Breeder pellets	Chick mash	Breeder pellets	Chick mash
pH	6.67	6.10	6.48	4.5
Lactic acid, %	0.22	0.18	0.84	2.83
Acetic acid, %	0.09	0.02	0.23	0.11
Dry matter, %	70	36	58	36

[1]From Bolton (1965)

Bjonnes et al. (1987) found that 17 to 20 day embryos and 0 to 7 day old chicks responded differently to low O_2 density. Embryos reduced their metabolism while chicks resorted to anaerobic glycolysis. However, both the embryo and chick showed a marked increase in blood lactic acid concentrations after a 20–minute exposure to low oxygen.

DIETARY NITROGEN AND SULPHUR CONTENT

Dietary sulphur

While no specific biochemical factor has been shown directly involved in SDS or ascites, from the complexity of both conditions one may surmise that a number of factors could be involved. When considering the practical conditions affecting acid–base balance of the bird, more attention should be paid to dietary factors. Bedford and Summers (1988), Whitehead et al. (1985; Table 4) and many others have demonstrated the detrimental effects that can occur in the presence of higher dietary protein levels and(or) imbalance in the ratio of essential to non-essential amino acids. High protein diets usually contain excessive levels of sulphur. High dietary sulphur intake can result in increased acid excretion (Whiting and Draper, 1980) which is accompanied by increased calcium excretion, depending on the level and source (organic versus inorganic) of sulphur fed as well as the presence of other salts that can alter or modify acid–base balance (Whiting and Cole, 1986).

Table 4. Performance of broiler breeders from 26 to 60 weeks of age when fed diets of different protein content[1].

Protein content, %	13.7	16.8
Hen day egg production, %	60.3	57.8*
Mean egg weight, g	63.4	63.0
Fertility, %	93.1	92.4
Hatchability of fertile eggs, %	88.6	85.5*
Saleable chicks of fertile eggs, %	84.5	80.5*

[1]From Whitehead et al. (1985)

The detrimental effects of high levels of sulphur in canola meal have been reported (Summers et al., 1989, 1990). Canola meal contains around 1.15% sulphur as compared with 0.45% for soybean meal. Diets containing appreciable quantities of canola meal have, on occasion, been reported to reduce weight gain, increase leg problems, and result in a higher incidence of SDS in some broiler flocks. Summers et al. (1992) presented data to show that soybean meal diets supplemented with sulphur resulted in weight gain similar to that of a canola meal basal diet containing the same level of sulphur (Table 5). In a recent study involving four levels of supplemental sulphur and chloride added to a soybean meal basal diet it was shown that weight gain was correlated to dietary mEq [Na + K − (Cl + S)] (Summers, 1993; Figure 3). The effect of diet mEq on broiler performance was further confirmed where sulphur, calcium and supplemental dietary mEq (0, 10, and 20 via equal additions of sodium and potassium carbonate) were added to a canola meal diet in a factorial arrangement (Summers, 1993; Table 6). Plotting weight gain against dietary mEq [Na + K + Ca − (Cl + S)] resulted in the plot shown in Figure 4. There is no question but that the high level of sulphur in canola meal affects bird performance and probably influences acid–base balance to a significant extent.

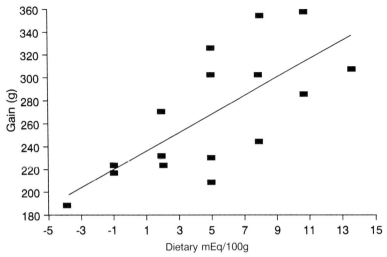

Figure 3. The relationship between dietary mEq and weight gain of male broilers (Summers, 1993)

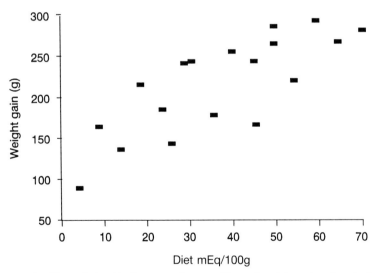

Figure 4. The relationship between dietary mEq and weight gain of male broilers (Summers, 1993)

While little mortality took place in the above studies and incidence of SDS was not recorded, the data do demonstrate that under what would be considered normal commercial feeding practices differences in dietary mEq are encountered which could significantly alter acid–base balance and hence bird performance.

Table 5. Interaction of sulphur and calcium in canola and soybean meal diets when fed to broiler cockerels from 7 to 21 days of age[1].

Protein source	Sulphur Supplement, %	Sulphur Diet level %	Diet Ca level, %	Average weight, g
Canola	–	0.46	0.37	424[f]
	0.26	0.72	0.37	371[g]
	–	0.46	1.32	560[bc]
Soybean	–	0.14	0.37	524[c]
	0.13	0.27	0.37	519[cd]
	0.26	0.40	0.37	479[de]
	0.39	0.53	0.37	373[g]
	–	0.14	1.32	635[a]
	0.13	0.27	1.32	598[ab]
	0.26	0.40	1.32	559[bc]
	0.39	0.53	1.32	451[ef]
				**
SD				27.5

[1]From Summers et al. (1992)

Table 6. Interaction of dietary mEq with sulphur and calcium[1].

	Supplemental diet mEq		
	0	10	20
Sulphur levels, %	Weight gain, g		
0.48	231	253	260
0.72	132	176	228
Calcium level, %	Feed:gain		
0.4	3.09	2.58	2.18
0.8	2.48	2.37	2.13
1.2	2.23	2.18	2.19

[1]From Summers, 1993

If increased dietary sulphur can result in altered acid–base balance leading to reduced bird performance, how widespread is this problem in the field? Recently Han and Baker (1993) compared the toxic effects of excess methionine and lysine when fed to chicks. Lysine at a supplemental level of 3.2% resulted in a moderate growth depression; however 4% DL-methionine gave a severe depression. While the authors do not mention the sulphur content of methionine as a factor influencing the toxicity of high supplemental methionine, the weight gain depression was similar to that reported by Summers (1993) with excess dietary sulphur.

Ammonia binding and ascites: Yucca extract

If excess dietary sulphur is a problem, then a number of practical considerations may interact to influence performance. Water quality, whether birds are fast or slow feathering, and other dietary salts, etc. must be considered. Of recent interest have been reports of high blood ammonia levels as a factor predisposing the bird to ascites. Several recent abstracts have reported reduced ascites mortality by feeding yucca extract (Arce et al., 1994; Balog et al., 1993; Straudinger et al., 1993; Walker, 1993). Products based on yucca extract have been marketed as an aid to reducing ammonia odours in confinement facilities. Yucca saponins have surfactant properties, cannot be absorbed from the digestive tract, and thus must act within the gut (Johnston et al., 1982). Until recently it was the generally held view that yucca extract exerts its effect by inhibition of the urease enzyme reaction shown in Figure 5. Poultry excrete nitrogen in the form of uric acid, not urea, and urease does not directly act on uric acid to produce ammonia, yet there is no question but that yucca extract will reduce poultry house levels of ammonia. Headon and Dawson (1990) demonstrated in a series of experiments that yucca extract reduced the level of ammonia by physically binding or complexing it, not through the inhibition of urease.

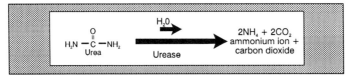

Figure 5. Hydrolysis of urea catalyzed by the enzyme unease.

Reduction in ascites mortality when ammonia-binding additives are added to the diet was initially thought to be an indirect effect of reducing respiratory damage and stress by lowering atmospheric ammonia. However, peak ascites losses have not always been correlated with maximum ammonia reduction. In a study in Mexico designed to enhance potential for ascites by restricting ventilation, addition of DeOdorase, a product based on *Yucca schidigera* extract, to broiler diets significantly reduced ascites and general mortality despite the fact that ammonia levels were both comparatively low and did not differ widely (Arce et al., 1994; Table 7).

Table 7. Effect of supplemental De-Odorase broiler mortality (to 56 days of age)[1].

Treatment	Average weight, g	Feed:gain	Mortality (%)	
			Overall	Ascites
Basal	2427	2.08	27.1[b]	18.9[b]
Basal + DeOdorase	2447	2.07	20.9[a]	13.8[a]

[a,b]Means differ, P<0.05
[1]Arce et al., 1994

207

If, as suggested earlier, altered acid–base balance is a contributing factor triggering ascites, could the marked increase in the use of synthetic lysine in broiler diets during the past few years have added to this problem? High supplemental levels of lysine HCl will often, along with similar dietary increases of methionine, significantly alter anion–cation balance. This imbalance is usually further aggravated as associated reductions in the level of soybean meal used reduce dietary potassium concentration.

Recently an experiment was undertaken in our laboratory to study effects of high levels of methionine and lysine supplementation on performance of male broilers. The basal diet was a commercial type formula consisting mainly of corn and soybean meal containing 21.4% protein, 0.86% total sulphur amino acids, and 1.12% lysine. To this diet supplements of methionine and lysine were made as shown in Table 8. While only small differences were noted for weight gain and feed:gain ratio, mortality values were markedly increased for birds fed the amino acid-supplemented diets. After two weeks of age the birds were stressed by keeping the pen temperatures noticeably cool. This may account for the higher than normal mortality even with the control birds.

Table 8. Performance of broiler males to six weeks of age when fed corn/soy diets supplemented with methionine and lysine.

Treatments	Average weight g,	Feed intake, g	Feed:gain	Mortality, %
Corn, soya				
+ 0.2% DL-methionine	2552	4520	1.77	10.0
+ 0.3% L-lysine	2569	4517	1.76	15.5
+ 0.2% DL-methionine	2507	4367	1.74	16.7
+ 0.3% L-lysine	2507	4387	1.75	21.1

Conclusions

While a definite answer to the ascites problem has not been suggested, an attempt has been made to focus more on the biochemical aspects of the disease and to consider how closely if any, ascites is related to similar problems like SDS and FLKS which are metabolic in nature and have been experienced in many areas of the world.

There is no question but that ascites and SDS are complex metabolic problems. It is time emphasis is shifted toward the biochemical aspects of these conditions as the clinical and pathological parameters have already been exhaustively studied.

References

Ahrens, F.A. 1967. Histamine, lactic acid and hypertoxicity as factors in the development of ruminitis in cattle. Amer. J. Vet. Res. 28:1334.

Asheim, A., O. Knudsen, A. Lindholm, C. Rulcker, and B. Saltin. 1974. Heart rates and blood lactate concentrations of standardbred horses during training and racing. JAVMA, 157:304.

Arce, J., E. Avila and C. Lopez-Coello. 1994. Effect of DeOdorase on performance and mortality due to ascites. Southern Poultry Science Association Meetings. January 16–18, Atlanta, Georgia.

Balnave, D., M.N. Berry, and R.B. Cumming. 1977. Clinical signs of fatty liver and kidney syndrome in broilers and their alleviation by the short-term use of biotin or animal tallow. Br. Poult. Sci. 18:749.

Balnave, D., and J. Pearce. 1979. Lactate administration and fatty liver and kidney syndrome development in biotin-deficient chicks. British Poult. Sci. 20:109.

Balog, J.M., N.C. Rath, W.E. Huff, N.B. Anthony, C.D. Wall, R.D. Walker, and R.O. Asplund. 1993. Effect of a urease inhibitor and air mixing on ascites in broilers. 2. Blood parameters, ascites scores and body and organ weights. Poult. Sci. 72: (Suppl. 1), 3 (Abst.)

Bedford, M.R., and J.D. Summers. 1988. The effect of the essential to non-essential amino acid ratio on turkey performance and carcass composition. Can. J. Anim. Sci. 68:899.

Bjonnes, P.O., A. Aulie and M. Hoiby. 1987. Effects of hypoxia on the metabolism of embryos and chicks of domestic fowl. J. Exp. Zool. Suppl. 1: 209.

Bolton, W. 1965. Digestion in the crop of the fowl. Br. Poult. Sci. 6:97.

Chung, H.C., W. Guenter, R.G. Rotter, G.H. Crow, and N.E. Stranger. 1993. Effects of dietary fat source on sudden death syndrome reticular calcium transport in broiler chickens. Poult. Sci. 72:310.

Eyssen, H., V. DePrins and DeSomer. 1962. The growth promoting action of virginiamycin and its influence on the crop flora in chickens. Poult. Sci. 41:227.

Garner, H.E., J.R. Coffman, A.W. Hahn, D.P. Hutcheson, and M.E. Tumbeson. 1975. Equine laminitis of alimentary origin: an experimental model. Amer. J. Vet. Res. 36: 41.

Garner, H.E., D.P. Hutcheson, J.R. Coffman, A.W. Hahn, and C. Salem. 1977. Lactic acidosis. A factor associated with equine laminitis. J. Anim. Sci. 45:1037.

Han, Y. and D.H. Baker. 1993. Effects of excess methionine or lysine for broilers fed a corn-soybean meal diet. Poult. Sci. 72:1070.

Headon, D.R. and K.A. Dawson. 1990. Yucca extract controls atmospheric ammonia levels. Feedstuffs, July 16.

Hopkinson, W.I., G.L. Griffiths, D. Jessop, and W. Williams. 1983. Sudden death syndrome in broiler breeders. Australian Vet. J. 60:192.

Hopkinson, W.I., W. Williams, G.L. Griffiths, D. Jessop, and W.M. Peters. 1984. Dietary induction of sudden death syndrome in broiler breeders. Avian Dis. 28:352.

Hulan, H.W., F.G. Proudfoot, and K.B. McRae. 1980. Effect of vitamins on the incidence of mortality and acute death syndrome ("flip-over") in broiler chickens. Poult. Sci. 59:927.

Hungate, R.E., R.W. Dougherty, M.P. Bryant, and R.M. Cello. 1952. Microbiological and physiological changes associated with acute indigestion in sheep. Cornell Veterinary 42:423.
Johnston, N.L., C.L. Quarles, and D.J. Fagerberg. 1982. Broiler performance with DSS40 yucca saponin in combination with monensin. Poult. Sci. 61:1052.
Kratzer, F.H., J.L. Buenrostro, and B.A. Watkins. 1985. Biotin-related abnormal fat metabolism in chickens and its consequences. Annals of the New York Academy of Sciences. 447:401.
Odom, T.W., B.M. Hargis, C.C. Lopez, M.J. Arce, Y. Ono, and G.E. Avila. 1991. Use of electrocardiographic analysis for investigation of ascites syndrome in broiler chickens. Avian Disease 35:738.
Riddell, C. and J.P. Orr. 1980. Chemical studies of the blood and histological studies of the heart of broiler chickens dying from acute death syndrome. Avian Disease 24:751.
Rotter, B.A., W. Guenter, and B.R. Boycott. 1985. Sudden death syndrome in broilers: dietary fat supplementation and its effect on tissue composition. Poult. Sci. 64:1128.
Staudinger, F.B., N.B. Anthony, G.C. Harris, J.M. Balog, W.E. Huff, and R.D. Walker. 1993. Effect of a urease inhibitor and air mixing on ascites in broilers 1. Incidence of ascites and environmental conditions. Poult. Sci. 72: (Supplement 1, 4) (Abst.).
Summers, J.D., M. Bedford, and D. Spratt. 1987. Sudden death syndrome: Is it a metabolic disease? Feedstuffs, Jan. 26, p. 20.
Summers, J.D., M. Bedford, and D. Spratt. 1989. Amino acid supplementation of canola meal. Can. J. Anim. Sci. 69:469.
Summers, J.D., M. Bedford, and D. Spratt. 1990. Interaction of calcium and sulphur in canola and soybean meal diets fed to broiler chicks. Can. J. Anim. Sci. 70:685.
Summers, J.D., D. Spratt, and M. Bedford. 1992. Sulphur and calcium supplementation of soybean and canola meal diets. Can. J. Anim. Sci. 72:127.
Summers, J.D. 1993. Canola meal and acid–base balance. Proceedings 29th Nutrition Conference for Feed Manufacturers, University of Guelph. pp. 24–33.
Walker, R.D. 1993. The effects of urease inhibitor on ascites mortality. Poult. Sci. 72: (Suppl. 1), 4 (Abst.).
Whitehead, C.C. and C.J. Randell. 1982. Interrelationships between biotin, choline and other B-vitamins and the occurrence of fatty liver and kidney syndrome and sudden death syndrome in broiler chickens. Brit. J. Nutr. 48:177.
Whitehead, C.C., R.A. Pearson, and K.M. Harron. 1985. Biotin requirements of broiler breeders fed diets of different protein content and effect of insufficient biotin on the viability of progeny. Br. Poult. Sci. 26: 73.
Whiting, S.J. and H.H. Draper. 1980. The role of sulphate in the calciuria of high protein diets in adult rats. J. Nutr. 110: 212.
Whiting, S.J., and D.E.C. Cole. 1986. Effect of dietary anion composition on acid-induced hyper calciuria in the adult rat. J. Nutr. 110:212.

WATER – FORGOTTEN NUTRIENT AND NOVEL DELIVERY SYSTEM

PETER H. BROOKS

Seale-Hayne Faculty of Agriculture, Food and Land Use, University of Plymouth, Devon, UK

Introduction

Water is involved in virtually all body functions. In addition it comprises almost 70% of the adult animal's body mass. An animal can lose practically all its fat and over half its protein and yet live, while a loss of one-tenth of its body water will result in its death. Furthermore, the rate of turnover of water within the body is greater than that of any other substance. Over the last fifty years there has been a massive investment in research on the nutrient requirement of the pig but very little investment in research on the animal's requirement for water. Consequently, we are still making important discoveries about the way in which water quality and quantity affects the productivity, health, welfare and pollutant output of the pig.

Over the last fifteen years, my colleagues John Carpenter, Pinder Gill, John Barber and I have been finding out just how important water is in the operation of the modern pig unit. This paper will summarize some of our findings and identify the areas which we identify as still needing further research and providing further potential for development. This paper will not attempt to provide a comprehensive review of all the recent work on water. In earlier papers we have reviewed the theoretical and practical basis for determining the pig's water requirement (Brooks et al., 1989; Brooks and Carpenter, 1990) and the effects of water quality on swine welfare and productivity (Brooks et al., 1992). The reader is directed to these for a much more detailed discussion of those factors than can be afforded in this paper. The aim of this paper is to provide an overview of four areas which we consider to be important and where increasing knowledge is facilitating the adoption of radically different management practices. These are:

- Water consumption and wastage: minimizing water wastage without compromising pig productivity and(or) welfare.
- Water quality and quantity: the way in which water supply affects production.
- Water treatment and water medication: water as a vehicle.

- Water and liquid feeding systems: their current advantages and future potential.

Water consumption and wastage – minimizing water wastage without compromising pig productivity and/or welfare

In 1985 my colleague John Carpenter and I became concerned about the volume of effluent being produced from pig units. John Carpenter calculated that the additional cost of storage, transport and spreading of effluent resulting from water wastage averaged £4.50–£7.00 per cubic metre. This translated to over 4 million cubic metres of water being wasted each year on UK pig farms alone at a cost of almost £25 million. These calculations gave us cause for concern on three counts. First, water is now an expensive commodity. In many developed as well as developing countries potable water is becoming a limited and limiting resource. Secondly, wastage of water increases the volume of pig effluent that has to be dealt with. Thus water wastage increases the costs of effluent processing, storage, transportation and application. Thirdly, increasing the volume of effluent produced by the pig unit increases the problems of disposing of it in an environmentally acceptable manner. In some parts of the world these problems are reaching almost unmanageable proportions. In the UK it is no longer possible to obtain planning permission to build a pig unit unless you are able to demonstrate that you have sufficient, suitable land available on which to dispose of the waste without causing environmental problems or causing nuisance to your neighbours from smell pollution (Humberside County Council, 1990; MAFF, 1991; MAFF, 1992). In the Netherlands the problem is even worse. Units are granted effluent quotas which control the amount of effluent which they may produce. Despite quotas they still have more effluent than they can dispose of on their available land. In order to overcome these problems they now have to transport pig effluent to central collection points and export it, on barges down the Rhine to areas in Germany where there is suitable land to facilitate its safe disposal.

The amount of effluent produced on a pig unit can be reduced considerably in two ways. First, by reducing water wastage due to spillage, and secondly, by engineering the diet to reduce unnecessary water consumption.

REDUCING WATER WASTAGE DUE TO SPILLAGE

Two factors have a profound effect on water wastage:
- Drinker design.
- Pig drinking behaviour and drinker operation.

Drinker design can have a dramatic effect on the amount of water wasted both while the pig is drinking and through leakage from poorly designed drinkers between drinking events. As an example Gill (1989) demonstrated that weaner pigs watered using five different types of drinker could differ in water to feed ratio by as much as 83% (Table 1). Similarly, the water to feed ratio recorded for growing-finishing pigs differed by as much as 60% (Table 2). It is worth noting that the water use per kilogram liveweight is very similar on the better designs of drinker (Tables 1 and 2). Using Gill's data it is possible to estimate the effect of drinker design on the amount of water used to produce a pig from weaning to 100 kg (Table 3). The difference between the 'Best' and 'Worst' drinker type used in these trials was 378 litres per pig produced. In the absence of significant differences in biological performance of the pigs it must be assumed that this represents increased water wastage and hence a proportional and unnecessary increase in effluent production.

European manufacturers of drinkers have responded very well to this type of information and in recent years have made significant

Table 1. Water use by weaned piglets from 3 to 6 weeks of age, provided with water from five different drinker types (After Gill 1989).

Drinker type	Daily gain (g)	Water to feed ratio (l/kg)	Water to weight ratio (l/kg LW)
Mono-flo nipple	199[b]	5.32	0.23
Arato 76 nipple	260[a]	3.23	0.13
Lubing bite type I	213[b]	3.68	0.12
Lubing bite type II	221[b]	2.90	0.13
Alvin bowl	224[ab]	3.49	0.14

[a,b] means with the same superscript did not differ at $P>0.05$

Table 2. Water use by growing pigs fed *ad libitum*, provided with water from four different drinker types (After Gill 1989).

Drinker type	Daily gain (g)	Water to feed ratio (l/kg)	Water to weight ratio (l/kg LW)
Mono-flo nipple	820	4.02	0.17
Arato 76 nipple	790	2.54	0.11
Lubing bite type I	820	2.50	0.10
Lubing bite type II	790	2.53	0.11

Table 3. Difference in water use by pigs watered with best and worst drinker types from weaning to 100kg (Extrapolated from the data of Gill 1989).

	Best	Worst	Difference
Daily water use (5–10 kg) (l)	0.91	1.59	0.68 (+74%)
Daily water use (11–100 kg) (l)	4.98	8.32	3.34 (+68%)
Total water used 5–10 kg (21 d) (l)	11.11	33.39	14.28
Total water used 11–100 kg (109 d) (l)	542.82	906.88	364.06
Water wasted by worst drinker (l/pig)			378.34

improvements to the design and engineering quality of their drinkers. The only benefit that we have ever heard suggested for having leaking drinkers is that newly weaned pigs may find the water source more readily if the drinkers leak! However, Ogunbameru and Kornegay (1991) found that 'pig-trainer' nipple waterers, which were designed to drip in order to attract piglets, did not result in any improvement in pig performance. Therefore, we must conclude that every effort must be made to design drinkers which reduce water wastage without inhibiting water consumption.

The way in which the pig uses the drinker also has a marked effect on the extent of water wastage from it. Gill (1989) observed the behaviour of pigs in relation to drinkers and found that a considerable amount of the wastage that was occurring was a result of the pig using the drinker in a manner which the designer had not intended. Some examples are given in Figure 1. Drinking pigs were frequently observed to take the drinker into their mouths in a manner which resulted in spillage from their mouth during drinking. In addition pigs were found to use the drinkers in quite unexpected manners, for example, operating a nipple drinker with their

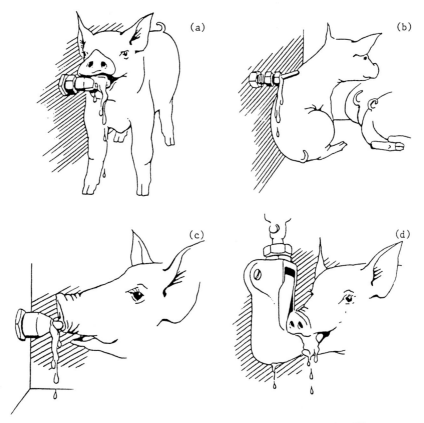

Figure 1. Water wastage by weaned piglets and growing pigs from the different types of drinker (Gill 1989).

nose, allowing water to drop onto the floor, then attempting to drink the water off the floor. Pigs were also observed leaning against drinkers to provide themselves with a shower for cooling purposes. Therefore, in attempting to design the less wasteful drinker it is very important to conduct behavioural observations to ensure that the pig operates and uses the drinker in the way that the design engineer intended.

ENGINEERING THE DIET TO REDUCE UNNECESSARY WATER CONSUMPTION

The quantity and composition of food consumed has a profound effect on the water requirement of the animal. The pig uses water to maintain homeostatic balance. Two factors have a particular influence on water consumption namely, the mineral content of the diet and the quality and quantity of protein. Hagsten and Perry (1976) showed how the water consumption of the pig increased in response to salt additions to the diet (Figure 2). More recently, Wrigley, Brooks and Morgan (1992, unpublished data) have shown that this increase in water intake results more from the pigs need to maintain homeostatic balance of chloride than sodium (Figure 3 and 4). These are not the only mineral ions of importance. The potassium content of the diet also has a dramatic effect on the water demand as Gill (1989) demonstrated (Figure 5).

It is not only the mineral content of the diet which is of importance in determining the pig's water requirement. Protein is another good

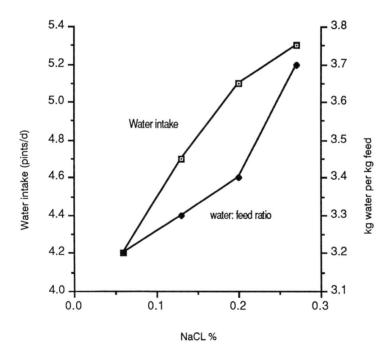

Figure 2. Effect of dietary salt on water intake and water:feed ratio.

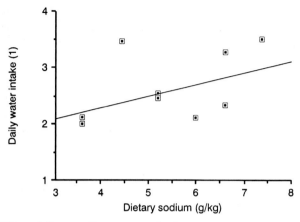

Figure 3. Effect of dietary sodium on water intake.

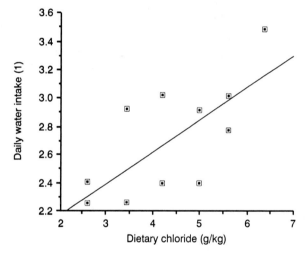

Figure 4. Effect of dietary chloride on water intake.

example. Any amino acids which are supplied in excess of the animal's protein requirement are deaminated and the nitrogen has to be excreted in the urine. As the pig has a limited ability to concentrate nitrogen in its urine, having extra nitrogen to get rid of means that it has to increase water intake. Consequently, if the pig is fed more protein than it can utilize for productive purposes, or an unbalanced protein, it has to increase water consumption. For example, increasing the protein content of diets for pigs between three and six weeks of age from 12 to 16% increased water intake from 3.90 to 5.26 litres per pig per day (35%) (Wahlstrom et al., 1970). A recent study at the University of Plymouth (Thompson et al., 1992) illustrates that the response to increasing protein in the diet is actually biphasic. The data (Figure 6) show water intake increasing only slowly as the protein supply increases

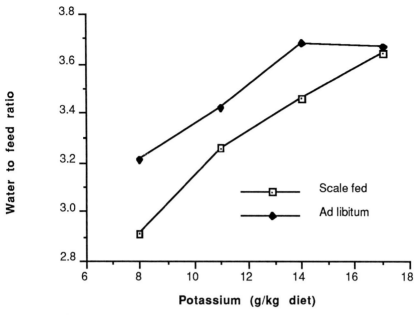

Figure 5. Effect of potassium on water to feed ratio.

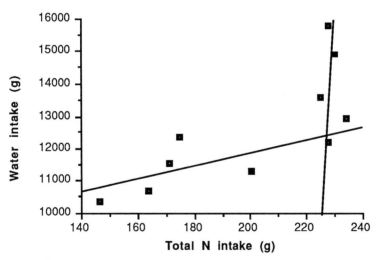

Figure 6. Effect of total N intake on water intake.

until the protein requirement for lean deposition is reached. Once this point is reached all the additional protein ingested has to be deaminated and the nitrogen excreted. Consequently, from this point onwards the water demand increases dramatically. This results in an increase in both urine output and the output of urinary nitrogen and hence the potential pollutant output of the pig.

Thus by manipulating the mineral and protein content of the diet it is possible to have a very significant effect on both the biological performance, the water demand and the potential pollutant output of the pig.

Water quality and quantity: the way in which water supply affects production

WATER QUALITY

The effect of water quality on pig performance has not received a great deal of attention. We can demonstrate quite clearly that certain aspects of water quality have a marked effect on performance. However, in a number of other cases we believe that there may be an effect but to date there is no objective, experimental, evidence to substantiate our belief.

Water quality is usually described in terms of three factors:

- Dissolved solids, i.e. the mineral content of the water.
- Suspended solids, i.e. the amount of organic matter carried in the water.
- Microbiological content. Usually thought of in terms of bacteria, but algae, invertebrates, insect larvae and various life cycle stages of parasites should also be considered in this context.

Therefore water quality is determined by two things, where it has come from and what has happened to it both before and after it arrived on the pig unit.

Dissolved solids

Depending upon where in the world the pig unit resides, its water supply may be derived from a lake or stream, from a bore hole or well or from a domestic or industrial water supply. The water may have its origin in and have flowed through very different rock strata. These different rock types will each impart their own, varied and variable mineral content to the water. This is generally referred to as the 'hardness' of the water (Table 4). Hardness is often expressed as the sum of the calcium and magnesium (reported in equivalent amounts of calcium carbonate). Generally, water that contains less than 1000 mg/litre of dissolved solids should present no serious problem to any class of stock.

When the dissolved solids level exceeds 1000 mg/litre the palatability

Table 4. Classification of water hardness.

Hardness range (mg/litre)	Description
0–60	Soft
61–120	Moderately hard
121–180	Hard
>180	Very hard

of the water declines. In addition the minerals which are present in solution may form deposits on the internal surfaces of the pipes thereby reducing effective diameter flow rate. It must also be remembered that the corrosive nature of some minerals may reduce the life of metal piping; and the interaction with the pipes can produce further dissolved solids. The low rates of flow created by these processes may themselves result in delivery rates to the animal which are less than optimal and therefore have an adverse effect on performance (see next section). Until recently elaborate and expensive equipment has been required to overcome the problem of 'furring up' of pipes. The problem may now be overcome by relatively inexpensive devices which fit around the incoming water main and subject the water to an intense electro-magnetic field. These devices are claimed to charge the particles in the water in such a way that they prevent them accumulating on the internal surfaces of pipes. At present there is no objective information to substantiate these claims. As importantly, there is no information whether water treated in this way has any effect on the performance of the animals which subsequently drink it. An alternative approach which circumvents the problem to a large extent is to use non-metallic pipes. These can often help prevent the problem occurring.

Some minerals can cause problems at levels well below 1000 mg/litre. Recommended limits suggested by CSIRO (1987) and NAC (1974) are given in Table 5.

Table 5. Recommended limits of concentration (mg/ml) of some potentially toxic substances in water.

Item	CSIRO 1987	NAC 1974	Item	CSIRO 1987	NAC 1974
Total dissolved salt	6000	5000	Chromium	1.0	1.0
Bicarbonate (HCO_3^-)	1000	–	Cobalt	–	1.0
Calcium (Ca_2^+)	1000	–	Copper	0.5	0.1
Fluoride (F^-)	2	2	Lead	0.5	0.1
Magnesium (Mg^{2+})	400	–	Mercury	0.002	0.01
Nitrate (NO_3^-)	500	100	Nickel	5.0	1.0
Nitrate (NO_2^-)	100	10	Vanadium	–	0.1
Sulphate (SO^{24-})	1000	1000	Zinc	20.0	25.0
Arsenic	0.5	0.2	Selenium	0.02	–
Cadmium	0.01	0.05			

Suspended solids

Suspended solids are a major cause of contamination and blockage of supply systems. They are often easier to deal with than dissolved solids because, as the name implies, they are suspended and therefore can be removed by filtration. Suspended solids are more often a problem where the water is being supplied from a stream, lake or bore hole. These 'on farm' supplies often lack a satisfactory filtration system. Organic matter can also proliferate in the water supply system in the form of slime moulds. Routine sanitization of water supplies within buildings should be a regular part of normal management. However, we have been made aware of problems arising *from* sanitization. Debris may become detached from the pipeline as a result of the cleansing process. This debris may collect behind or in drinkers rendering them inoperable. In extreme cases this can leave the animals without any supply of water. Consequently, when sanitization is undertaken care must be taken to remove drinkers and flush out the pipeline to remove all debris and prevent subsequent blockage of the system.

Generally water derived from domestic or 'mains' supplies contains less suspended solids. However, even otherwise high quality mains water may contain a certain amount of silt. This is generally derived from corrosion of water supply pipes and is more likely to be a problem when the pig unit receiving the supply is situated at the end of a distribution line.

Microbiological contamination
Water can be a vehicle whereby pathogenic organisms are transmitted to the animal. Water from domestic supplies usually has good microbiological quality when it arrives on the unit because it has been chlorinated. Many farm supplies would also benefit from a sterilization system of some kind. Chlorination can be achieved by using a proportioner to introduce the sterilant into the supply. However, this should only be used on the main incoming supply line where the flow rate is high as proportioners can be very inefficient and inaccurate when the flow rate is low and intermittent. Most proportioners used these days do not require an electricity supply so their use is feasible even in remote areas.

As an alternative to chlorination, ultra violet (UV) sterilization may be used. UV sterilization equipment is now readily available but does necessitate an electrical supply. This technique may have an advantage over chlorination in that it removes the possibility of accidental over-chlorination. Excessive chlorination will certainly have an adverse effect on water consumption and hence on performance. What we are not sure about at present is the effect that chlorination at 'normal' levels has on the palatability of water for the pig. Certainly some humans are very conscious of chlorine in the water, as evidenced by the massive growth in mineral water sales and of domestic water filters.

Neither water filtration nor sterilization overcome one problem, namely the contamination of the supply with toxins produced by microbial organisms. Of particular significance in this context are the toxins produced by blue-green algae, which can be lethal. These are a particular problem where water is derived from ponds or reservoirs where algal blooms form readily (particularly in areas where the water has a high nitrate content). The rapid decomposition of these blooms can cause particular problems. Blue-green algae can also proliferate in on-farm reservoirs and storage tanks if these are open to the light. Algal toxins probably occur more frequently than realized in troughs used for drinking by outdoor or 'free-range' animals.

Not all algae and slime moulds growing in water supplies are directly harmful, but it appears that many of them can impart 'off' taints to the water which reduce its acceptability and hence intake.

Quality at the point of delivery

Even some pig producers (who really should know better) seem to think that the pig is not concerned about the cleanliness of the water it is given. The data in Figure 7 (Barber, 1992) show just how wrong that is! The pigs were given a choice between drinking from a water bowl or from a drinker. As long as the water bowl stayed clean the pigs chose to drink from it. As soon as it became fouled the pigs rejected it and drank from the nozzle drinker instead. As far as the pig was concerned fouling did not mean getting dung or urine in the water. Food particles from around its mouth left in the water following drinking also counted as 'fouling'.

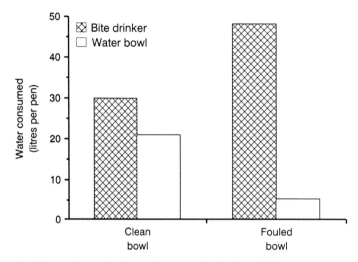

Figure 7. Effect of water bowl cleanliness on pigs choice of drinker.

Pigs prefer to drink from a free water surface provided that the water is clean. Sadly, this is rarely the case. As well as fouling with dung, urine and feed, the supply is often contaminated by bacterial and slime mould growth in the bowl and around the filling mechanism. Some producers claim that their pigs prefer water bowls to drinkers. Generally they prove to be the producers whose units take their water supply directly from the main supply (illegal in the United Kingdom) and have nozzle drinkers delivering water to the pig at a very high velocity. Given a choice of drinking from a nipple or bite drinker delivering water at a high rate of flow, but a low velocity (which implies a low pressure system) or a less than scrupulously clean water bowl, the pig will usually chose the clean supply from the drinker.

Overcoming problems of poor water flavour

Pigs like things that taste nice and that includes water. As noted earlier, if the water which the pig has available to it is contaminated with food, urine or faeces it will only drink enough water to keep itself alive, not the amount needed to maximize its productivity. Similarly, water which has a particularly high mineral content will be less well accepted. However, it is not only water that should really be regarded as 'undrinkable' which causes problems. Some sources of water which have a perfectly acceptable mineral content and a good microbiological quality still have a poor flavour, while others are particularly palatable. This is demonstrated by the massive growth in demand for bottled water by human consumers. Clearly it is not feasible to provide all pigs with 'Perrier Water'. However, it may be feasible to improve the palatability of water by including sweeteners, flavours, or flavour enhancers. There may be benefits from adjusting the pH of the water too. It is not possible to predict the result that will be achieved on every unit, for as shown above, water is not a single universal product. However, it is an approach well worth trying once you have assured yourself that there is not a major underlying problem of water quality which should be tackled first.

Water quantity

If water were like other nutrients then it would be relatively easy to derive factorial estimations of water requirement. However as discussed above water is required for other functions as well:

- maintenance of mineral homeostasis
- excretion of the end products of digestion (particularly protein digestion).

In addition the pig uses water to:

- adjust/control of body temperature
- achieve satiety (gut fill)
- satisfy behavioural drives.

Two important points need to be emphasized. First, the functions listed above create needs or requirements which may be additive and additional to the requirements for maintenance and production. Thus, the classic nutritional requirement expressed in nutritional terms represents the minimum value which is only operative if the requirements for the other functions have fortuitously been met within a given situation. Secondly, some of the functions have a higher priority than production. Hence, in a situation where water supply is inadequate to support all the animals needs, prioritization of these functions will result in an undersupply of water to support production and therefore a reduction in animal performance.

One of the more interesting discoveries that we have made is that the rate at which water is supplied to the young pig influences performance. It has been assumed in the past that the pig consumes food and then drinks sufficient water to metabolize its food and detoxify itself. Results from one of our studies (Barber et al., 1989), presented in Figure 8, suggest that the reverse may be the case, namely that the availability of water influences the amount of water the pig consumes and that this in turn affects its voluntary feed intake and subsequent performance. A surprise in this study was the relatively limited amount of time that these young pigs were prepared to devote to drinking behaviour each day. The pigs on the most restricted water delivery rate were not prepared

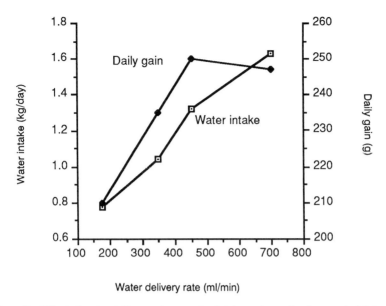

Figure 8. Effect of water delivery rate on water intake and growth of weaned piglets.

to increase their drinking time to compensate for the reduced availability of water (Figure 9). This behavioral limitation to water consumption may be a result of the conditioning that the pig has received during lactation when milk is only available for very short periods of time, at regular intervals, and where the whole group suckles together. We have not yet elucidated whether the provision of greater numbers of drinkers in pens for newly weaned pigs will overcome the problem. However, it is clear that when water delivery rates are restricted, older pigs will

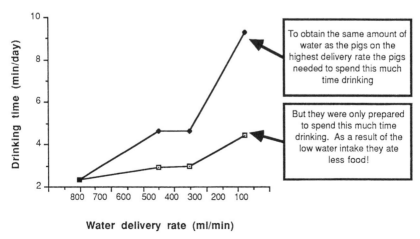

Figure 9. Time spent drinking by 3–6 week old weaned piglets from drinkers with different delivery rates.

increase drinking time to compensate and growth rate is not necessarily compromised (Barber, 1992).

Some years ago Yang (1981) suggested that the pig had a requirement for total volumetric intake and that the water to feed ratio would be minimised when the pig was fed ad libitum. That is to say the pig would limit its water intake to the minimum needed to detoxify itself and would maximize feed intake. Yang's studies suggested that when feed intake was restricted the pig would increase water intake to satisfy its demand for gut fill. In his studies with young pigs the total volumetric limit appeared to be around 19% of liveweight. In our studies (Barber, 1992) using somewhat older pigs the limit appears to be approximately 12%. We are now in a position to propose a tentative model of the relationship between total volumetric intake, water intake and feed intake (Figure 10) which accommodates the various effects that we have observed as follows.

- Where water availability is limited (irrespective of the reason for the limitation) feed intake is restricted. The animal will not consume food, the waste products from the metabolism of which it is unable to excrete.
- As water availability increases voluntary feed intake will increase

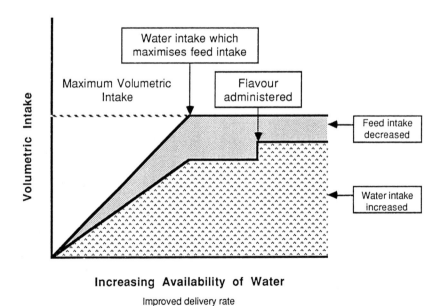

Figure 10. A schematic model of the relationship between volumetric intake, water intake and feed intake in the pig.

proportionally up to a maximum set by the total volumetric intake limit.
- Where feed intake is limited and water availability is not, the pig will increase water intake to satisfy its demand for gut fill.
- Where water availability (consumption) is adversely affected by palatability the problem may be rectified by flavouring the water.
- Where water flavourings are particularly attractive, consumption may be stimulated beyond the normal intake. This can result in water comprising an abnormal proportion of the total volumetric fill. In such cases feed intake will be decreased.

Water treatment and water medication – water as a vehicle

PROBLEMS WITH FEED MEDICATION
For many years 'in-feed' medication has been the principal route by which prophylactic or therapeutic products have been administered to large numbers of pigs or poultry. Despite the widespread use of this method of administration it is an approach which has a number of important deficiencies:

Target animals have reduced food intake
The animal which we wish to treat, the 'target' animal, is the sick one. Generally animals which are suffering from disease have reduced feed intake compared with their healthy contemporaries. In extreme cases they may fail to compete successfully with their healthy pen mates at the feed trough. Consequently, those animals at which the product is targeted are the least likely to consume it. Conversely, healthy animals consume more of the ration and receive medicinal products unnecessarily.

Treatment may be delayed by the use of the in-feed route
With diseases of confined animals rapid treatment is vital. Feed medication may fail to achieve this. Unless the feed compounder happens to keep stocks of feed containing the required medicinal product there will be an inevitable delay in treatment while a diet containing the required product is manufactured and delivered. At busy times and over holiday periods delays can be extremely costly to the producer.

Even when the medicated food has been delivered to the unit delays are still possible. When pigs are fed *ad libitum* producers are unwilling to empty out feed hoppers or bulk bins in order to refill them with medicated food.

Bulk bins cause big problems
On intensive livestock units feed delivery is frequently automated and food is usually supplied to the unit in bulk. Many units find it extremely difficult to arrange separate feeding for specific pens of pigs. Feed troughs are not arranged for manual filling and there is often nowhere to store 'special' diets and insufficient labour available to undertake the task. Consequently, medicated food, delivered in bulk, is generally

added to unmedicated food already present in the bin. This has three very deleterious effects. If the bin emptied in a sequential manner (most don't) drug administration would be delayed until the unmedicated food already present in the bin has been used. The more usual situation is that the medicated food will tend to mix with the unmedicated food as it is drawn from the bin. Consequently, sick pigs will receive medication at a lower rate than intended and often one which is insufficient to deal with the problem. Similar problems occur when a new load of unmedicated feed is added to the bin. Once again mixing of the two diets will occur during normal bin emptying. Not only does this reduce the effective dose given to the animal but also increases the risk of disease organisms developing resistance to the drug.

The other major, and very worrying, problem posed by the use of medicated feed in automated systems lies in the failure of producers to observe statutory withdrawal periods. Regrettably, many producers fail to obey the legislation on withdrawal periods. Concern about drug residues in meat caused by this problem has already led to a ban on feed medication in some European countries and may well result in a ban on feed medication throughout Europe in the next few years.

Feed medication is expensive
There is a widespread belief amongst pig producers that feed medication is cheap, or at least cheaper than, alternatives such as water medication. This is a false picture. Many producers are only asked to pay slightly more than the drug cost for the production of 'special' mixes. This does not reflect the true cost of medication. In the UK the additional expense of making 'special' mixes in a modern high performance mill is estimated to be between £30 and £40 per tonne. However, the competitiveness of the feed industry is such that no one is prepared to pass on the full cost direct to the customer. Consequently the costs incurred in producing 'special' mixes are absorbed in the overall overheads which means that **all** compound feed users pay!

Nuisance value to the feed compounder
Most people in the feed compounding industry would be quite happy to see feed medication banned. Feed medication produces massive operational problems. Modern mills are designed for long production runs of standard products and not for the stop, start production of small batches of 'special' diets. There is also a large risk element in the manufacture of special diets. Accidental additions of incorrect quantities or the cross-contamination of diets intended for other species can be very expensive and damaging to the reputation of the feed compounder.

POTENTIAL ADVANTAGES OF WATER MEDICATION

Given the problems outlined above it may seem strange that the major alternative route for medication, that is, through addition to drinking water, has not gained more support. Water medication could overcome many of the problems listed above, for the following reasons:

Sick animals drink even if they do not eat
The sick animal that does not want to eat or cannot compete for food will continue to drink. Indeed it will often increase its fluid intake. In one observation in our unit, pigs suffering from post-weaning scour increased their water intake four-fold compared with their healthy contemporaries. Therefore, the water route for medication increases the chances of the 'target' (sick) animal receiving the medication.

No delay in the administration of drugs with water medication
One of the greatest benefits of water medication is that there need be no delay in starting treatment. If the attending veterinarian has appropriate products in stock, water medication can commence within minutes of diagnosis. The timeliness of treatment can be a vital factor in containing disease spread in a large confinement unit.

Statutory withdrawal periods can be satisfied easily
Water medication makes it much easier for the veterinarian to ensure that his/her clients observe the statutory withdrawal periods prior to slaughter. Providing a separate supply of unmedicated water is much easier than providing unmedicated food in many large units.

Water medication need not be more expensive
Cost comparisons between feed and water medication are difficult because of the feed pricing distortion noted above. However, the water-additive versions of drugs are sometimes more expensive than the feed-additive forms. In some cases at least, this is a reflection of the relative volume of sales. Increasing use of water medication will result in increased sales volume and reduced unit cost of many water soluble products. There are also direct savings to be made from water medication, merely by restricting treatment to those animals (or pens of pigs) which need it rather than applying blanket treatment of an entire house via the feed. However, it is important to ensure that the water delivery system does not result in water wastage as this will result in unnecessary loss of product.

In addition to the benefits listed above the water medication has particular benefits for the veterinarian. First, the veterinarian can exercise more control over both the products used and which animals are treated. With water medication, blanket treatment of all the pigs in the house is not obligatory. Therefore, problems of disease resistance may be reduced. Secondly, by ensuring that statutory withdrawal periods are observed human health can be better protected. Thirdly, water medication does not incur the risks of cross-contamination which can occur with the feed medication.

PROBLEMS WITH WATER MEDICATION

Given that water medication appears to have so many advantages over food medication, why is it not practised more widely? The answer lies partly in inertia within the industry and partly in the problems of administering the drugs. There are three main ways in which water medication can take place on the livestock unit and all have their problems.

Water medication via separate drinkers
For small animals and birds the provision of medicated water in fountain drinkers of various types offers one method of administration. These have to be tended manually, however, and are prone to spillage and to damage. In addition, the existing water supply to the pen has to be cut off. For larger animals; free standing tanks with their own drinker attached can be used. These are laborious to use, heavy to move and take up significant amounts of space in the pen.

Water medication via a header tank or reservoir
Water medication via the header break tank has all the problems of feed medication and then some! The first problem is that not all units have header tanks (it is a legal requirement to have an air break between the incoming water main and the delivery system to animals in the UK). The ball valve has to be tied up to stop it operating or the water supply turned off. The volume of the tank has to be estimated and the correct amount of drug added and dispersed. This may cause more problems than it solves, as stirring in the drug stirs up the sediment in the tank. On a number of units we have seen this result in every drinker in the house becoming blocked with sediment, and in at least one case a large number of pigs died of water deprivation as a consequence. The same problem can occur if no one remembers to untie the ball valve or turn the water back on after treatment.

Water medication via proportioner
The third method is to use a proportioning device which will introduce the medication into the water at the correct dilution rates. A number of such devices have appeared on the market over the years and, sadly, many have disappeared as quickly as they came. The problem has tended to be that the devices would only work effectively if supplying a constant and large demand for water provided from a good pressure head. Unfortunately, these conditions are rarely met in livestock units. The demand for water is variable and intermittent and the available head is often small. Users should be very careful to read the manufacturer's instructions before using water proportioners. Most proportioners work tolerably well when operated strictly within the manufacturer's recommended conditions. Many are almost unbelievably inaccurate when operated outside the conditions specified by the manufacturers.

Solubility and taste of medications
There is one major problem with water medication which should not be overlooked and that is the drug itself. Some materials are very unstable once mixed with water and quickly lose their efficacy. Others are not easily dispersed. We have found that many so-called water soluble products are not very soluble! We have found products sold as water soluble that are only miscible rather than soluble and others which are

not soluble at all but are very fine suspensions. This may not a problem if you are mixing them by hand and delivering them to the pig in a trough, but can lead to all manner of problems if and when you attempt to deliver them through a water proportioner and a supply pipeline. Finally, it must be noted that not all drugs are pleasant tasting. Therefore, care needs to be taken to ascertain that the animals are not reducing their liquid intake, and as a consequence their feed intake, as a result of the medication being used in the water.

WATER AND LIQUID FEEDING SYSTEMS: CURRENT ADVANTAGES AND FUTURE POTENTIAL

Liquid feed delivery systems tend to be more common where:

- liquid by-products are readily available
- pigs are normally rationed in order to reduce carcass fatness
- finishing units are large in size and hence the capital expenditure on a computerized liquid feed delivery system is more easily justified.

Although the availability of liquid by-products may be an incentive to consider a liquid feed system, many European producers install such systems even when they are not available. Furthermore, the concept of liquid feeding now extends beyond the finishing pig. Some European units are fast moving to a situation where the entire unit, pregnant and lactating sows, weaners, growers and finishers, are all being fed on liquid diets.

There are a number of perceived advantages in liquid feeding, namely:

- reduction of food loss, as dust, during handling and feeding
- improvement in the pig's environment and health due to the reduction of dust in the atmosphere
- improved pig performance and FCR
- flexibility in raw material use (opportunity to utilize more economic food sources and reduce cost per kg gain)
- improved materials handling (system can act as both a feed mixing and distribution system)
- increased accuracy of rationing (computer control brings a degree of accuracy to the system that it is difficult to emulate with dry feeding systems)
- improved dry matter intake in problem groups (e.g. weaners and lactating sows)
- increased eating speed in pregnant sows fed using Electronic Sow Feeders and consequent reduction in capital cost per sow of such equipment

Improved performance in liquid fed pigs

There are several reasons why liquid feeding, particularly if the feed is soaked for a period prior to feeding, may have beneficial effects on performance, namely:

- Reduction in feed wastage as dust.
- Increased acidity of the diet.

- Increased phosphorus availability.
- Improved accessibility of the substrate to digestive enzyme action.

Reduction of wastage

When making comparisons between feeding systems, differences in wastage are always difficult to quantify. Where wastage occurs it is difficult to ascertain whether this is an intrinsic difference between the forms of feed under test, or a result of differences in the design of feeders. However, wet feeding does reduce the opportunities for wastage of feed as dust. Dust losses may be of much greater significance than often realized. A finishing pig consuming 2.0 kg/d needs only to waste 4 g feed per hour to produce an apparent reduction of FCR of 5%. Wastage, as dust, contributes to the atmospheric dust load. This in turn increases the prevalence of respiratory disease and further reduces productivity.

Increased acidity

Wet feeding also alters the composition of the feed. Smith (1976) showed that *Lactobacillus* species, which occur naturally on cereal grains, will proliferate in a wet feed and reduce the pH. In his study adding water to the meal at feeding time produced a pH of 5.81. Soaking the mixture for 24 h resulted in a massive proliferation of *Lactobacilli* which produced lactic acid and reduced the pH to 4.10. Increasing the soaking time to 84 h did not further increase acidity. Because pipeline wet feeding systems are not sterilized between feeds they are microbiologically active and act as a fermenter. Hansen and Mortenson (1989) conducted a large survey in Denmark and found that it took 3–5 days for the *Lactobacillus* levels to elevate and stabilize in pipeline feeding systems. In their studies they found that it was detrimental to sterilize pipeline feeding systems

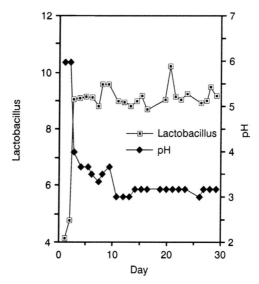

Figure 11. Relationship between lactobacillus and pH.

as this removed the *Lactobacilli* and increased the feed pH by 1.5–2.0 units. This in turn allowed *coliform* bacteria to proliferate for 1–5 days until the *Lactobacilli* re-established themselves and lowered the pH. They found that sterilization of pipeline feeding systems was actually disadvantageous as outbreaks of diarrhoea often resulted from the coliform 'bloom' which followed the sterilization of pipeline systems. Recent studies at the University of Plymouth (Russell et al., unpublished data) have shown a similar development of a *Lactobacillus* population in an *ad libitum* wet feeding system for weaner pigs, accompanied by a reduction of pH and *Escherichia coli*. (Figure 11). However, we must also ask whether the *Lactobacillus* spp. that are developing in the systems are the best organisms or whether we should be 'seeding' the system with specific organisms. These are areas of study in which we are actively involved at present.

Increased phosphorus availability
In addition to activating the natural flora of the diet, soaking also activates enzymes present in the diet ingredients. In some preliminary studies (Segier et al., unpublished data) we have shown that phytases, which occur naturally in the pericarp of cereals, are activated in a wet medium and increase the dissociation of phosphorus from phytate-phosphorus. We are interested to find out the extent to which this can be utilized in commercial feeding practice.

Improved enzyme activity
Providing the pig with its feed in a wet form also appears to improve its digestibility. Studies at University of Plymouth, (Gill et al., 1986) showed that increasing the water to feed ratio had a beneficial effect on growth rate and feed conversion ratio (Figure 12). Subsequently, Barber et al. (1991) showed that increasing the water to feed ratio improved the

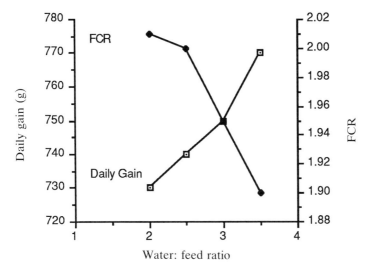

Figure 12. Effect of water to feed ratio on pig performance.

Table 6. Effect of water to feed ratio on diet digestibility (Barber 1992).

	Water to feed ratio			
	2:1	2.67:1	3.33:1	4:1
Dry matter digestibility (%)	79.12[a]	77.78[a]	80.34[a,b]	82.93[b]
Estimate DE (MJ/kg/DM)	15.16	14.96	15.41	15.80
Nitrogen retention (g/kg $W^{0.75}$/d)	1.49	1.40	1.63	1.74

[a,b]means with the same superscript do not differ significantly at $P>0.05$

digestibility of both dry matter and digestible energy (Table 6). These studies suggest that the simple expedient of changing the water content of the digesta influences the utilization of nutrients. We postulate that this results from more effective permeation of the digesta by the digestive enzymes.

Ad libitum wet feeding of weaners

Feeding young pigs on liquid diets has been attempted on a number of occasions in the past. The major problems have been devising suitable feeding equipment and maintaining the feed in a hygienic and palatable state. Kornegay et al. (1981) compared *ad libitum* feeding of dry and wet (water to feed ratio 2:1) diets and found no advantage from wet feeding. This appears to be in contrast to the results that are reported from commercial units in the UK who are now using *ad libitum* wet feeding. Preliminary results from our unit show a quite staggering difference in the growth rate of pigs on dry and liquid diets (Figure 13). The difference may stem from the combined effects of different diets and different feeding equipment. Our studies are using contemporary commercial diets of very high digestibility and palatability whereas Kornegay et al. used corn-soya or corn-soya-whey diets. Another important feature of our studies is that the wet feeding system is showing a rapid and dramatic development of *Lactobacillus sp.* which are producing acid conditions in the feed system and as a consequence maintaining the palatability of the feed. A significant research programme is now under way in our Faculty to investigate optimum strategies for the use of this feeding system.

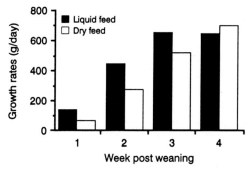

Figure 13. Effect of liquid or dry feed on growth rate of weaners.

THE FUTURE FOR LIQUID FEEDING SYSTEMS

In recent years there has been considerable interest both in the possibility of using probiotics instead of antibiotics and in the use of enzymes in the diet to improve diet digestibility. To a large extent attempts to pursue both these paths are limited by their application to dry feed systems. These do not provide environments in which probiotics are easily retained or in which enzymes have the opportunity to permeate the substrate. The likelihood of success is further undermined when feed compounders subject diets to high temperatures and pressures in order to sterilize and (or) pellet diets. The liquid medium provides opportunities for biologically engineering diets for the pig by bacterial or enzymic action. It is clear from our studies that fermentation processes are a normal component of liquid feed systems. The challenge is to learn how to modify and control these processes to obtain the outcomes that we desire.

There is considerable potential for enzymic modification of liquid diets. At present the approaches being taken are very crude. Enzyme 'cocktails' are being added to complete diets. This is a wasteful and inappropriate strategy. In applying the technology to liquid diets it would be much more appropriate to concentrate on the treatment of individual raw materials before they were incorporated into the final diet. We should also be more adventurous in our thinking. Merely using enzymes to enhance processes already present in the pig may yield useful results, but how much more could be achieved if we developed enzymic processes that removed anti-nutritional factors from raw materials, or as in the case of phytase, achieved dissociations that the animal cannot accomplish? This is a major focus of our current research.

In conclusion, I would suggest that for too long nutritionists have thought only about the dry matter component of the pig's diet. We have given scant attention to the other major component, water. Almost everyone who has ever published a review making reference to water refers to it as the forgotten nutrient. I would argue that it has not so much been forgotten as ignored. To continue to ignore both its current importance and future potential in production systems would be a tragedy.

References

Barber, J. 1992. The rationalisation of drinking water supplies for pig housing. Ph.D., Polytechnic South West.
Barber, J., P.H. Brooks and J.L. Carpenter. 1989. The effects of water delivery rate on the voluntary food intake, water use and performance of early-weaned pigs from 3 to 6 weeks of age. *In:* The Voluntary Feed Intake of Pigs. British Society of Animal Production, Edinburgh. 103–4.

Barber, J., P.H. Brooks and J.L. Carpenter. 1991. The effects of water to food ratio on the digestibility, digestible energy and nitrogen retention of a grower ration. Anim. Prod. 52: 601 (Abstr.).

Brooks, P.H. and J.L. Carpenter. 1990. The water requirement of growing-finishing pigs; theoretical and practical considerations. *In:* Haresign, W. and D.J.A.Cole, Recent Advances in Animal Nutrition 1990. Butterworths, London. 115-136.

Brooks, P.H., J.L. Carpenter and J. Barber. 1992. Impact of water quality on swine welfare and productivity. In: Foxcroft, G.R. Advances in Pork Production Volume 3. University of Alberta, Edmonton. 1-21.

Brooks, P.H., J.L. Carpenter, J. Barber and B.P. Gill. 1989. Production and welfare problems relating to the supply of water to growing-finishing pigs. Pig Vet. J. 23: 51-66.

CSIRO 1987. Feeding Standards for Australian Livestock: Pigs. Commonwealth Scientific and Industrial Research Organisation, East Melbourne.

Gill, B.P. 1989. Water use by pigs managed under various conditions of housing, feeding, and nutrition. Ph.D. Thesis, Polytechnic South West, Plymouth.

Gill, B.P., P.H. Brooks and J.L. Carpenter. 1986. Voluntary water intake by growing pigs offered a liquid feed of differing water to meal ratios. In: Pig Housing and the Environment. British Society of Animal Production, Edinburgh. 1-2.

Hagsten, I. and T.W. Perry. 1976. Evaluation of dietary salt levels for swine. I. Effect on gain, water consumption and efficiency of food conversion. J. Anim. Sci. 42: 1887-1190.

Hansen, I.D. and B. Mortenson. 1989. Pipe-cleaners beware. Pig Int. 19: (11), 8-10.

Humberside County Council. 1990. Intensive livestock units local plan. Humberside County Council, Beverley.

Kornegay, E.T., H.R. Thomas, D.L. Handlin, P.R. Noland and D.K. Burbank. 1981. Wet versus dry diets for weaned pigs. J. Anim. Sci. 52: 14-17.

MAFF. 1991. Code of good agricultural practice for the protection of water. MAFF Publications, London.

MAFF. 1992. Code of good agricultural practice for the protection of air. MAFF Publications, London.

NAC. 1974. Nutrients and Toxic Substances in Water for Livestock and Poultry. National Academy of Sciences, Washington DC.

Ogunbameru, B.O., E.T. Kornegay and C.M. Wood. 1991. A comparison of drip and non-drip nipple waterers used by weanling pigs. Can. J. Anim. Sci. 71: 581-583.

Smith, P. 1976. A comparison of dry, wet and soaked meal for fattening bacon pigs. Expt. Husb. 30: 87-94.

Wahlstrom, C., A.R. Taylor and R.W. Seerley. 1970. Effects of lysine in the drinking water of growing swine. J. Anim. Sci. 30: 368-373.

Yang, T.S., B. Howard and W.V. McFarlane. 1981. Effects of food on drinking behaviour of growing pigs. Appl. Anim. Eth. 7: 259-270.

A BIOLOGICAL APPROACH TO COUNTERACT AFLATOXICOSIS IN BROILER CHICKENS AND DUCKLINGS BY THE USE OF *SACCHAROMYCES CEREVISIAE* CULTURES ADDED TO FEED

G. DEVEGOWDA, B.I.R. ARAVIND, K. RAJENDRA,
M.G. MORTON, A. BABURATHNA and C. SUDARSHAN
*Department of Poultry Science, University of Agricultural Sciences
Bangalore, India*

Summary

Two broiler trials, one trial with ducklings and an in vitro study were conducted to evaluate efficacy of a *Saccharomyces cerevisiae* culture feed additive in counteracting aflatoxicosis. The first experiment involved three levels of yeast culture (0, 0.1 or 0.2% Yea-Sacc[1026], Alltech Inc.) added to broiler diets containing either 500 or 1000 ppb aflatoxin. Body weight, feed efficiency, serum total protein, albumin and HI titre against Newcastle disease were significantly reduced and mortality significantly increased with addition of aflatoxin. This trend was reversed upon yeast culture (YS1026) supplementation of diets containing aflatoxin.

A second experiment was undertaken to confirm results of the first trial using larger numbers of birds. The three treatments consisted of Control (no aflatoxin, no YS1026), 500 ppb aflatoxin, and 500 ppb aflatoxin plus 0.1% YS1026. Body weight and feed efficiency in groups given aflatoxin were depressed ($P< 0.05$), mortality increased, relative weight of liver and activity of gamma glutamyl transferase were increased, and relative weight of the bursa of Fabricius was decreased. The toxic effects of aflatoxin were alleviated by supplementation with YS1026.

A third experiment involved 160 male ducklings assigned to five treatments: Control, 100 or 200 ppb aflatoxin with either 0 or 0.1% YS1026. Toxicity of aflatoxin was characterized by significant reductions in body weight, relative weight of the bursa of Fabricius, serum concentrations of albumin and total protein, red blood cell count, % hemoglobin and packed cell volume, and by significant increases in mortality, aflatoxin concentration in liver tissue, relative weight of the liver, activity of gamma glutamyl transferase and alanine transaminase. The addition of 0.1% YS1026 significantly alleviated the effects of the mycotoxin.

A fourth experiment, an in vitro study of aflatoxin biodegradability, was conducted using three levels of YS1026 (0, 0.5 and 1 g) and two levels of aflatoxin (250 and 500 ppb) in Sabouraud's culture broth at three incubation periods (48, 72 and 96 hours). Aflatoxin was found to be degraded to the extent of 88%.

Introduction

Aflatoxins are toxic metabolites produced by the storage fungi *Aspergillus flavus* and *Aspergillus parasiticus* which are frequently encountered in a variety of foods and feedstuffs. The adverse effects of aflatoxin on poultry and other livestock are well documented (Hamilton, 1987; Huff, 1988). Ducklings were found to be more sensitive than other species of poultry (Muller et al., 1970).

Contamination of corn and other grains with significant levels of aflatoxin continues to be a major problem in many parts of the world. Aflatoxins have been found in stored corn (Shotwell et al., 1975), in the field, and in ensiled high moisture corn (Meronuck, 1993).

The aflatoxicosis problem is generally more severe in developing countries than in developed countries due to tropical and subtropical climatic conditions coupled with poor handling during harvest and post-harvest leading to great economic losses (Devegowda, 1989; Ramakrishna et al., 1992).

Numerous strategies for the detoxification of aflatoxin in feeds and feedstuffs have been tried including chemical methods, physical separation, irradiation and thermal inactivation (Goldblatt, 1971; Goldblatt and Dollear, 1979). Despite these efforts, practical and cost effective methods of detoxification beyond addition of hydrated sodium calcium aluminosilicate or sodium aluminosilicate clays to the diet are not available (Phillips et al., 1988; Kubena et al., 1990; 1993; Barmase and Devegowda, 1990). However, a rapid explosion in biotechnology has opened an avenue for biological means of detoxification. Ceigler et al. (1986) screened various fungi, bacteria and yeast for their ability to detoxify aflatoxin. They reported that *Flavobacterium aurantiacum* was able

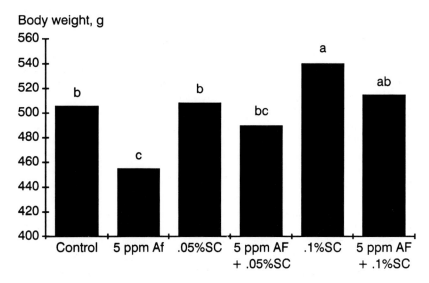

Figure 1. Effects of aflatoxin (AF) and 0.05 or 0.1% *Saccharomyces cerevisiae* culture (SC) on body weight of broilers. Adapted from Stanley et al., 1993.

to destroy aflatoxin in vitro. A biological trial conducted by Stanley et al. (1993) with inclusion of *Saccharomyces cerevisiae* in broiler diets containing 5 ppm aflatoxin resulted in improvement in body weight and feed efficiency (Figure 1). Further, it has been suggested that yeast cells can influence health of animals by adsorbing toxins and pathogenic bacteria to the cell wall (Gedek, cited by Dawson, 1993).

Published data on effects of yeast culture in diets containing aflatoxin are lacking. Therefore, the present research evaluated the effects of a commercial yeast culture in diets fed broilers and ducklings. Additionally, biodegradability of aflatoxin in the presence of yeast culture was evaluated in vitro.

Experiment 1: Effect of 0.1 and 0.2% Yea-Sacc^{1026} in diets containing 500 or 1000 ppb aflatoxin

MATERIALS AND METHODS

Two hundred and ten day-old, sexed commercial broiler chicks were randomly assigned to seven treatment groups. Each treatment had three replicates of ten chicks. Dietary treatments consisted of two levels of aflatoxin (500, 1000 ppb), three levels of yeast culture (0, 0.1 or 0.2% Yea-Sacc^{1026}) and a negative control.

Aflatoxin was produced on rice using *Aspergillus parasiticus* NRRL 2999 as described by Shotwell et al. (1966) and was added to the basal corn/soy/fishmeal diet to obtain the desired aflatoxin level. Standard management practices were followed with feed and water available ad libitum. Body weight and feed consumption were measured weekly and mortality noted upon occurrence.

Blood samples were drawn from the wing vein of 12 randomly selected birds from each group. Serum was separated to determine levels of hemagglutination inhibition (HI) titre against Newcastle disease, total protein and albumin using an Hitachi 705 autoanalyzer. The birds were killed by cervical dislocation and liver samples taken and preserved in 10% neutral buffered formalin for subsequent histopathology. Liver sections were stained with hematoxylin and eosin.

Data were subjected to analysis of variance (Snedecor and Cochran, 1967). Means were compared with Duncan's Multiple Range test (Duncan, 1955).

RESULTS

Body weight at six weeks, total serum protein and HI titre against Newcastle disease were significantly reduced and mortality significantly increased in groups fed diets containing aflatoxin (Table 1). Supplementation with YS1026 counteracted the adverse effects of aflatoxin on all the above parameters with the exception of body weight in the group given 1000 ppb plus 0.1% YS1026. Feed efficiency in groups given YS1026 was improved.

Table 1. Effect of Yea-Sacc[1026] on body weight, feed conversion, mortality, organ weights, serum protein, enzyme activity and HI titre against Newcastle disease in broilers fed diets containing aflatoxin.

Aflatoxin, ppb	Yea-Sacc, %	Body weight, g	Feed:gain	Mortality, %	Total protein, g/dl	HI titre, log 2 values
Control	0	1099[c]	2.21[c]	3	2.75[d]	2.41[b]
500	0	975[b]	2.25[c]	13	2.38[c]	1.30[a]
500	0.1	1105[c]	2.00[a]	6	2.51[c]	2.10[b]
500	0.2	1106[c]	2.03[a]	6	2.78[d]	2.00[b]
1000	0	830[a]	2.21[c]	20	1.52[a]	1.00[a]
1000	0.1	854[a]	2.13[b]	16	2.10[b]	2.00[b]
1000	0.2	945[b]	2.07[ab]	10	2.46[c]	2.10[b]

[a,b]Means in a column with different superscripts differ $P < 0.05$.

On gross examination livers of birds fed diets containing aflatoxin without YS1026 were enlarged, pale and friable. The microscopic hepatic lesions in groups fed aflatoxin showed uncharacteristic fatty changes, massive heterophillic infiltration and severe biliary hyperplasia. Addition of YS1026 resulted in lesions of a milder degree indicating less liver damage.

Experiment 2: Effects of 0.1% Yea-Sacc[1026] in broiler diets containing 500 ppb aflatoxin

MATERIALS AND METHODS

Four hundred and eighty day-old, sexed, commercial broilers were assigned to one of three treatment groups. Each treatment had four replications of 40 chicks. Treatments were control, 500 ppb aflatoxin, and 500 ppb aflatoxin plus 0.1% YS1026. Birds were reared in floor pens on rice hull litter. Production of aflatoxin, management and statistical analysis were as in Experiment 1. Measurements and sampling were also as in the previous experiment with the addition of obtaining the bursa of Fabricius for histopathological examination.

RESULTS

Aflatoxin depressed body weight, reduced feed efficiency and increased mortality significantly (Table 2). Inclusion of YS1026 to diets containing aflatoxin significantly improved the body weight, feed efficiency and reduced mortality. Relative liver weight was significantly increased and bursa weight decreased in birds fed aflatoxin. This trend was reversed on supplementation with YS1026. Serum parameters showed a significant decrease in total protein, gamma glutamyl transferase and HI titre to Newcastle disease in groups fed aflatoxin. This effect was nullified on the addition to YS1026 to the diets which demonstrated a protective effect to liver and enhanced immune response. Gross and microscopic hepatic lesions were similar to those seen in Experiment 1.

Table 2. Effect of Yea-Sacc^{1026} on body weight, feed conversion, mortality, total protein and HI titre level against Newcastle disease in broilers fed diets containing aflatoxin.

Aflatoxin, ppb	0	500	500
Yea-Sacc^{1026}, %	0	0	.1
Body weight, g	1409^a	912^b	1308^c
Feed:gain	2.27^a	3.00^b	2.36^a
Mortality, %	3.87^a	38.40^c	3.30^a
Liver, g/100g BW	2.65^a	3.40^b	2.87^a
Bursa of Fabricius, g/100g BW	0.32^a	0.20^b	0.29^a
Total protein, g/dl	2.68^a	2.12^b	2.56^a
Gamma glutamyl transferase, IU/l	19.25^a	15.25^b	19.77^a
HI titre, log 2 values	2.46^a	1.74^b	2.62^a

^{a,b}Means in a row with different superscripts differ, P< 0.05

Experiment 3: Effects of aflatoxin and Yea-Sacc^{1026} in diets fed to ducklings

MATERIALS AND METHODS

One hundred sixty day-old male Khaki Campbell ducklings were randomly assigned to five treatment groups. Dietary treatments consisted of Control, 100 ppb aflatoxin, 200 ppb aflatoxin, 100 ppb aflatoxin plus 0.1% YS1026, and 200 ppb plus 0.1% YS1026. Ducklings were raised in battery brooders to four weeks of age. Production of aflatoxin, management and statistical analysis were as in Experiment 1.

Blood samples drawn from the wing vein were used for hematological examination. Parameters included red blood cell counts, % hemoglobin and packed cell volume measured using an ERMA clinical blood cell counter. Serum was separated to determine levels of total protein and albumin as well as the activity of the gamma glutamyl transferase using an Hitachi 705 autoanalyzer. Following cervical dislocation liver and the bursa of Fabricius were removed and weighed. Aflatoxin was extracted from liver according to AOAC procedure (1990).

RESULTS

As in the studies with chicks, body weights decreased and mortality increased on addition of aflatoxin to feed (Table 3). However, addition of YS1026 to the diet significantly diminished the adverse effects of aflatoxicosis.

The concentrations of total serum protein and albumin were significantly decreased while activity of gamma glutamyl transferase and concentration of aflatoxin in the liver were increased in groups fed

Table 3. Effect of Yea-Sacc[1026] on body weight and mortality of ducklings fed diets containing aflatoxin.

Aflatoxin, ppb	Yea-Sacc[1026], %	Body weight, g		Mortality, %
		Day 14	Day 28	
0	0	212[c]	548[b]	6.3
100	0	142[a]	257[a]	25.0
200	0	139[a]	251[a]	37.5
100	0.1	198[bc]	544[b]	6.3
200	0.1	195[b]	537[b]	6.3

Means in a column with different superscripts differ, $P< 0.05$

diets containing aflatoxin. On supplementation of YS1026 the above parameters were comparable to control values. The relative weight of liver and bursa of Fabricius were improved and comparable to those of control by addition of YS1026 when aflatoxin was in the diet.

Feeding aflatoxin decreased red blood cell counts, hemoglobin and packed cell volume of ducklings (Table 4). Addition of YS1026 essentially reversed aflatoxin effects on hematological parameters. Gross examination revealed liver enlargement and greenish discoloration and distention of the gall bladder of ducklings fed aflatoxin. Ducklings additionally given YS1026 did not have these lesions.

Table 4. Effect of Yea-Sacc[1026] on organ weights, concentration of aflatoxin in liver, total serum protein, albumin and enzyme activity in ducklings fed diets containing aflatoxin.

	Control	Alatoxin, 100 ppb		Aflatoxin, 200 ppb	
	0 % YS1026	.1% YS1026	0% YS1026	.1% YS1026	
Relative liver weight, g/100 g BW	4.57[a]	5.36[b]	4.61[a]	6.11[c]	4.50[b]
Relative weight of bursa of Fabricius, g/100 g BW	0.25[c]	0.19[b]	0.24[c]	0.11[a]	0.23[c]
Aflatoxin in liver, ppb	ND[1]	1.41	0.99	2.50	1.20
Total protein, g/dl	1.94[d]	1.11[b]	1.92[d]	0.81[a]	1.81[c]
Albumin, g/dl	0.49[c]	0.37[b]	0.51[d]	0.29[a]	0.47[c]
Gamma glutamyl transferase, IU/l	2.83[a]	5.83[c]	3.00[a]	7.00[d]	4.50[b]

[ab]Means in a row with different superscripts differ $P< 0.05$.
[1]Not detectable.

Experiment 4: In vitro biodegradability of aflatoxin in presence of Yea-Sacc[1026]

MATERIALS AND METHODS

The data in the three in vivo experiments clearly indicated that supplementation of diets contaminated with aflatoxin with YS1026 improved growth performance and immune response and markedly reduced the deleterious effects of the mycotoxin. The following in vitro trial tested whether the viable yeast culture might act to reduce aflatoxicosis via degradation of the toxin.

Aflatoxin was produced on rice using the method described by Shotwell et al. (1966) and quantified using the Romer minicolumn method as per AOAC (1990). Sabouraud's culture broth with pH adjusted to 5.5 was used at 25 ml per sample to grow yeast from YS1026. Samples of YS1026 at three levels (0, 0.5 and 1 g) and aflatoxin at 250 and 500 ppb were incubated in triplicate flasks at 37°C. The samples containing no YS1026 were considered the control to exclude the possibility of aflatoxin degradation by the culture broth. After 48 hours the viability of YS1026 was determined, aflatoxin was extracted and quantified using the CB method (AOAC, 1990). This method was adopted after trials showed 92% recovery of aflatoxin from media alone compared with 86% recovery by the BF method and 81% recovery by Romer minicolumn as per AOAC (1990). Similar trials were conducted after 72 and 96 hours incubation.

RESULTS

Viability of YS1026 was unaffected by levels of aflatoxin or time of incubation in all the samples based on growth in Sabouraud's media after incubation for 48 hours (Figures 2 and 3).

Regardless of aflatoxin level the pattern of degradation was uniform with only minor variations. There was a gradual increase in the amount of aflatoxin degraded as the time of incubation was increased (50%

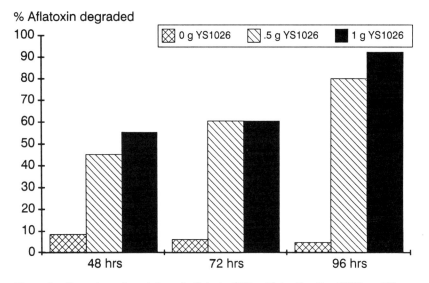

Figure 2. Percentage degradation of aflatoxin (250 ppb) by Yea-Sacc[YS1026] at different periods of incubation.

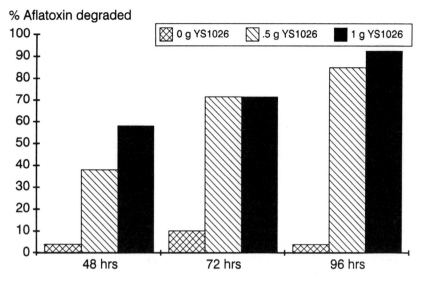

Figure 3. Percentage degradation of aflatoxin (500 ppb) by Yea-Sacc^{YS1026} at different periods of incubation.

degradation at 48 hours to 88% degradation at 96 hours). Aflatoxin degradation was not significantly affected by the two levels of YS1026 used for 72 hours of incubation whereas for 48 and 96 hours the percentage of aflatoxin degradation increased with the higher level of YS1026. Losses during extraction ranged from 5 to 10% as shown in controls. The above results correlate with in vivo studies. These results suggest that YS1026 can degrade aflatoxin to the extent of 88%.

Discussion

Effects of aflatoxin on body weight, mortality and feed efficiency agree with the findings of Huff et al. (1986), Umesh and Devegowda (1990) and Kubena et al. (1993) in broiler chickens and Muller et al. (1970) in ducklings. Addition of 0.1% Yea-Sacc[1026] yeast culture reduced the adverse effects of the toxin. These data confirm the findings of Stanley et al. (1993) who noted beneficial effects of yeast culture on body weight when broiler diets contained aflatoxin.

Use of yeast culture to improve performance and efficiency of broilers (Taklimi et al., 1993), hatchability in broiler breeders (McDaniel, 1991; Khan, 1993) and reduced depletion rate in broiler breeder chicks (Ganpule, 1993) has been documented; however little if any information on mode of action exists. One possible explanation for the ability of Yea-Sacc[1026] to counteract aflatoxicosis is the biodegradation of the toxin. Additionally, it is possible that toxins are adsorbed by the yeast cell wall.

The relative weight of the liver was significantly increased by feeding diets containing aflatoxin. Similar findings were reported by Phillips et al. (1988); Kubena et al. (1990) and Umesh and Devegowda (1990). The liver is considered the main target for aflatoxin toxicity. These data show the protective effect of Yea-Sacc[1026] as indicated by enlargement of the liver to a lesser degree in groups fed diets containing both aflatoxin and Yea-Sacc[1026]. Addition of Yea-Sacc[1026] reduced the level of aflatoxin in liver tissue. This finding has important implications since liver tissue enters the human food chain.

The bursa of Fabricius is also considered to be a target for aflatoxin in young birds leading to lower production of B-lymphocytes. In the present study, a protective effect of Yea-Sacc[1026] was also observed.

The activity of gamma glutamyl transferase was increased by feeding diets contaminated with aflatoxin. This trend was reversed upon addition of the Yea-Sacc[1026]. Serum gamma glutamyl transferase is a sensitive indicator or liver disease, whether the disorder involves liver inflammation or obstruction of the biliary tract (Kubena et al., 1990). Further, Kubena et al. (1990) reported an increase in the activity of gamma glutamyl transferase by feeding aflatoxin-contaminated diets.

Total serum protein and albumin levels were reduced when aflatoxin was in the diet. The reduction in concentrations of total protein and albumin due to inhibition of DNA and protein synthesis (Thaxton et al., 1974) serve as an indicator of aflatoxicosis (Tung et al., 1975). Supplementation of Yea-Sacc[1026] to contaminated diets improved the total serum protein and albumin level significantly. Reversal of aflatoxicosis may be due to normal functioning of liver and normal serum biochemical values including hematological parameters reported elsewhere in the present study by addition of Yea-Sacc[1026].

HI titre values against Newcastle disease were reduced in birds fed aflatoxin. The lower antibody level may be due to regression of bursa of Fabricius. Immunosuppression caused by aflatoxin has been demonstrated in broiler chickens (Thaxton et al., 1974; Chang and Hamilton, 1982; Barmase and Devegowda, 1990). Yea-Sacc[1026] inclusion in the contaminated diets has shown protective effects on HI titre values which may be due to optimum production of B-lymphocytes, since the bursa of Fabricius was near normal in size.

The results of the present investigation clearly demonstrate that aflatoxicosis in broilers and ducklings can be counteracted by dietary supplementation of Yea-Sacc[1026]. Inclusion of 0.1% Yea-Sacc[1026] in the diet almost totally negated the effects of aflatoxin in broiler chickens and in ducklings. Additionally in vitro study has revealed the degradation of aflatoxin to the extent of 88% by addition of Yea-Sacc[1026]. Further research work is in progress to elucidate the mechanism of action of Yea-Sacc[1026].

References

AOAC. 1990. Official methods of analysis, 15th ed. Association of Official Analytical Chemists, Washington, D.C.

Barmase, B.S. and G. Devegowda. 1990. Reversal of aflatoxicosis through dietary adsorbents in broiler chicks. Proc. 13th Annual Poultry Science Conference and Symposium, Bombay, India.

Ceigler, A., E.B. Lillehoj, R.E. Peterson and H.H. Hall. 1966. Microbial detoxification of aflatoxin. Appl. Microbiol. 14(6):934–939.

Chang, C. and P. B. Hamilton. 1982. Increased severity and new symptoms of infectious bursal disease during aflatoxicosis in broiler chickens. Poultry Sci. 61:1061–1068.

Dawson, K.A. 1993. Current and future role of yeast culture in animal production: A review of research over the last seven years. In: Biotechnology in the Feed Industry. Proc. 9th annual symposium. Alltech Technical Publications, Nicholasville, Kentucky.

Devegowda, G. 1989. Aflatoxins, aflatoxicosis and animal production. In: Proc. of symposium on aflatoxin. The Compound Livestock Feed Manufacturer's Association of India, Bombay, India.

Duncan, D.B. 1955. Multiple range and multiple F test. Biometrics. 11:1–42.

Ganpule, S.P. 1993. Usefulness of Lacto-Sacc as a performance enhancer in broiler breeder ration. Vetcare Update Bulletin 1(1).

Goldblatt, L.A. 1971. Control and removal of aflatoxin. J. Am. Oil. Chem. Soc. 48:605–610.

Goldblatt, L.A., and F.G. Dollear. 1979. Modifying mycotoxin contamination in feeds – use of mold inhibitors, ammoniation, roasting. In: Interactions of mycotoxins in animal production. National Academy of Sciences, Washington, DC. Pages 167–184

Hamilton, P.B. 1987. Why the animal industry worries about mycotoxins. Proc.of Symposium on Recent Developments in the study of mycotoxins. Kaiser Chemicals, Cleveland, Ohio.

Huff, W.E., L.F. Kubena, R.B. Harvey, D.E. Corrier and H.H. Mollenhaur. 1986. Progression of aflatoxicosis in broiler chickens. Poultry Sci. 64:1891–1899.

Huff, W.E., 1988. Mycotoxin: toxicity and control. US Feed Grains Council. Grain handling and quality workshop. New Delhi, India.

Khan, S.A. 1993. Improving hatchability of broiler parents by supplementing Lacto-Sacc. Vetcare Update Bulletin 1(1).

Kubena, L.F., R.B. Harvey, T.D. Phillips, D.E. Corrier and W.E. Huff. 1990. Diminution of aflatoxicosis in growing chickens by the dietary addition of a hydrated sodium calcium aluminosilicate. Poultry Sci. 69:727–735.

Kubena, L.F., R.B. Harvey, W.E. Huff, M.H. Elissalde, A.G. Yersin, T.D. Phillips and G.E. Rottinghaus. 1993. Efficiency of a hydrated sodium calcium aluminosilicate to reduce the toxicity of aflatoxin and diacetoxyscirpenol. Poultry Sci. 72:51–59.

McDaniel, G. 1991. Effect of Yea-Sacc[1026] on reproductive performance of broiler breeder males and females. In: Biotechnology in the Feed Industry. Proc. 7th annual symposium. Alltech Technical Publications, Nicholasville, Kentucky.

Meronuck, G. 1993. Mycotoxins in feed. Feedstuffs 65(30):150–153.

Muller, R.D., C.W. Carlson, G.Semeniuk and G.S. Harshfield. 1970.

The response of chicks, ducklings, goslings, pheasants and poults to graded levels of aflatoxins. Poultry Sci. 49:1346–1350.
Phillips, T.D., L.F. Kubena, R.B. Harvey, D.S. Taylor and N.D. Heidelbaugh. 1988. Hydrated sodium calcium aluminosilicate: a high affinity adsorbent for aflatoxin. Poultry Sci. 67:243–247.
Ramakrishna, G.T., G. Devegowda, D. Umesh and M.R. Gajendragad. 1992. Evaluation of mold inhibitors in broiler diets and their influence on the performance of broilers. India J. of Poultry. Sci. 27(2):91–94.
Shotwell, O.L., C.W. Hesseltine, R.D. Staublefield and W.G. Soreson. 1966. Production of aflatoxin on rice. Appl. Microbiol. 14:425–428.
Shotwell, O.L., R. J. Gaulden, R.J. Battrast and C.W. Hesseltine. 1975. Mycotoxins in hot spots in grains. 1. Aflatoxin and zearalenone occurrence in stored corn. Cereal Chem. 49:458–465.
Snedecor, G.W. and W.G. Cochran. 1967. Statistical Methods, 6th Ed. Iowa State University Press, Ames, IA.
Stanley, V.G., R. Ojo, S. Woldesenbet and D.H. Hutchinson. 1993. The use of *Saccharomyces cerevisiae* to suppress the effects of aflatoxicosis in broiler chicks. Poultry Sci. 72:1867–1872.
Taklimi, S.M., B.I.R. Aravind, C.V. Gowdh and G. Devegowda. 1993. Effect of live yeast culture (Yea-Sacc[1026]) on performance of broilers fed varying energy and protein levels. Proc. 6th Anim. Nutr. Res. Worker's Conf., Bhubaneswar, India.
Thaxton, J.P., H.T. Tung and P. B. Hamilton. 1974. Immunosupression in chickens by aflatoxin. Poultry. Sci. 53:721–725.
Tung, H.T., D.D. Wyatt, J.P. Thaxton and P.B. Hamilton. 1975. Concentrations of serum proteins during aflatoxicosis. Toxicol. Appl. Pharmacol. 34:320–326.
Umesh, D. and G. Devegowda. 1990. Reducing adverse effects of aflatoxin through nutrition in broiler chickens. Proc. 13th Annual Poultry Sci. Conf. and Symposium, Bombay, India.

SHRIMP FARMING: A BREAKTHROUGH IN CONTROLLING NITROGEN METABOLISM AND MINIMIZING WATER POLLUTION

CHAI WACHARONKE

Animal Health Department, Diethelm Trading Co., Ltd.
Bangkok, Thailand

Presented at Alltech's 9th Annual Symposium on Biotechnology in the Feed Industry, April 26–28, 1993

Introduction

Prawn culture performance, mortality and growth are intimately influenced by environmental conditions in this complex and dynamic ecosystem. Factors controlling the composition of pond water, which determine to a great extent the success or failure of prawn culture operations, are extremely varied and include physical, chemical and biological processes. The parameters most concerned with shrimp life include ammonia (NH_3), temperature, dissolved oxygen (DO), acidity, salinity, suspended solids and other potentially toxic substances such as hydrogen sulfide, heavy metals and other contaminants.

On Thailand shrimp farms a serious ecosystem imbalance problem due to ammonia typically occurs at least once every prawn culture cycle. During this imbalance plankton populations collapse, NH_3 rises to toxic levels and DO falls to subsystem levels. Feed intake and growth are depressed and mortality rate sharply increases during such crises. Immediate attention must focus on correcting environmental conditions, especially decreasing ammonia and increasing dissolved oxygen as these are the most critical factors threatening survival. Addition of *Yucca schidigera* extract (De-Odorase) containing glycocomponents which bind ammonia has shown considerable benefit both under experimental and practical conditions. The following paper describes the results of studies and field trials with the ammonia-binding agent in controlling water quality problems.

Experiment 1:
Efficacy of De-Odorase at low level usage in reducing or controlling un-ionized ammonia in sea water without shrimp.

Experiment 2:
Efficacy of De-Odorase used at low levels on un-ionized ammonia concentrations and shrimp mortality in sea water at 5 × normal stocking density (high ammonia conditions)

Field Trials:
Prevention Program:
Effects of De-Odorase at 0.3 ppm applied in a long term prevention program until harvest on growth and survival rate of the prawn culture in commercial prawn production.

Treatment Program:
To study the role of De-Odorase during a crisis in a prawn culture production cycle.

Experiment 1: Effect of De-Odorase on un-ionized ammonia in sea water

MATERIALS AND METHODS

Thirteen plastic nursery ponds with circulating seawater were used for this experiment. One pond to which nothing was added served as a control. In the other 12 ponds ammonia in the form of ammonium chloride was added at either 0.5, 1.0 or 2.0 ppm (Table 1). At each ammonia level De-Odorase was added at either 0.3, 0.5 or 1.0 ppm. Ammonia was measured by titration at 1, 6, 12 and 24 hrs.

Table 1. Experimental design for the efficacy of De-Odorase in controlling un-ionized ammonia in sea water.

Pond	Sea water (litres)	Ammonia (ppm)	De-Odorase (ppm)	Designation
1	150	–	–	Control 1
2	150	0.5	–	Control 2
3	150	0.5	0.3	Treatment 1
4	150	0.5	0.5	Treatment 2
5	150	0.5	1.0	Treatment 3
6	150	1.0	–	Control 3
7	150	1.0	0.3	Treatment 4
8	150	1.0	0.5	Treatment 5
9	150	1.0	1.0	Treatment 6
10	150	2.0	–	Control 4
11	150	2.0	0.3	Treatment 7
12	150	2.0	0.5	Treatment 8
13	150	2.0	1.0	Treatment 9

Note: Salinity maintained at 25–30 ppt.

RESULTS

Comparing the levels of un-ionized ammonia at 1 and 24 hours of the control (Controls 2, 3, 4) and the treatment groups (Treatments 1, 2, 3, 4, 5, 6, 7, 8, 9), it was clear that De-Odorase added at low rates (0.3, 0.5 and 1.0 ppm) could reduce the NH_3 in the sea water significantly (Table 2). Reductions ranged from 15 to 55%. Response followed dosage in that the higher the dosage of De-Odorase, the lower the level of un-ionized ammonia at 24 hrs.

Table 2. Effect of De-Odorase in controlling un-ionized ammonia in sea water in the absence of prawns.

Treatment	Un-ionized NH_3, mg/l				
	1 hr	6 hr	12 hr	18 hr	24 hr
0 NH_3					
Control 1	0.0628	0.0125	nil	nil	nil
0.5 ppm NH_3					
Control 2	0.2236	0.1633	0.1884	0.2386	0.2286
0.3 ppm DeO	0.2236	0.1382	0.1382	0.2135	0.1306
0.5 ppm DcO	0.1884	0.1909	0.1080	0.2085	0.1130
1.0 ppm DeO	0.2236	0.1633	0.1633	0.1432	0.1030
1.0 ppm NH_3					
Control 3	0.3743	0.4899	0.3844	0.4648	0.4019
0.3 ppm DeO	0.5150	0.4137	0.2386	0.3618	0.3769
0.5 ppm DeO	0.3266	0.2889	0.2889	0.3668	0.3266
1.0 ppm DeO	0.4497	0.4145	0.3140	0.3140	0.2763
2.0 ppm NH_3					
Control 4	0.9044	0.7663	0.7411	0.7436	0.5778
0.3 ppm DeO	0.8291	0.7411	0.7160	0.7160	0.6808
0.5 ppm DeO	0.6030	0.6406	0.6281	0.4145	0.4019
1.0 ppm DeO	0.5773	0.6030	0.5753	0.3894	0.3769

Experiment 2: Effect of De-Odorase on un-ionized ammonia and consequent effect on survival of 90 day old tiger prawns

MATERIALS AND METHODS

Plastic ponds (200 liter capacity) containing recirculating seawater were stocked with 30 tiger prawn (90 days of age) each in order to study effect of four different De-Odorase treatments on water quality. Initial ammonia concentration was 2.0 ppm added as ammonium chloride. De-Odorase was added at either 0.5 or 1.0 ppm (one treatment). These were compared with 0.5 ppm given either twice at 0 and 6 hrs or three times (0, 6 and 12 hrs). Treatments were compared under aerated or non-aerated conditions (Table 3). Salinity was maintained at 25–39 ppt. Un-ionized ammonia was measured by titration at 1,6 and 24 hrs. Cummulative mortality was also measured at 1, 6 and 24 hrs.

RESULTS

Some un-ionized ammonia was present due to excretion by the shrimp in the control and treatment 1 ponds even without adding any NH_3 (Table 4). Un-ionized ammonia was higher in aerated than in non-aerated water which implied that shrimp would excrete more NH_3 when aerated due to greater metabolism.

In the treatments where ammonia was at 2.0 ppm, De-Odorase reduced ammonia under non-aerated conditions by 18 to 22%.

Table 3. Treatment structure of Experiment 2.

Pond	Sea water (liter)	Prawns (number)	Ammonia (ppm)	Aerated[1]	De-Odorase (ppm)	Design
1	150	30	–	No	–	Control 1
2	150	30	–	Yes	–	Trt 1
3	150	30	2.0	No	–	Trt 2
4	150	30	2.0	Yes	–	Trt 3
5	150	30	2.0	No	1.0	Trt 4
6	150	30	2.0	Yes	1.0	Trt 5
7	150	30	2.0	No	0.5	Trt 6
8	150	30	2.0	Yes	0.5	Trt 7
9	150	30	2.0	No	0.5 (2×)	Trt 8
10	150	30	2.0	Yes	0.5 (2×)	Trt 9
11	150	30	2.0	No	0.5 (3×)	Trt 10
12	150	30	2.0	Yes	0.5 (3×)	Trt 11

[1] Non-aerated: dissolved oxygen 1.5 ppm; Aerated, dissolved oxygen 7.0 ppm.

Table 4. Effect of De-Odorase on ammonia levels and mortality rates of prawn.

Aerated	De-Odorase (ppm)	Treatment	Un-ionized NH_3, mg/l			Remarks
			1 hr	6 hr	24 hr	
No	-	Control	0.5076	0.4812	0.4431	all dead in 12 hr.
Yes	-	Trt 1	0.6406	0.7114	0.8199	
No	-	Trt 2	0.6916	0.7125	0.7313	all dead in 12 hr.
Yes	-	Trt 3	0.6273	0.8578	0.9974	
No	1	Trt 4	0.6738	0.6021	0.5983	all dead in 12 hr.
Yes	1	Trt 5	0.6184	0.8203	0.953	
No	0.5	Trt 6	0.8046	0.8417	0.8643	all dead in 12 hr.
Yes	0.5	Trt 7	0.7581	0.9134	1.1304	
No	0.5 (2×)	Trt 8	0.5363	0.5505	0.5761	all dead in 12 hr.
Yes	0.5 (2×)	Trt 9	0.5273	0.8687	0.9974	
No	0.5 (3×)	Tr 10	0.5519	0.5645	0.5761	all dead in 12 hr.
Yes	0.5 (3×)	Tr 11	0.9399	0.9986	1.0861	

The NH_3 level at 24 hrs in treatment 6 which was higher than the control might have resulted from titration error.

Under non-aerated condition where DO was extremely low, all shrimp were dead within 12 hrs. In treatment 1 (no added ammonia), even with aeration there was 16.67% mortality (Table 5). This was attributed to stocking rates five times higher than normal culture density.

Mortality rate in treatment 3 (2 ppm NH_3) at 24 hrs was about twice that in treatment 1 (no added ammonia) at 24 hrs; most probably due to the fact that NH_3 was 20 per cent higher in treatment 3 than in treatment 1.

It was surprising that survival rates in treatments 7, 9 and 11 (non-aerated, 2 ppm + 0.5 ppm De-Odorase 1×, 2× or 3×) were higher than in treatment 3 even though the NH_3 level at 24 hrs was lower.

The most effective application program for De-Odorase was 0.5 ppm De-Odorase applied three consecutive times at 6–hr intervals.

Table 5. Survival rate of tiger prawn at 24 hours.

NH_3	Aerated	De-Odorase (ppm)	Treatment	Survived (n)	Dead (n)	Survival (%)
–	No	–	Control 1	–	30	0
–	Yes	–	Trt 1	25	5	83.33
2 ppm	No	–	Trt 2	–	30	0
2 ppm	Yes	–	Trt 3	20	10	66.66
2 ppm	No	1	Trt 4	–	30	0
2 ppm	Yes	1	Trt 5	20	10	66.66
2 ppm	No	0.5	Trt 6	–	30	0
2 ppm	Yes	0.5	Trt 7	24	6	80.00
2 ppm	No	0.5 (2×)	Trt 8	–	30	0
2 ppm	Yes	0.5 (2×)	Trt 9	26	4	86.66
2 ppm	No	0.5 (3×)	Trt 10	–	30	0
2 ppm	Yes	0.5 (3×)	Trt 11	29	1	96.66

Experiment 3: Field trials with De-Odorase in water quality crisis situations

MATERIALS AND METHODS

Field trials with De-Odorase were conducted in both "Crisis Prevention" and "Crisis Treatment" situations in commercial ponds. Ten ponds, each 6,400 m³, were used to test a prevention program. The prevention program consisted of spraying De-Odorase onto the water surface at a rate to supply 0.3 ppm on Day 1 and every 15 days until harvest. Five ponds were treated and five served as controls. Control and treatment ponds had the same stocking density (40 fry/m³) and identical water management. Prawn size and survival rate were measured.

Treatment of a water quality crisis with De-Odorase was investigated during plankton collapse. Shrimp showed obvious clinical signs of crisis by swimming near the surface of water. There were five different cases in different farms and areas. In each case one pond served as a control while a second was treated with De-Odorase.

RESULTS

Water quality crises were successfully prevented with 0.3 ppm De-Odorase. The control groups had to be harvested prior to reaching desired market weight due to continuous problems with disease and persistantly high mortality rates (Table 6). Control groups experienced water quality crises (plankton collapse and water color change) two to three times during the production cycle while no crises were experienced in the ponds treated with De-Odorase at 0.3 ppm. Treated ponds were harvested following a full production cycle.

Table 6. Effect of De-Odorase treatment on production parameters in commercial prawn production.

	Control groups	Treatment groups
Average prawn size, #/kg	92	38
Average survival rate, %	28	65
Days to harvest	72	120

Table 7. Effect of De-Odorase on recovery time from water quality crisis and mortality.

	NH_3-N, ppm		Mortality rate during crisis	Hours to recovery[2]	Feed intake[3]
	Initial	After[1]			
Treated pond	8.0	5.0	Low	24	+
Control pond	8.0	10.0	High	No recovery	−
Treated pond	3.0	3.0	Low	12	+
Control pond	3.0	5.0	Moderate	No recovery	−
Treated pond	3.0	1.0	Low	12	+
Control pond	5.0	5.0	High	No recovery	0
Treated pond	2.0	0.5	Low	8	+
Control pond	3.0	5.0	Moderate	No recovery	−
Treated pond	5.0	3.0	Moderate	12	+
Control pond	5.0	5.0	High	No recovery	0

[1]After treatment with De-Odorase.
[2]Recovery after treatment was determined by whether shrimp returned to the bottom of the pond.
[3]+ means intake increased; 0 means intake maintained, − means intake decreased

In general, treating ponds undergoing water quality crisis reduced ammonia levels and mortality rates (Table 7). Recovery, defined by the shrimp returning to the bottom of the pond, required 8 to 24 hrs after treatment.

Conclusions

1. De-Odorase, even at low usage levels, significantly reduced NH_3 in sea water.
2. De-Odorase, even under high NH_3 conditions, increased survival rate of the shrimp. The most effective treatment was 0.5 ppm applied three times at 6-hour intervals.
3. When given De-Odorase as a long term application program at 0.3 ppm starting from Day 1 and repeated once every 15 days until harvest in the prawn culture system, growth and survival rate were improved significantly.
4. De-Odorase was one of the essential factors for reducing shrimp mortality during the crisis of plankton collapse.
5. When dissolved oxygen was extremely low De-Odorase could not prevent mortality.

TRACE MINERALS: THE ROLE OF MINERAL PROTEINATES IN IMMUNITY, REPRODUCTION AND PERFORMANCE

UNDERSTANDING STRESS IN CATTLE

CHERYL F. NOCKELS
*Professor of Animal Sciences, Colorado State University,
Fort Collins, Colorado 80523*, USA

Introduction

Animals alter their metabolism in response to noxious stimuli or stress. This change in metabolism is primarily to provide the animal with a continuing source of energy, amino acids for protein synthesis derived from its own resources, and minerals. Very little stress effect may be noted during an acute short-term stress, but metabolic changes in the animal may become substantial during long-term chronic stress. The animal begins to use its own tissues for energy production, and routes the energy sources to specific tissues while decreasing it to others. This coordination of energy production, distribution and utilization is a tightly controlled process, regulated by specific hormones. The metabolic control exerted by some hormones has been known for some time, but newly recognized hormones and their functions are just now beginning to be discerned. Hormones orchestrate the activity of the metabolic pathways through which compounds pass by altering the activity of the rate regulating enzyme(s) of that pathway. In stress, while many metabolic pathways, such as those involving fatty acids, glucose and certain proteins, may be catabolic in supplying a continued energy source; others, such as acute phase protein synthesis, are anabolic as long as energy is sufficient. In this redistribution of resources, mineral composition of tissues may also be changed with many elements being excreted. It is very important to elucidate and understand the metabolic changes occurring in stress so we can assist the animal in controlling those that could be harmful.

In the following thesis I have attempted to integrate some of the metabolic changes in the bovidae, when possible, as it moves from a fed to fasted state while subjected to increasing stress and trauma. In order to illustrate the dynamic alterations in metabolism when stress and trauma intensify, I will use data obtained from cattle arriving at the feedlot (Bennett *et al.*, 1989). Generally, as stress intensity and length increased, so did blood levels of cortisol, glucose, creatine phosphokinase (CPK), serum aspartate amino transaminase (AST), urea-nitrogen, creatinine, and fibrinogen. Decreases in albumin (A) and increases in globulin (G) reduced the A/G ratio, although total protein remained constant.

Stress hormones

Before trying to explain the metabolic cause of the aforementioned biochemical changes, the different noxious stimuli producing the stress response need to be identified. These stimuli are defined in part as handling, transport, physical trauma, fasting, fatigue and unfamiliar environment. Any one or all of these elicit the following hormonal changes (Roth and Kaeberle, 1982; Rulofson et al., 1988). Adrenocorticotropic hormone (ACTH) release from the pituitary increases, which augments adrenal cortex release of cortisol and aldosterone (Ganong, 1987). Additionally, through neural stimulation, the adrenal medulla releases the catecholamines, epinephrine and norepinephrine to the circulation. Furthermore, beta-endorphin is cosecreted with ACTH from the pituitary and enkephalins are cosecreted with epinephrine from the adrenal medulla. Both beta-endorphins and enkephalins are also produced by activated macrophages and lymphocytes. These hormones may in part regulate inflammation, immune function and temperature control in infection (Braezile, 1988). Norepinephrine may also be released at most sympathetic postganglionic nerve endings. Additional blood hormonal changes result from alterations in blood glucose levels (Mayes, 1988).

Hormones alter metabolism

When hormones respond to stress, they concomitantly affect metabolism. Blood glucose levels are maintained at a fairly constant concentration by regulating the amount entering the circulation from the liver and kidney or its uptake by glucose-utilizing tissues. These changes in entry and removal are hormonally regulated. Blood glucose rises rapidly when catecholamines initiate hepatic glycogenolysis. This increase in blood glucose in turn increases insulin secretion, which enhances the extra-hepatic uptake of glucose. When blood glucose decreases during fasting, insulin declines and glucagon and cortisol secretion are enhanced. These latter two hormones are gluconeogenic, and promote glucose production by the kidney and liver in order to maintain blood glucose levels. These hormones act to provide carbon compounds from body tissues from which glucose is ultimately synthesized. The metabolic pathways of glycogenolysis, proteolysis and lipolysis are catabolic and are increased in activity by these hormones. While all tissues can use glucose for energy synthesis, the red blood cells (RBC) and portions of the central nervous system (CNS) have an absolute requirement for it. In order to partition and protect the low level of glucose produced during fasting for the RBC and CNS, cortisol prevents glucose entry into other tissues. Insulin, which is low at this time, is needed for glucose entry into all tissues except the liver, RBC and CNS. When glucose is kept pooled in the blood, it may reach quite high levels, as found in chronically stressed cattle (Bennett et al., 1989). The source of carbon compounds from which glucose is synthesized is also ensured by cortisol. During stress and when energy is limiting, protein synthesis in most tissues is largely curtailed with certain proteins being preferentially degraded in some tissues to furnish amino acids for glucose synthesis and energy and for the production of other proteins. These changes in protein metabolism will be discussed later.

Energy regulation

The primary energy source for tissues other than the RBC and the CNS is fatty acids arising from fat stored in adipose tissue. Enzymes, responsible for fat catabolism (lipolysis), enabling the fatty acids to exit adipose tissue are responsive to ACTH, thyroid-stimulating hormone (TSH), epinephrine, norepinephrine and glucagon. Most of these lipolytic events require the presence of glucocorticoids and thyroid hormone for an optimal effect. Fatty acids leaving the adipocyte combine with albumin for transport as free fatty acids (FFA), which rise in the circulation. Large quantities of FFA are removed by the liver. Generally, in the fed animal, some of these fatty acids contribute to energy production of the liver via oxidative phosphorylation, with the majority converted into very low density lipoproteins (VLDL) and introduced back into the blood. However, both energy and VLDL production are greatly diminished in the fasted animal so that the FFA remain in the liver. In an effort to reduce fatty livers and export an energy source elsewhere, the FFA are converted to ketone bodies which are returned to the circulation. The two primary ketone bodies are 3–hydroxybutyric acid and acetoacetic acid. These ketoacids may lead to a metabolic acidosis. Cattle arriving at the feedlot may have a metabolic acidosis from ketoacidosis and lactic acid production. These stressed cattle may suffer from serious mineral deficiencies of calcium, phosphorous, potassium, magnesium, zinc, and copper; and, as a result of acidosis, high aldosterone, cortisol production and tissue catabolism (Nockels, 1990).

Acidosis may also inhibit the synthesis of the active form of vitamin D, 1,25 dihydroxycholecalciferol ($1,25D_3$) (Ching *et al.*, 1989). This hormone-vitamin is needed for regulating calcium and phosphorous homeostasis and the differentiation of promonocytes and monocytes to macrophages, and macrophage functions (Reinhardt and Hustmyer, 1987; Manolagas *et al*, 1989; Koeffler *et al.*, 1989). Acidosis has been shown to reduce renal synthesis of $1,25D_3$. Within the promoter region of the osteocalcin gene there is a binding site for $1,25D_3$ as well as a separate binding site for dexamethasone, a synthetic glucocorticoid, (Morrison *et al.*, 1989). The osteocalcin promoter was repressed by dexamethasone, which slowed or inhibited transcription of the osteocalcin gene. Genes or gene products regulated by $1,25D_3$ might similarly be repressed by glucocorticoids (Minghetti and Norman, 1988). The production of macrophages is very important as the macrophage is one of several cell types which are sources of small protein hormones know as cytokines which will be discussed later.

Protein and amino acid metabolism

In the initial stages of fasting, cellular proteins are degraded first in the liver, kidney, and intestine, whereas decreases in skeletal muscle occur much later (Mortimore, 1986). The freed amino acids may be catabolized to provide energy within the cell or released to be used elsewhere. When amino acids are catabolized in the cell, the amino group is captured by α-keto acids with a net production of alanine and glutamine, the relative proportion of these two released amino acids may be regulated by hormones. Glucagon increases efflux from skeletal muscle of alanine relative to glutamine, which is a better substrate for hepatic gluconeogenesis (Smith, 1986). All forms of acidosis in rats, including ketoacidosis, produced significant increases in glutamine synthetase in skeletal muscle. This stimulates glutamine synthesis and efflux, which may result in muscle atrophy (Falduto et al., 1989).

Glutamine has several important roles in stress (Lacey and Wilmore, 1990). Glutamine may serve as a preferred energy source for rapidly proliferating cells such as enterocytes and lymphocytes. It acts as a regulator of acid base balance through the production of urinary ammonia. Glutamine transports nitrogen between tissues; and perhaps its most important function is as a precursor for nucleic acids, nucleotides and proteins necessary for cell proliferation. Furthermore, during stress insufficient amounts of glutamine may be produced as its supplementation increases host survival, improves nitrogen balance and attenuates skeletal muscle proteolysis (Lacey and Wilmore, 1990).

Under neutral pH conditions most of the circulating alanine and glutamine, along with other amino acids arising from muscle proteolysis, enters the liver with various subsequent fates. Following removal of the amino group by transaminases, which activity is increased by cortisol, the carbon skeleton enters the glucose synthetic pathway as directed by the gluconeogenic hormones. The ammonia is combined with another waste product, carbon dioxide, in the urea cycle. Activity of the urea cycle enzymes is increased by glucagon and glucocorticoids (Anonymous, 1988). When catabolism of protein is high, as in stressed cattle, blood urea level rises. As muscle protein is degraded, creatinine level rises as it is released from the tissue. Another important fate of amino acids arising from proteolysis is resynthesis into different vital proteins which may occur in the liver and other organs and tissues.

Interorgan transport of amino acids arising in stress and (or) disease is very important in providing the substrate for synthesis of acute-phase proteins in the liver, clonal expansion of immune cells, immunoglobulins and cytokines (Johnstone and Klasing, 1990). Macrophage production and release of the cytokines interleukin-1,-2,-6, and tumor necrosis factor (TNF) occurs following tissue trauma, infection or inflammation. Induction of hepatic acute-phase protein synthesis occurs in response to interleukin-6 (IL-6), which is synergistically aided by IL-1, (Mizel, 1989) and is induced in vitro by either IL-1 or tumor necrosis factor (TNF) (Akira et al., 1990). A few of the acute-phase proteins produced are C-reactive protein and serum amyloid A (both are immunosuppressive), haptoglobin, ceruloplasmin (increases serum copper, scavenges free radicals), alpha$_1$ acid glycoprotein (potentiates clotting), fibrinogen (fibrin precursor) and C3 (protective-part of the complement pathway) (Tizard, 1988). Increases in acute-phase proteins produce hypoferremia, hypozincemia and hypercupremia in animals in response to cytokines

(Grimble, 1989; Moldawer *et al.*, 1989; Johnstone and Klasing, 1990). Induction of procoagulant activity on endothelial cells by TNF and Il-1 (Le and Vilcek, 1987) may be one reason for the difficulty noted in obtaining blood from highly stressed cattle due to rapid blood clotting. (Bennett *et al.*, 1989). Proteins which are decreased in the acute-phase response are albumin, prealbumin (transthyretin), retinol binding protein and transferrin (Fleck, 1989).

Acute-phase response effects on plasma proteins are demonstrated following bacterial or parasitic infection, mechanical or thermal trauma, malignant growth or ischemic necrosis (Fleck, 1989). Injury, but not protein-energy depletion, reduces albumin levels (Fleck, 1989). The decline in plasma albumin and increase in fibrinogin in highly stressed cattle (Bennett *et al.*, 1989) demonstrates that a functional acute-phase response is occuring. Part of the stress these cattle have received may be trauma which results in cell and tissue damage. Products from the damaged cells probably increase chemotaxis of macrophages and neutrophils to the injured site to begin recovery from the injury. Arriving leukocytes, macrophages and mast cells then release IL-1, IL-2, IL-6 and TNFα. These cytokines initiate hormonal changes which activate proteolysis of destroyed proteins so that new tissue rebuilding may occur. Evidence that serum TNF is increased with severe tissue damage was reported by Reuter *et al.*, 1988. Serum TNF values increased in patients suffering from burns or multiple injuries with and without sepsis relative to controls. Furthermore, the increase in TNF was directly correlated to the severity of the pathology. These polypeptide cytokines perform various functions, including alterations in acute-phase protein levels in plasma (Fleck, 1989). Since stressed cattle (Bennett *et al.*, 1989) evidence increased acute-phase protein changes which are augmented by cytokines, this is indirect evidence that cytokine quantities may be increased in stressed, traumatized cattle. The interactions between immune cells and inflammatory cells are conducted largely by the twelve IL and TNF. Their production, cell receptors, immune function, and toxic effects are beyond the scope of this paper and have been recently reviewed (Le and Vilcek, 1987; Smith, 1988; Mizel, 1989; Kunkel *et al.*, 1989; Sisson and Dinarello, 1989; Akira *et al.*, 1990). Instead, the influence of IL and TNF on metabolism will be described.

EFFECTS OF CYTOKINES

Receptors to the cytokines are present in many or most tissues of the body so that they may have not only paracrine, but endocrine function. They are very much involved in enhancing immune function and recovery of injured tissue, which requires a source of energy and amino acids. Much of what is known regarding metabolic changes in infection, inflammation and repair processes is directed by IL-1 and(or) TNF (Pomposelli *et al.*, 1988). IL-1 has been shown to stimulate ACTH secretion with consequent adrenal release of glucocorticoids while TNF directly causes adrenal catecholamine release. The secretion of ACTH in response to IL-1 increases glucocorticoid release, which acts by a feedback mechanism to control the immune response through immunosuppression by reducing further production of IL-1 (Lumpkin, 1987). Besides causing anterior pituitary secretion of ACTH, IL-1 also promotes growth hormone. Leutinizing hormone and TSH secretion (Bernton et al., 1987) IL-1 was also reported to stimulate pancreatic secretion of insulin and glucagon (Pomposelli et al., 1988). Cytokine-induced production of these hormones would then initiate the metabolic strategies previously described. Metabolic rate would be expected to increase due to enhanced TSH secretion stimulating thyroid hormone release (Granner, 1988) as well as from the fever caused by the cytokines (Grimble, 1989). An initial hyperglycemia due to epinephrine-induced hepatic glycogenolysis and glucose release may be followed by hypoglycemia due to ensuing insulin-induced glucose entry into extrahepatic tissues. Hypertriglyceridemia may occur as free fatty acids from adipose tissue are converted by the liver to triglyceride-containing very low density lipoproteins, followed by a mild ketoacidosis. Proteolysis may result in increases in blood urea and amino acids which would provide the substrate for increased production of some of the acute-phase and immune proteins. Evidence for these metabolic changes induced by the cytokines is increasing and will be subsequently reported.

Differences in glucose metabolism occur following burns, trauma and sepsis. Nonseptic burn patients show increased glucose production which is depressed with sepsis (Pomposelli et al., 1988). These findings suggest that different forms of injury produce different mediators which do not produce uniform changes in carbohydrate metabolism. Administering IL-1 to mice was temporarily associated with increased insulin, corticosterone and glucagon and decreased glucose blood levels (Del Rey and Besedovsky, 1987). These authors indicated that cytokine-induced hypoglycemia was entirely due to the hyperinsulinemia. TNF injections into mice did not elicit a change in blood glucose, insulin or corticosterone. IL-1 administered to rats induced a febrile response, increased blood insulin levels and increased glucose utilization in skeletal muscle, diaphragm and in macrophage-rich tissues including the lung, spleen, liver and skin (Lang and Dobrescu, 1989). However, these authors believed that the increase in tissue glucose utilization occurred by insulin-dependent mechanisms. Dogs given TNF responded with a hypoglycemia associated with increased blood levels of ACTH, cortisol, glucagon and epinephrine (Evans et al., 1989). Glucagon production increased, as did blood glucose disappearance and clearance. As blood glucose declined, hindlimb glucose uptake and clearance increased markedly despite a 50% fall in mean serum-insulin level. Hindlimb release of lactate and pyruvate rose two- to three-fold after TNF administration. These authors state that the increased hindlimb glucose uptake was not mediated by insulin, nor due to the increased body temperature. IL-1 and TNF have both been implicated in producing lactic acidosis and reduced oxygen consumption by shifting glucose utilization to anaerobic metabolism (Johnstone and Klasing, 1990).

Lipid metabolism is also affected by cytokines. Lipoprotein lipase (LPL) is an enzyme which releases fatty acids in the blood from their lipoprotein carrier, such as chylomicrons and VLDL. This then allows entry of the fatty acids into the tissue. Lipoprotein lipase is attached to the endothelial capillaries adjacent to most tissues, including heart, liver, adipose and muscle tissue. TNF, IL-1 and IL-2 have been shown to reduce lipoprotein lipase activity in adipose tissue, but not elsewhere (Pomposelli *et al.*, 1988). This inhibition of LPL activity in adipose tissue results in a continual loss from this depot with no re-entry of fatty acids. An increased concentration of VLDL in plasma has been observed following decreased LPL activity (Porat, 1989; Gruenfeld *et al.*, 1989). This repartitioning of the fatty acids through LPL regulation is energetically feasible as the fatty acids are needed for energy production elsewhere.

Chronic TNF production causes cachexia, a potentially lethal syndrome which involves lipid and protein wasting (Porat, 1989). Earlier work reported that IL-1 produced skeletal muscle proteolysis similar to that caused by endotoxin (Pomposelli *et al.*, 1988). However, more recent research has shown that purified IL-1 did not stimulate whole body leucine flux, oxidation and muscle breakdown, although it induced fever, increased plasma acute-phase protein quantities and decreased serum iron and zinc (Pomposelli *et al.*, 1988). TNF, a possible contaminant of early IL-1 preparations, has been shown to increase total leucine oxidation and muscle catabolism. Coadministration of IL-1 and TNF had a synergistic effect on increasing skeletal muscle breakdown and reducing the percentage of protein in muscle (Pomposelli *et al.*, 1988).

TNF administration in rats has profound effects on individual amino acid utilization (Argiles and Lopez-Soriano, 1990). This cytokine increases both the total hepatic amino acid uptake and the individual uptake of gluconeogenic amino acids while decreasing the uptake of leucine, isoleucine and phenylalanine. Mealy *et al.* (1990) sought to determine if effects exerted by TNF were mediated by glucocorticoids in rats. While either TNF or corticosterone decreased nitrogen balance and carcass nitrogen content, only TNF induced increased liver DNA and protein content and diminished jejunal mucosal DNA and protein level which suggested that TNF may have effects independent of glucocorticoid.

Cytokines, while promoting peripheral protein wasting and nitrogen excretion, are essential for supplying amino acids needed for hepatic acute-phase protein synthesis, immune cell proliferation and peptide modulators of the neural and endocrine systems. Another common manifestation of cytokine production is a decrease in food intake followed by a depression in body weight. TNF and IL-1 both have been found to produce anorexia (Moldawer *et al.*, 1988), another acute-phase response. TNF is instrumental in the pathogeneses of cachexia (Beutler and Cerami, 1988). Continuous administration of TNF into rats or mice produced changes in nitrogen balance, a slight reduction in body weight followed by a gradual increase to' normal levels which was associated with fluid retention (Johnstone and Klasing, 1990).

Many effects of administering recombinant bovine TNF to cattle were reported by Ohmann *et al.*, 1989. Depending on the amount and length of TNF administration, some of the responses found were hyperthermia, leukopenia, decreased serum iron and zinc, increased serum copper and urea, depression, anorexia, cachexia, diarrhea, atrophy of the thymus, heart and skeletal muscle and loss of body fat. Several different intracellular pathways have been proposed to account for TNF-mediated cellular injury and cytotoxicity (Larrick and Wright, 1990).

Many of the changes precipitated by cytokine administration are classical signs observed in highly stressed cattle arriving at the feedlot. Cytokines are multifunctional in the regulation and integration of metabolism and immunity. It is important to realize that the initial insult producing a stress response may of itself be short-lived or acute, but the metabolic changes arising from cytokine release may be of sufficient magnitude to cause a chronic stress response. While repartitioning of nutrients during stress may be advantageous to survival, prolonged endocrine-induced effects may prove pernicious.

Production of ROM

In addition to all the forgoing effects of stress on metabolism it would be remiss if the production and effects of the highly reactive oxygen-derived molecules (ROM), peroxides, and free radicals (Freeman and Crapo, 1982) were not mentioned. Hydroperoxide, peroxide and peroxy radicals may be increased in stress and prove injurious to animal health. These ROM are produced in all cells within membranes, in the cytosol and in activated phagocytes. Cells may be destroyed if these ROM are not controlled by antioxidants or degraded by specific metalloenzymes (Freeman and Crapo, 1982; Machlin and Bendich, 1987; Nockels, 1988). Free radicals generated in active phagocytes as a means of destroying ingested particles may cause self-destruction and damage to adjacent tissues if antioxidant systems are inadequate (Weiss and Buglio, 1982). Because stress may also increase loss of metals and proteins necessary for synthesis of enzymes which may deactivate ROM (Nockels, 1990), ROM levels may potentially remain higher in tissues for a longer time and produce more damage.

One of the means by which ROM may damage cells is through destruction of its membranes via lipoperoxidation (Machlin and Bendich, 1987). When this occurs, cellular enzymes penetrate the membrane and enter the circulation where their high concentration is diagnostic of tissue injury. Circulating levels of CPK and AST were found to increase to very high levels in stressed cattle (Bennett et al., 1989), but it was not known if this was due to ROM levels or tissue trauma.

Summary

The determination of metabolic changes occuring in animals as a result of stress still requires intensive investigation. As our knowledge in this area expands, opportunities to prevent some of the deleterious effects of stress and trauma may be realized.

References

Akira, S., T. Hirano, T. Taga and T. Kishimoto. 1990. Biology of multifunctional cytokines: IL6 and related molecules (IL-1 and TNF). FASEB J. 4:2860–2867.
Anonymous. 1988. Regulation of urea cycle enzymes. Nutr. Rev. 46:326–327.
Argiles, J.M. and F.J. Lopez-Soriano. 1990. The effects of tumour necrosis factor- (cachectin) and tumour growth on hepatic amino acid utilization in the rat. Biochem J. 266:123–126.
Bennett, B.W., R.P. Kerschen and C.F. Nockels. 1989. Stress induced hematological changes in feedlot cattle. Agri-Practice 10:16–28.
Bernton, E.W., J.E. Beach, J.W. Holaday, R.C. Smallridge and H.G. Fein. 1987. Release of multiple hormones by a direct action of interleukin-1 on pituitary cells. Science 238:529–521.
Beutler, B. and A. Cerami. 1988. Tumor necrosis, cachexia, shock and inflammation: a common mediator. Ann. Rev. Biochem. 57:505–518.
Braezile, J.E. 1988. The physiology of stress and its relationship to mechanisms of disease and therapeutics. In: J.L. Howard, ed. The Veterinary Clinics of North America: Food Animal Practice. W.B. Saunders Co., Philadelphia 4:441–480.
Ching, S.V., M.J., Fettman, D.W. Hamar, L.A. Nagode and K.R. Smith. 1989. The effect of chronic dietary acidification using ammonium chloride on acid-base and mineral metabolism in the adult cat. J. Nutr. 119:902–915.
Del Rey, A. and H. Besedovsky. 1987. Interleukin-1 affects glucose homeostasis. Am. J. Physio. 253:R794–R798.
Evans, D.A., D.O. Jacobs, and D.W. Wilmore. 1989. Tumor necrosis factor enhances glucose uptake by peripheral tissues. Am. J. Physiol. 257:R1182–R1189.
Falduto, M.T., R.C. Hickson and A.P. Young. 1989. Antagonism by glucocorticoids and exercise on expression of glutamine synthetase in skeletal muscle. FASEB J. 3:2623–2628.
Fleck, A. 1989. Clinical and nutritional aspects of changes in acute-phase proteins during inflammation. Proc. Nutr. Soc. 48:347–354.
Freeman, B.A. and J.D. Crapo. 1982. Biology of disease free radicals and tissue injury. Lab. Invest. 47:412–426.
Ganong, W.F. 1987. Hormonal control of calcium metabolism and the physiology of bone. In: Review of Medical Physiology, Thirteenth edition. Appleton and Lange, Norwalk, Connecticut. p. 331.
Granner, D.K. 1988. Thyroid hormones. In: R.K. Murray, D.K. Granner, P.A. Mayes and V.W. Rodwell, eds. Harper's Biochemistry, Twenty-first edition. Appleton and Lange, Norwalk, Connecticut. pp. 496–501.
Grimble, R.F. 1989. Cytokines: their relevance to nutrition. Eur. J. Clin. Nutr. 43: 217–230.
Gruenfeld, C., R. Gulli, A.H. Moser, L.A. Gavin, and K.R. Feingold. 1989. Effect of tumor necrosis administration in vivo on lipoprotein lipase activity in various tissues of the rat. J. Lipid Res. 30:579–585.

Johnstone, B.J. and K.C. Klasing. 1990. Nutritional aspects of leukocytic cytokines. Nutr. Clin. Metabol. 4:7–27.

Koeffler, H.P., H. Reichel, A. Tobler and A.W. Norman. 1989. Macrophages and vitamin D_3. In: M. Zembala and G.L. Asherson, eds. Human Monocytes. Academic Press, New York. pp. 345–351.

Kunkel, S.L., D.G. Remick, R.M. Strieter and J.W. Larrick. 1989. Mechanisms that regulate the production and effects of tumor necrosis factor-α. Critical Rev. Immun. 9:93–117.

Lacey, J.M. and D.W. Wilmore. 1990. Is glutamine a conditionally essential amino acid? Nutr. Rev. 48:297–309.

Lang, C.H. and C. Dobrescu. 1989. Interleukin-1 induced increases in glucose utilization are insulin mediated. Life Sci. 45:2127–2134.

Larrick, J.W. and S.C. Wright. 1990. Cytotoxic mechanism of tumor necrosis factor-α. FASEB J. 4:3215–3223.

Le, J. and J. Vilcek. 1987. Biology of disease. Tumor necrosis factor and interleukin-1: Cytokines with multiple overlapping biological activities. Lab. Invest. 56:234–248.

Lumpkin, M.D. 1987. The regulation of ACTH secretion of IL-1. Science 238:452–454.

Machlin, L.J. and A. Bendich. 1987. Free radical tissue damage: protective role of antioxidant nutrients. FASEB J. 1:444–445.

Manolagas, S.C., F.G. Hustmyer and X. Yu. 1989. 1,25–dihydroxyvitamin D_3 and the immune system. S.E.B.M. 192:238–245.

Mayes, P.A. 1988. Regulation of carbohydrate metabolism. In: R.K. Murray, D.K. Granner, P.A. Mayes and V.W. Rodwell, eds. Harper's Biochemistry, Twenty-first edition. Appleton and Lange, San Mateo, CA. pp. 100–107.

Mealy, K., J. van Lanschot, B.G. Robinson, J. Rounds and D.W. Wilmore. 1990. Are the catabolic effects of tumor necrosis factor mediated by glucocorticoids? Arch. Surg. 125:42–48.

Minghetti, P.P. and A.W. Norman. 1988. 1,25–$(OH)_2$-Vitamin D_3 receptors: gene regulation and genetic circuitry. FASEB J. 2:3043–3053.

Mizel, S.B. 1989. The interleukins. FASEB J. 3:2379–2388.

Moldawer, L.L., C. Anderson, J. Gelin and K.G. Lundholm. 1988. Regulation of food intake and hepatic protein synthesis by recombinant-derived cytokines. Am. J. Physiol. 254:G450–G456.

Moldawer, L.L., M.A. Marano, H. Wei, Y. Fong, M.L. Silen, G. Kuo, K.R. Manogue, H. Vlassara, H. Cohen, A. Cerami and S.F. Lowry. 1989. Cachectin/tumor necrosis factor-alters red blood cell kinetics and induces anemia in vivo. FASEB J. 3:1637–1643.

Morrison, N.A., J. Shine, J. Fragonas, V. Verkest, M.L. McMenemy and J.A. Eisman. 1989. 1,25–Dihydroxyvitamin D-responsive element and glucocorticoid repression in the osteocalcin gene. Science 246: 1158–1161.

Mortimore, G.E. 1986. Regulation of hepatic protein degradation by circulatory amino acids. Fed. Proc. 45:2169–2172.

Nockels, C.F. 1988. The role of vitamins in modulating disease resistance. In: J.L. Howard, ed. The Veterinary Clinics of North America:

Food Animal Practice. Vol. 4, W.B. Sanders Co., Philadelphia. pp. 531–542.
Nockels, C.F. 1990. Mineral alterations associated with stress, trauma, and infection and the effect on immunity. The Compendium 12: 1133–1139.
Ohmann, H.B., M. Campos, M. Snider, N. Rapin, T. Beskorwayne, Y. Popowych, M.J.P. Lawman, A. Rossi and L.A. Babiuk. 1989. Effect of chronic administration of recombinant bovine tumor necrosis factor to cattle. Vet. Pathol. 26:462–472.
Pomposelli, J.J., E.A. Flores and B.R. Bistrain. 1988. Role of biochemical mediators in clinical nutrition and surgical metabolism. J. Parenteral and Enteral Nutr. 12:212–217.
Porat, O. 1989. The effect of tumor necrosis factor α on the activity of lipoprotein lipase in adipose tissue. Lymphokine Res. 8:459–469.
Reinhardt, T.A. and F.G. Hustmyer. 1987. Role of vitamin D in the immune system. J. Dairy Sci. 70:952–962.
Reuter, A., J. Benier, P. Gysen, Y. Gebaert, R. Gathy, M. Lopez, G. Dupont, P. Damas and P. Franchimont. 1988. A RIA for tumor necrosis factor (TNFα) and interleukin 1B (IL-1B) and their direct determination in serum. *In*: M.C. Powanda, J.J. Oppenhein, J.J. Kluger and C.A. Dinarello, eds. Monokines and other Nonlymphocytic Cytokines. Alan R. Liss, Inc., New York. pp. 377–381.
Roth, J.A. and M.L. Kaeberle. 1982. Effect of glucocorticoids on the bovine immune system. JAVMA 18:894–901.
Rulofson, F.S., D.E. Brown and R.A. Bjur. 1988. Effect of blood sampling and shipment to slaughter on plasma catecholamine concentration in bulls. J. Anim. Sci. 66:1223–1229.
Sisson, S.D. and C.A. Dinarello. 1989. Interleukin-1. *In*: M. Zemballa, ed. Human Monocytes. Academic press, New York, pp. 183–215.
Smith, K.A. 1988. Interleukin-2. Inception, impact and implications. Science 240:1169–1176.
Smith, R.J. 1986. Role of skeletal muscle in interorgan amino acid exchange. Fed. Proc. 45:2172–2176.
Tizard, I.R. 1988. Regulation of the immune response. *In*: Immunology: An Introduction. Twenty-second edition. Saunders College Publishing, New York. pp 261–283.
Weiss, S.J. and A.F. Buglio. 1982. Biology of disease phagocyte-generated oxygen metabolites and cellular injury. Lab. Invest. 47:5–18.

STRESS EFFECTS ON CHROMIUM NUTRITION OF HUMANS AND FARM ANIMALS

RICHARD A. ANDERSON

Vitamin and Mineral Nutrition Laboratory, Beltsville Human Nutrition Research Center, U.S. Department of Agriculture, ARS Beltsville, Maryland, USA

Introduction

Stress, which can be defined as "abnormal pressure, strain, constraining force or influence", can occur in biological systems in various forms from very mild dietary stresses to severe physical trauma or injury. While the degree of stress can be approximated, a similar amount of stress can have very different effects on different people or animals. For example, a mild amount of exercise for one individual may be a rather extreme form of stress for another.

Chronic stresses may also alter nutrient requirements. If dietary intake is suboptimal, a small stress-induced alteration may have significant effects on the signs and symptoms of deficiency. One metal that is suboptimal in the diet and strongly influenced by stress in humans and farm animals is chromium (Cr).

The essentiality of chromium was demonstrated in rats whose impaired glucose tolerance was improved following Cr supplementation (Schwarz and Mertz, 1959). Chromium has subsequently been shown to affect glucose and(or) lipid metabolism in mice, squirrel monkeys, guinea pigs, rabbits, chickens, turkeys, pigs, cattle and humans (Anderson, 1988). Documentation of the essential role of Cr in humans was first reported in 1977 (Jeejeebhoy et al., 1977) for a woman on total parenteral nutrition. The patient was on total parenteral nutrition for five years. She became diabetic with severe glucose intolerance, weight loss and impaired nerve conduction. Symptoms were refractory to the daily addition of 45 units of insulin. Following the addition of 250 µg of Cr as chromium chloride for two weeks, diabetic symptoms were alleviated and exogenous insulin requirement dropped from 45 units/d to zero. Intravenous glucose tolerance and respiratory quotient returned to normal. This work has subsequently been confirmed in numerous laboratories (Freund et al., 1979; Brown et al., 1986; Anderson, 1989). Chromium is now routinely added to total parenteral nutrition solutions (Anderson, 1994).

However, not all patients on total parenteral nutrition may require supplemental Cr (Anderson, 1994). Chromium is often a contaminant in parenteral nutrition fluids and extra chromium may not be required. Chromium content of parenteral nutrition fluids should be monitored routinely to prevent chromium accumulation especially in patients with impaired kidney function.

Signs of Cr deficiency in humans are not limited to subjects on total parenteral nutrition. Improvements in glucose and(or) lipid concentrations have been observed in children with protein calorie malnutrition, adults and the elderly. Adults with low blood sugar, marginal to impaired glucose tolerance and diabetics have all been shown to respond to supplemental Cr (see review, Anderson, 1993).

The signs of Cr deficiency in humans and experimental and farm animals shown to improve following supplemental Cr are listed in Table 1. Improvements following Cr supplementation may not be reversing signs of deficiency but rather therapeutic effects. For example, some early studies involving turkey poults (Steele and Rosebrough, 1979) and pigs (Steele et al., 1982) involved Cr added to the diet at 20-200 µg/kg of diet. Recent studies show beneficial effects in the nutritional range (100 to 1000 times lower) (Burton et al., 1992; Chang and Mowat, 1992; Page et al., 1993; Evock-Clover et al., 1993). However, true nutritional and therapeutic effects still need to be ascertained. Glucose, lipids and body composition of animals consuming stock diets containing presumably ample amounts of Cr (greater than 300 µg/kg) often improve when supplemented with Cr. However, the form of Cr used may also be very important. Anderson et al. (1993) supplemented rats with nine different forms of Cr and demonstrated that some forms

Table 1. Signs and symptoms of chromium deficiency.

Function	Animals
Impaired glucose tolerance	Human, rat, mouse, squirrel monkey, guinea pig
Elevated circulating insulin	Human, rat, pig
Glycosuria	Human, rat
Fasting hyperglycemia	Human, rat, mouse
Impaired growth	Human, rat, mouse, turkey
Hypoglycemia	Human
Elevated serum cholesterol and triglycerides	Human, rat, mouse, cattle, pig
Increased incidence of aortic plaques	Rabbit, rat, mouse
Increased aortic intimal plaque area	Rabbit
Neuropathy	Human
Encephalopathy	Human
Corneal lesions	Rat, squirrel monkey
Ocular eye pressure	Human
Decreased fertility and sperm count	Rat
Decreased longevity	Rat, mouse
Decreased insulin binding	Human
Decreased insulin receptor number	Human
Decreased lean body mass	Human, pig, rat
Elevated per cent body fat	Human, pig
Humoral immune response	Cattle
Morbidity	Cattle

Source: Adapted from Anderson, 1988.

of Cr including inorganic chromium chloride, were incorporated poorly into tissues. Chromium incorporation into rat tissues varied several-fold among the Cr compounds tested (Anderson et al., 1993). Therefore, total dietary Cr may be an inaccurate reflection of bioavailable Cr and a poor reflection of Cr nutritional status.

Stress effects on human chromium nutrition

Stresses shown to alter Cr nutrition are glucose loading, high simple sugar diets, lactation, infection, acute exercise, chronic exercise and physical trauma (Table 2, Anderson, 1988). Urinary Cr losses can be used as a measure of the Cr mobilized and lost since Cr is not reabsorbed in the kidney but is excreted via the urine (Doisy et al., 1971).

Various forms of stress increase Cr losses and the degree of stress is roughly related to the amount of Cr lost in the urine. For example, mild aerobic exercise at 50% of V_{O_2max} does not result in significant increases in urinary Cr losses while exercise at 90% of V_{O_2max} does result in significant urinary Cr losses (Anderson et al., 1988). Running a 10 km race at near maximal capacity resulted in nearing a doubling of daily urinary Cr losses compared with a nonexercise day (Table 2, Anderson et al., 1982). Urinary Cr concentration was almost five-fold higher two hours following exercise compared with pre-exercise values. Anaerobic exercise also leads to increased Cr losses (Vincient et al., 1994). While acute strenuous aerobic and anaerobic exercise results in significantly enhanced Cr losses, aerobic training leads to decreased basal Cr losses (Table 2; Anderson et al., 1988).

Chromium values listed in Table 2 were all determined in our laboratory at Beltsville by the same methods and can be compared directly. Basal urinary Cr concentrations varied greatly prior to 1980 but accepted values are 0.12–0.22 µg/d for normal sedentary subjects. Decreased Cr losses observed in response to aerobic training may be an indication of decreased body Cr stores in response to repeated bouts of exercise. Lower basal losses in trained subjects may also be an adaptive response. If aerobic training led to increased Cr deficiency then people who are well-trained would be expected to display increased signs of Cr deficiency including impaired glucose tolerance and insulin resistance. However, this is not the case and aerobically trained individuals have improved glucose tolerance and insulin sensitivity (Rodnick et al., 1987). It is

Table 2. Stress effects on urinary Cr losses of humans.

Stress	Urinary Cr, µg/day	Reference
Basal	0.16 ± 0.02	Anderson et al., 1982
Acute exercise	0.30 ± 0.07	Anderson et al., 1982
Chronic exercise (training)	0.09 ± 0.01	Anderson et al., 1988
Lactation	0.37 ± 0.02	Anderson et al., 1993
High sugar diet	0.28 ± 0.01	Kozlovsky et al., 1986
Physical trauma	10.8 ± 2.1	Borel et al., 1984

likely that training which leads to obvious adaptive increases in strength, endurance, heart stroke volume etc. leads to increased conservation of body Cr stores. Adaptive changes may involve a redistribution of Cr in specific tissues which is supported by animal studies. Vallerand et al. (1984) reported that exercise-trained rats displayed significantly higher Cr concentrations in the heart and kidneys compared with respective tissues of controls. Humans who train regularly may also compensate for increased Cr losses associated with acute exercise with increased caloric intake, resulting in higher Cr intake. Improved dietary habits including decreased consumption of simple sugars (which leads to enhanced Cr losses) would also lead to improved dietary Cr status. However, people who exercise strenuously but sporadically to lose weight would not have increased dietary intake, would not have the adaptive mechanisms of training, and would not have improved Cr nutrition, but would have low intake coupled with high Cr losses due to acute exercise. This could obviously lead to decreased Cr stores.

A direct correlation of Cr losses with stress was shown by Anderson et al. (1991; Figure 1) who reported a correlation of serum cortisol and urinary Cr losses associated with strenuous head out immersion exercise. Serum cortisol is related to the degree of exercise intensity and is therefore a measure of exercise stress (Kuoppasalmi et al., 1980). Increased exercise capacity associated with carbohydrate loading was less stressful based on serum cortisol and also less stressful based on decreased Cr losses (Anderson et al., 1991).

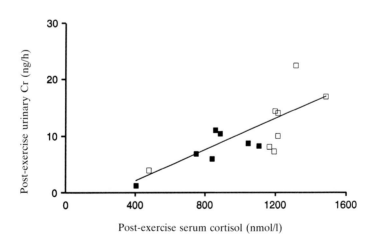

Figure 1. Correlation of urinary chromium losses and post-exercise serum cortisol. Blood samples for serum cortisol were taken immediately following head out immersion exercise at 25°C. Urine samples were collected during the four hours of exercise and two hours following. (□) Control diet period; (■) following carbohydrate loading.(Source: Anderson et al., 1991)

Severe forms of stress such as those associated with physical trauma severe enough to be taken to a shock trauma unit resulted in several-fold increases in Cr losses (Table 2; Borel et al., 1984). Chromium losses decreased as the patients improved and the degree of stress decreased.

Stress effects in farm animals

The effects of stress on Cr metabolism observed in humans are also observed in farm animals. Chang and Mowat (1992) reported that Cr in the form of high Cr yeast increased average daily gain by 30% and feed efficiency by 27% in steer calves following the stress of shipping. During the growing period, Cr had no effect on weight gain or feed efficiency. Chromium also decreased serum cortisol and increased immunoglobulin M and total immunoglobins in calves fed diets with soybean meal but had no effect in calves fed urea–corn supplementation. Humoral immune responses of periparturient and early lactating dairy cows were also improved by supplemental Cr (Burton et al., 1993). These data suggest that stressed animals fed normal diets may be showing signs of Cr deficiency including decreased feed efficiency, increased stress hormone, morbidity, and impaired immune function.

These signs of Cr deficiency are only observed in the stressed animals. The stress associated with shipping appears to increase the Cr requirement of the steers. Long-acting oxytetracycline which is given to combat stress has effects similar to Cr but when Cr and tetracycline were administered simultaneously there were no additive effects. The authors postulated that by reducing stress, long-acting injectable oxytetracycline alleviated Cr losses and prevented signs of marginal Cr deficiency. Klasing and Roura (1991) demonstrated that administration of antibiotic to chicks decreased immunologic stress by reducing interleukin-1 levels resulting in improved growth. Chromium supplementation to unstressed growing-finishing steers had no effect on carcass composition and tissue mineral concentrations or any performance characteristics. The benefical effects of Cr in stressed animals demonstrates that stress increases Cr requirements.

Chromium has also been shown to have beneficial effects on body composition, glucose, lipids and related parameters in swine. Steele et al. (1982) reported that high levels of insulin-potentiating or biologically active Cr increased growth and feed efficiency in pigs. Page et al. (1993) reported increased daily gain, longissimus muscle area, increased percentage of muscling and decreased tenth rib fat in pigs following Cr supplementation in the form of Cr picolinate but not with chromium chloride. Evock-Clover et al. (1993) reported improved HDL-cholesterol, total cholesterol: HDL ratio, glucose, insulin and insulin:glucose ratio in both control and growth hormone treated pigs following supplementation with Cr in the form of Cr picolinate. Supplemental Cr counteracted the negative effects of growth hormone on glucose and insulin variables. Chromium effects were greater at 60 kg than at 45 kg. Chromium also increased serum growth hormone in the growth hormone-treated pigs.

Page et al. (1993) reported variable results on blood variables. For example, "serum cholesterol was reduced by Cr picolinate in Experiments 1 and 2 but not affected by Cr picolinate in Experiment 3". Effects on body composition were also somewhat variable. In a follow-up study, Evock-Clover et al. also did not observe as large or consistent effects of Cr as in the initial study (unpublished observations). Other investigators have also found some inconsistencies in their studies involving Cr nutrition in humans and animals. One important factor that has been poorly controlled in all studies is stress. Variable responses to Cr may be related to stress levels.

Summary

Dietary and physical stresses have profound effects on Cr nutrition of humans and farm animals. Stress leads to increased Cr losses in humans and may lead to higher Cr requirements. Stress-induced higher Cr requirements have been documented in farm animals. While dietary stress can be controlled quite closely, environmental stresses including temperature, humidity and pathogens often vary considerably. External pathogens as well as pathogens and toxins found in foods could lead to altered levels of stress. Any or all of these stresses may alter Cr metabolism and nutritional Cr requirements and function.

References

Anderson, R.A. 1988. Chromium. In: Trace Minerals in Foods, K.Smith (ed.) Marcel Dekker, Inc., New York.. p. 231.
Anderson, R.A. Chromium and parenteral nutrition. Nutrition (in press).
Anderson, R.A. 1989. Essentiality of chromium in humans. Sci. Total Environ. 86:75.
Anderson, R.A. 1993. Recent advances in the clinical and biochemical effects of chromium deficiency. In: Essential and Toxic Trace Elements in Human Health and Disease, A.S. Prasad, ed., Wiley Liss Inc., New York. p. 221.
Anderson, R.A., N.A. Bryden, and M.M. Polansky. 1993. Form of chromium effects tissue chromium concentration. FASEB J. 7:A204.
Anderson, R.A., N.A. Bryden, M.M. Polansky, and P.A. Deuster. 1988. Exercise effects on chromium excretion of trained and untrained men consuming a constant diet. J. Appl. Physiol. 64:249.
Anderson, R.A., N.A. Bryden, M.M. Polansky, and J.W. Thorp. 1991. Effect of carbohydrate loading and underwater exercise on circulating cortisol, insulin and urinary losses of chromium and zinc. Eur. J. Appl. Physiol. 63:146.

Anderson, R.A., M.M. Polansky, N.A. Bryden, E.E. Roginski, K.Y. Patterson, and D.C. Reamer. 1982. Effect of exercise (running) on serum glucose, insulin, glucagon and chromium excretion. Diabetes 31:212.
Borel, J.S., T.C. Majerus, M.M. Polansky, P.B. Moser, and R.A. Anderson. 1984. Chromium intake and urinary chromium excretion of trauma patients. Biol. Trace Elem. Res. 6:317.
Brown, R.O., S. Forloines-Lynn, R.E. Cross, and W.D. Heizer. 1986. Chromium deficiency after long-term total parenteral nutrition. Dig. Dis. Sci. 31:661.
Burton, J.L., B.A. Mallard, and D.N. Mowat. 1993. Effects of supplemental chromium on immune responses of periparturient and early lactation dairy cows. J. Anim. Sci. 71:1532.
Chang, X., and D.N. Mowat. 1992. Supplemental chromium for stressed and growing feeder calves. J. Anim. Sci. 70:559.
Doisy, R.J., D.H.P. Streeten, M.L. Souma, M.E. Kalafer, S.L. Rekant, and T.G. Dalakos.1971. Metabolism of chromium 51 in human subjects. In: Newer Trace Elements in Nutrition, W. Mertz and W.E. Cornatzer, eds., Marcel Dekker, New York. p. 155.
Evock-Clover, C.M., M.M. Polansky, R.A. Anderson, and N.C. Steele. 1993. Dietary chromium supplementation with or without somatotropin treatment alters serum hormones and metabolites in growing pigs without affecting growth performance. J. Nutr. 123:1504.
Freund, H., S. Atamian, and J.E. Fischer. 1979. Chromium deficiency during total parenteral nutrition. J. Am. Med. Assoc. 241:496.
Jeejeebhoy, K.N., R.C. Chu, E.B. Marliss, G.R. Greenberg, and A. Bruce-Robertson. 1977. Chromium deficiency, glucose intolerance, and neuropathy reversed by chromium supplementation in a patient receiving longterm total parenteral nutrition. Am. J. Clin. Nutr. 30:531.
Klasing, K.C. and E. Roura.1991. Interactions between nutrition and immunity in chickens. Proc. Cornell Nutr. Conf. p. 94.
Kuoppasalmi, K., H. Naveri, M. Harkonen, and H. Aldercreutz. 1980. Plasma cortisol, androstinedione, testosterone and leutenizing hormone in running exercise of different intensities. Scand. J. Clin. Lab. Invest. 40:403.
Page, T.G., L.L. Southern, T.L. Ward, and D.L. Thompson, Jr. 1993. Effect of chromium picolinate on growth and serum and carcass traits of growing finishing pigs. J. Anim. Sci. 71:656.
Rodnick, K.J., W.L. Haskell, A.L.M. Swislocki, J.E. Foley, and G.M. Reaven. 1987. Improved insulin action in muscle, liver, and adipose tissue in physically trained human subjects. Am. J. Physiol. 253:E489.
Schwarz, K. and W. Mertz. 1959. Chromium (III) and the glucose tolerance factor. Arch. Biochem. Biophys. 85:292.
Steele, N.C., M.P. Richards, and R.W. Rosebrough.1982. Effect of dietary chromium and protein status of hepatic insulin binding characteristics of swine. J. Anim. Sci. 55 (Suppl. 1):300.
Steele, N.C. and R.W. Rosebrough. 1979. Trivalent chromium and nicotinic acid supplementation for the turkey poult. Poult. Sci. 58:983.

Vallerand, A.L., J.P. Cuerrier, D. Shapcott, R.J. Vallerand, and P.F. Gardiner. 1984. Influence of exercise training on tissue chromium concentrations in the rat. Am. J. Clin. Nutr. 39:402.

Vincient, K.R., P.M. Clarkson, P.S. Freedson, and R.A. Anderson. Weight training, carbohydrate feeding and chromium excretion. Am. Coll. Sports Med. (in press).

ORGANIC CHROMIUM: A NEW NUTRIENT FOR STRESSED ANIMALS

D.N. MOWAT
Department of Animal and Poultry Science
University of Guelph
Guelph, Ontario, Canada

Presented at Alltech's 9th Annual Symposium on Biotechnology in the Feed Industry, April 26-28, 1993

Intensively managed livestock are prone or exposed to a host of stresses. In fact, periodic stress may be the norm rather than the exception. During parturition and early lactation, dairy cows are under great physical, psychosocial and metabolic stresses which are reflected in altered hormone profiles, increased disease susceptibility (eg. mastitis, ketosis) and possibly reduced production and reproduction. Moreover, beef and veal calves undergo a variety of psychosocial, physical, nutritional and environmental stresses during weaning, marketing, transport, mixing and adaptation to the feedlot. The result is a period of decreased feed intake and growth and increased susceptibility to bovine respiratory disease complex (BRD) or shipping fever. Despite improvements in vaccines, nutrition and management, BRD remains the major disease in many feedlots.

Nutrition and stress

Increased attention in preventative medicine programs is now being given to nutrition and its role in reducing stress and improving the immune response or effectiveness of vaccines. Recent research has shown beneficial effects of various nutrients on immunocompetence and consequent resistance to infectious agents in beef and dairy cattle (Nockels, 1991). Nutrients involved in the antioxidant system (selenium (Se), vitamin E, β–carotene, zinc (Zn), manganese (Mn) and copper (Cu)) can aid in reducing mastitis, probably by reducing free radical damage to phagocytic host cells (Weiss *et al.*, 1991). The microminerals Zn and Cu, for example, are required at elevated levels to improve performance and immune response of newly arrived feeder calves. Increased urinary excretion of Zn and Cu have been reported with market-transit stress, fasting and after IBR virus infection (Orr *et al.*, 1990). Studies with humans and mice have indicated that various stressors (eg, elevated intakes of glucose, low protein diets, infection, strenuous exercise, trauma) increased urinary excretion of chromium (Cr) (Pekarek *et al.*, 1975; Borel *et al.*, 1984; Anderson *et al.*, 1988). Excessive Cr losses in exercising individuals appear to place athletes at a greater risk of Cr depletion. Furthermore, Cr supplementation has been shown in mice to protect against stress-induced losses of several trace elements (Zn, Fe, Cu and Mn) (Schrauzer et al., 1986).

The nutrition or ration considerations for sick or stressed calves are substantially different than for healthy calves. However, most nutrient standards are based on information available for feeding normal animals. Nutrition–health interactions of newly arrived feeder cattle have been reviewed elsewhere (Cole, 1982; Matheson, 1992). Most forms of stress and disease result in some reduction in feed intake, creating ration formulation problems. The problems of formulating starter diets are further confounded by the low percentage of calves actually eating during the first few days (Hutcheson, 1990).

For Cr (as well as other nutrients) to exert beneficial effects for stressed animals, it is imperative that the nutrients be administered early in (or before) the stress period. In our studies so far with stressed feeder calves, Cr was typically fed at a level of 4 mg per head daily for the first 3–5 days after arrival at the feedlot, followed by a reduced level of .5 ppm in diet dry matter.

An essential nutrient

Chromium is well known as an essential trace element for humans and laboratory animals. Numerous reviews report on the role of Cr in human nutrition (Anderson, 1987; Anderson, 1988; Offenbacher and Pi-Sunyer, 1988). Chromium was first determined to be the active constituent of the glucose tolerance factor (GTF) isolated from brewers' yeast by Schwarz and Mertz in 1959. Although the chemical structure of GTF has not been completely identified, it is considered to be a nicotinic acid-trivalent chromium-nicotinic acid axis with ligands of glutamic acid, glycine and cysteine (Mertz et al., 1974). Chromium may have several biological functions, including roles in nuclear protein and RNA synthesis, but its predominant physiological role seems to be as an integral component of GTF to potentiate the action of insulin (Anderson and Mertz, 1977). The exact mechanism by which GTF improves glucose tolerance and insulin action is not known. However, it is suggested that GTF enhances the binding of insulin to its specific receptors, possibly through initiating disulphide linkages between insulin and cell membranes (Mooradian and Morley, 1987).

Inorganic trivalent Cr is absorbed at very low levels (0.4–3%); and absorption is inversely related to dietary intake (Anderson, 1988). When present in foods in the GTF as opposed to the inorganic form, Cr has been shown to be more readily absorbed and more biologically active. Conversion of chromium chloride to GTF may be slow or even entirely lacking in certain animal species (Ranhotra and Gebroth, 1986). Thus, the content of total Cr in a diet may bear little relationship to its effectiveness as biologically active Cr. Moreover, several synthetic GTF compounds have been found to have chemical and biological properties similar to naturally occurring GTF (Mooradian and Morley, 1987). Recently, however, Evans and Pouchnik (1993) showed that Cr coordination complexes with nicotinate or picolinate differ markedly in both chemical composition and biological action.

Insufficient dietary Cr in humans leads to signs and symptoms often associated with maturity-onset diabetes and(or) cardiovascular diseases (Anderson, 1988). The recommended safe and adequate adult human daily intake for Cr is 50–200µg. However, most reliable reports in the U.S. and other countries indicate that females and males are only consuming roughly 40 and 60% of the minimum suggested intake, respectively (Anderson, 1987). While Cr toxicity is usually limited to industrial settings and is due to hexavalent Cr, Cr deficiency is widespread and concerned with Cr in the trivalent state.

While data on the role and function of Cr in humans are accumulating more rapidly, studies with food-producing animals are just beginning. Most animal nutritionists fail to consider this element. One would hypothesize, however, that a deficiency of the element may exist for animals for which stress has depleted body stores. Stress often leads to increases in glucose metabolism. Increased utilization of glucose leads to increased mobilization of Cr. Once Cr is mobilized, it is not reabsorbed and is excreted in the urine (Anderson, 1988). Thus, the key to whether or not many diets provide sufficient bioavailable Cr is stress.

Benefits to animals

Since fall 1989, we have conducted four large Cr trials with recently weaned beef calves obtained at feeder sales. Studies during the first two years involved calves shipped a long distance (from Saskatchewan to Ontario). During the last two years local (Ontario) calves were used. All were fed high corn silage diets with or without supplemental organic Cr fed as high Cr yeast or chelated Cr. Results have been consistent and rather dramatic (Chang and Mowat, 1992; Moonsie-Shageer and Mowat, 1993; Mowat et al., 1993). Supplemental Cr increased weight gain (or prevented the depression in gain) during the initial 21–28 days in the feedlot by 24% when averaged over all four trials, with the greatest effect (82%) occurring in the bottom third (slowest gaining, more heavily stressed or prone to stress, Figure 1). Gain was not affected later during the relatively unstressed growing and finishing period (Chang and Mowat, 1992).

Probably the biggest benefit was the large reduction in morbidity or sickness due to BRD. The most effective Cr source (chelated) in one trial reduced morbidity to less than 1/3 of control (Mowat et al., 1993). For supplemental Cr to have maximum benefits, it would appear necessary to administer Cr (in the organically complexed form) early in the stress period. Ideally, this may involve feeding supplemental Cr in preconditioning (at the cow-calf farm) as well as in receiving diets.

The marked reduction in incidence of BRD with supplemental Cr means less need for preventative and therapeutic antibiotics. We have noted an interaction between Cr and an antibiotic. In our initial studies (Chang and Mowat, 1992), improvements in weight gain during the first 28 days after arrival were comparable between supplemental Cr

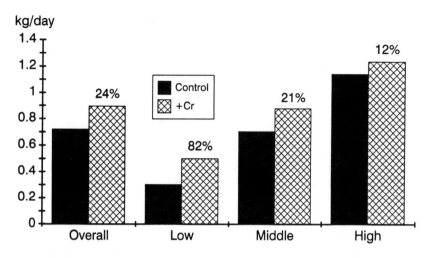

Figure 1. Initial gain of stressed calves (1989–1992).

and long-acting injectable oxytetracycline. In fact, in steers not treated with preventative or therapeutic antibiotic, Cr increased rate of gain by 52%. In the presence of antibiotic, Cr had no benefits on performance. In the study of Mowat et al. (1993), weight gain was increased 43% by chelated Cr during the initial 21 days. The main reason for this large improvement was the marked reduction in number of sick calves. However, during days 21–35 post arrival, a large compensatory gain occurred in the morbid or antibiotic-treated animals. Klasing and Roura (1991) indicated that chicks also undergo a period of compensatory growth subsequent to reducing immunological stress by administering an antibiotic. Furthermore, the lack of an additive effect of Cr with presence of antibiotic (long-acting oxytetracycline) suggests that antibiotic may decrease the Cr requirement of calves by reducing immunologic stress or immune response. Klasing and Roura (1991) showed that administration of antibiotic to chicks in a dirty environment decreased immunologic stress by reducing plasma interleukin-1 levels and improved growth. Thus, antibiotic may have decreased environmental or infection stress of calves, thereby reducing Cr losses.

Data on effects of supplemental Cr for high-producing dairy cows are more preliminary. In our initial trial, supplemental Cr increased milk production in primiparous cows but had no effect in multiparous cows (Subiyatno et al., 1993a). Supplemental Cr reduced days to first service in multiparous cows but had no effect with primiparous cows. During the trial (from six weeks prepartum to 16 weeks postpartum), Cr supplementation reduced body condition score, particularly with primiparous cows. Some evidence for a reduction in subclinical ketosis in multiparous cows supplemented with Cr was also obtained.

The effects of supplemental Cr on body composition need further clarification, but are receiving a good deal of attention, especially in the pig sector. Page et al. (1993) found that supplemental Cr, in the form of Cr picolinate (but not Cr chloride), increased loin eye area and percentage of muscling and decreased 10th rib fat thickness in growing-finishing pigs. In contrast, Swiss workers found that chromium chloride was comparable to or better than Cr yeast and Cr picolinate in improving growth rates and feed efficiency during the finish phase of Swiss landrace gilts. Effects on fat thickness and longissimus dorsi were not significant, however pigs given Cr in the form of Cr yeast had an 8.7% higher longissimus dorsi area (Table 1, Gerber and Wenk, personal communication). This confirmed earlier studies in humans with Cr picolinate (Evans, 1989). Recently, Evans and Pouchnik (1993) showed that Cr picolinate accelerated development of lean body mass in exercising humans while Cr nicotinate was ineffective. As mentioned, Subiyatno et al. (1993) observed a decrease in body condition score with late gestation/early lactation dairy cows fed chelated Cr. However, Chang et al. (1992) noted that continuous supplementation of high-Cr yeast to unstressed, growing-finishing steers had no effect on carcass composition. It should be emphasized that Cr is a nutrient, not a therapeutic agent or drug. Therefore, supplementation will benefit only when marginal Cr deficiency occurs.

Table 1. Effect of chromium and manganese supplements on performance, efficiency, backfat thickness and longissimus dorsi muscle area of gilts.

	Control	Cr chloride	Cr yeast	Cr picolinate	Mn proteinate
Number of animals	10	10	10	10	10
Initial weight, kgs	27.5	27.6	27.3	27.2	27.5
End weight, kgs	106.5	107.3	107.1	105.0	106.6
Days to finish	105.3	100.4	103.0	102.3	102.5
Daily gain, g					
growing[1]	667	671	673	653	656
finish[2]	829[a]	916[b]	875[ab]	878[ab]	887[ab]
Feed efficiency					
growing[1]	2.37	2.35	2.36	2.38	2.39
finishing[2]	3.05[b]	2.73[a]	2.86[ab]	2.88[ab]	2.83[ab]
DE per kg gain, MJ	39.0	36.2	36.6	37.3	37.8
Backfat thickness, cm	1.8	1.9	2.0	1.9	2.0
L. dorsi area, cm^2	49.5	51.2	53.8	48.6	53.5

[1] 27 to 60 kg
[2] 60 kg to finish

Effects on immune and endocrine systems

The University of Guelph has been the first to scientifically report on the ability of Cr to alter immune function, including response to vaccines. This could have profound effects on human as well as veterinary preventative medicine. In our initial study (Chang and Mowat,

1992), supplemental Cr increased serum immunoglobulin levels with soybean meal supplementation of corn silage during the growing period. Moonsie-Shageer and Mowat (1992) then showed that Cr during the stress period reduced rectal temperatures (0.5°C) and increased antibody titres to injected human erythrocytes. Chirase *et al.* (1991) indicated that calves receiving supplemental zinc methionine also had lower rectal temperatures during the stress period. Recently, in beef calves fed Cr, increased antibody titers were noted following vaccination with a commercial live attenuated IBR vaccine (Burton *et al.*, 1993a). Dairy cows receiving supplemental chelated Cr during early lactation also had higher antibody responses to a variety of test antigens and superior cell-mediated immunity following *in vitro* stimulation with a T-cell mitogen (Burton *et al.*, 1993b). It is clear that organically complexed Cr is a potent immunomodulator of specific immune responses in cattle.

In our initial study supplemental Cr from high-Cr yeast decreased serum cortisol of growing steers (Chang and Mowat, 1992). Various stressors are known to increase serum cortisol. Glucocorticoids generally are considered to be growth-inhibiting steroids. Furthermore, glucocorticoids are known to suppress the immune system (Munck *et al.*, 1984). Decreased cortisol may have caused the increased immunoglobulin levels noted in steers supplemented with Cr and soybean meal. In a subsequent study, supplemental Cr caused a linear decrease in serum cortisol at day 28 after arrival (Moonsie-Shageer and Mowat, 1993). However, it took time for Cr to exert its influence on cortisol with no effects noted at days 7 or 14 but a trend beginning to occur at day 21 after arrival.

Decreased serum cortisol with Cr supplementation could have particular implications for the high producing (metabolically stressed) dairy cow. Cortisol has been reported to be antagonistic to milk production (Sartin *et al.*, 1988). Insulin availability may limit the onset of ovarian activity leading to first ovulation in cows (Butler and Canfield, 1989). Moreover, increased cortisol concentrations associated with physiological stress may negatively affect estrus cyclicity in high producing dairy cows (Vighio and Liptrap, 1990). Recently, Subiyatno (1993b) noted reduced serum cortisol levels with supplemental Cr for prepartum primiparous and postpartum cows. In the latter study, Cr supplementation resulted in more rapid clearance of infused glucose in postpartum primiparous cows. This may possibly account in part for the increased milk production.

Summary

Chromium could well become a common micronutrient for the animal nutritionist in the near future, at least during periods or conditions of stress. In North America supplemental Cr sources are readily available at local human health food stores but have not been registered as yet for animal feed.

The biggest potential for Cr supplementation of stressed feeder (veal and dairy) calves is in reducing the incidence of BRD (including improving effectiveness of vaccines) and thus the need for antibiotics. It is far better to maintain health and prevent infection than to incur the costs of medication, labour, decreased productivity and potential antibiotic residue concerns.

The effects of malnutrition (Cr, Zn, Se, Vitamin E, etc) upon the immunity of farm animals are only now beginning to be clarified. Further research is in progress to clarify the extent of increased milk production, improved reproduction and possible reduction in certain metabolic diseases in dairy cattle as well as benefits in stressed swine and poultry.

Literature cited

Anderson, R.A. 1988. Chromium. In: K.T. Smith (Ed.) Trace Minerals in Foods. p. 231. Marcel Dekker Inc., New York.

Anderson, R.A., N.A. Bryden, M.M. Polansky and P.A. Deuster. 1988. Exercise effects on chromium excretion of trained and untrained men consuming a constant diet. J. Appl. Physiol. 64:249.

Anderson, R.A. 1987. Chromium. In: W. Mertz (Ed.) Trace Elements in Human and Animal Nutrition. Vol 1 p. 225. Academic Press, New York.

Anderson, R.A. and W. Mertz. 1977. Glucose tolerance factor: an essential dietary agent. Trends Biochem. Sci. 2:277.

Borel, J.S., T.C. Majerus, M.M. Polansky, P.B. Moser and R.A. Anderson. 1984. Chromium intake and urinary chromium excretion of trauma patients. Biol. Trace Elem. Res. 6:317.

Burton, J.L., B.A. Mallard and D.N. Mowat. 1993a. Effects of supplemental chromium on response of newly weaned beef calves to IBR/PI$_3$ vaccination. Can. J. Vet. Res. (In press).

Burton, J.L., B.A. Mallard and D.N. Mowat. 1993b. Effects of supplemental chromium on immune responses of perparturient and early lactation dairy cows. J. Anim. Sci. 71: (1532–1539).

Butler, W.R. and R.W. Canfield. 1989. Interrelationships between energy balance and postpartum reproduction. Proc. Cornell Nutr. Conf. pp. 66.

Chang, X. and D.N. Mowat. 1992. Supplemental chromium for stressed and growing feeder calves. J. Anim. Sci. 70:559.

Chang, X., D.N. Mowat and G.A. Spiers. 1992. Carcass characteristics and tissue-mineral contents of steers fed supplemental chromium. Can. J. Anim. Sci. 72:663–669.

Chirase, N.K., D.P. Hutcheson and G.B. Thompson. 1991. Feed intake, rectal temperature, and serum mineral concentrations of feedlot cattle fed zinc oxide or zinc methionine and challenged with infectious bovine rhinotracheitis virus. J. Anim. Sci. 69:4137.

Cole, N.A. 1982. Nutrition-health interactions of newly arrived feeder cattle. Proc. Symp. on Management of Food Producing Animals. Purdue Univ. Vol. 2:683.

Evans, G.W. 1989. The effect of chromium picolinate on insulin controlled parameters in humans. Int. J. Biosoc. Med. Res. 11:163–180.

Evans, G.W. and D.J. Pouchnik. 1993. Composition and biological activity of chromium-pipidine carbonylate complexes. J. Inorg. Biochem. 49:177–187.

Hutcheson, D.P. 1990. Nutrition critical in getting calves started right. Feedstuffs 62(11):14.

Klasing, K.C. and E. Roura. 1991. Interactions between nutrition and immunity in chickens. Proc. Cornell Nut. Conf. pp. 94.

Matheson, G.W. 1991. Starter feedlot cattle nutrition. Proc. Western (Canada) Nutr. Conf. pp. 232.
Mertz, W., E.W. Toepfer, E.E. Roginski and M.M. Polansky. 1974. Present knowledge of the role of chromium. Fed. Proc. 33:2275.
Moonsie-Shageer, S. and D.N. Mowat. 1993. Levels of supplemental chromium on performance, serum constituents and immune status of stressed feeder calves. J. Anim. Sci. 71:232–238.
Mooradian, A.D. and J.E. Morley. 1987. Micronutrient status in diabetes mellitus. Amer. J. Clin. Nutr. 45:877.
Mowat, D.N., X. Chang and W.Z. Yang. 1993. Chelated chromium for stressed feeder calves. Can. J. Anim. Sci. 73:49–56. .
Munck, A., P. Guyre and N. Holbrook. 1984. Physiological functions of glucocorticoids in stress and their relation to pharmacological actions. Endocr. Rev. 5:25.
Nockels, C.F. 1991. Impact of nutrition on immunological function:beef and dairy cattle. Minnesota Nutr. Conf. pp. 65.
Offenbacher, E.G. and G.X. Pi-Sunyer. 1988. Chromium in human nutrition. Ann. Rev. Nutr. 8:543.
Orr, C.L., D.P. Hutcheson, R.B. Grainger, J.M. Cummins and R.E. Mock. 1990. Serum copper, zinc, calcium and phosphorus concentrations of calves stressed by bovine respiratory disease and infectious bovine rhinotracheitis. J. Anim. Sci. 68:2893.
Page, T.G., L.L. Southern, T.L. Ward and D.L. Thompson. 1993. Effect of chromium picolinate on growth and serum and carcass traits of growing-finishing pigs. J. Anim. Sci. 71:656–662.
Pekarek, R.S., E.C. Hauer, E.J. Rayfied, R.W. Wannemacher and W.R. Beisel. 1975. Relationship between serum chromium concentrations and glucose utilization in normal and infected subjects. Diabetes 24:350.
Ranhotra, G.S. and J.A. Gebroth. 1986. Effects of high-chromium bakers' yeast in glucose tolerance and blood lipid in rats. J. Am. Assoc. Cereal Chem. 63:411.
Schrauzer, G.N., K.P. Shrestha, T.B. Molenaar and S. Mead. 1986. Effects of chromium supplementation on feed energy utilization and the trace element composition in the liver and heart of glucose-exposed young mice. Biol. Trace Elem. Res. 9:79.
Schwarz, K. and W. Mertz. 1959. Chromium (III) and the glucose tolerance factor. Arch. Biochem. Biophys. 85:292.
Subiyatno, A., D.N. Mowat and R.M. Liptrap. 1993a. The effect of supplemental chromium on early lactation performance of Holstein cows. Proc. Amer. Soc. Dairy Sci. Assoc. Ann. Mtg. (Abst.).
Subiyatno, A., W.Z. Yang, D.N. Mowat and G.A. Spiers. 1993b. Chelated chromium alters plasma metabolite responses to glucose infusion in dairy cows. Proc. Amer. Soc. Dairy Sci. Assoc. Ann. Mtg. (Abst.).
Vighio, G.H. and R.M. Liptrap. 1990. Plasma hormone concentrations after administration of dexamethasone during the middle of the luteal phase in the cow. Am. J. Vet. Res. 51:1711.
Weiss, W.P., J.S. Hogan and K.L. Smith. 1991. Managing mastitis in dairy cattle through nutrition. BASF Tech. Symp. Minnesota Nutr. Conf.

TRANSITION METALS, OXIDATIVE STATUS, AND ANIMAL HEALTH: DO ALTERATIONS IN PLASMA FAST-ACTING ANTIOXIDANTS LEAD TO DISEASE IN LIVESTOCK?

JAMES K. MILLER
Animal Science Department, The University of Tennessee, Knoxville, Tennessee, USA

FRED C. MADSEN
Suidae Technology, Greensburg, Indiana, USA

Summary

Performance of dairy cows may suffer when reactive oxygen metabolites (ROM) are generated faster than they can be safely neutralized by antioxidant mechanisms resulting in oxidative stress. Nutrients essential in manufacture or structure of known components of defense against ROM include copper (Cu), magnesium (Mg), manganese (Mn), zinc (Zn), vitamin E, vitamin A, and β-carotene in addition to adequate energy, nitrogen, and sulfur. Some elements (notably iron (Fe) and Cu) can, under some circumstances, also generate ROM. Protection therefore depends on orchestration of factors which influence ROM quenching, chain breaking, and reduction of site specific damage from catalytic metals such as Cu^{+1} and Fe^{+2}. Fast-acting antioxidants (FAA), an in vitro assay of ROM quenching capacity of blood plasma, may reflect an animal's ability to protect itself against oxidative stress. Research with periparturient dairy cows showed that incidence of retained fetal membranes (RFM) and udder edema were higher when plasma FAA were below average and that FAA could be elevated by the antioxidant vitamin E. In addition, incidence of RFM was reduced by vitamin E but increased by a potential prooxidant, excess dietary Fe. Control of oxidative stress by supplying all known nutrients in adequate and balanced amounts and by minimizing effects of substances that stimulate ROM is important to optimize performance of high producing cows. Under some conditions it may be necessary to feed extra chain-breaking antioxidants such as vitamin E and additional site-specific damage inhibitors such as Zn.

Introduction

Aerobic organisms have achieved greater energetic efficiency during metabolism than anaerobic organisms by using oxygen to accept electrons. However, partially reduced oxygen metabolites produced during stepwise reduction of oxygen to water can be potentially toxic. This property of oxygen has been investigated extensively in recent years (for reviews see Halliwell, 1987; Halliwell and Gutteridge, 1990; Levine and Kidd, 1985; Sies, 1985). When reactive forms of oxygen are produced faster than they can be safely neutralized by antioxidant mechanisms, oxidative stress results (Sies, 1985). Accumulating evidence suggests reactive forms of oxygen may contribute to health disorders in cattle (Miller et al., 1993).

Our purpose here is to discuss origin and propagation of reactive forms of oxygen, conditions which contribute to their increased production, possible effects on health and performance of livestock, and strategies for minimizing their adverse effects. In addition, we have evaluated plasma fast-acting antioxidants (FAA) as an index of the animal's ability to deal with oxidative stress. Our research has involved periparturient dairy cows so we have used these as examples. However, the same principles should also apply to other classes of livestock.

Sources of reactive forms of oxygen

SUPEROXIDE AND HYDROGEN PEROXIDE PRODUCTION

Stable, nonradical molecules contain paired electrons with opposite spins in their orbitals. A free radical, in contrast, has a single unpaired electron in an outer orbital which results in instability and reactivity (Levine and Kidd, 1985). Not all reactive metabolites of oxygen (e.g., hydrogen peroxide), however, are free radicals by this definition. For this reason, we will hereafter use the term "reactive oxygen metabolite" (ROM) (Powell, 1991). The initially-formed ROM, superoxide (O_2^-) results when a single electron is added to molecular oxygen (Table 1). Oxidative enzymes (e.g., aldehyde oxidase and xanthine oxidase) are examples of O_2^- generators. Another important generator of O_2^- is the electron transfer system. Oxidative phosphorylation is a process by which electrons are passed down a chain of enzymes in the mitochondria to generate ATP. Approximately 2 to 5% of all electrons fail to reach

Table 1. Possible sources of reactive oxygen metabolites[1].

Reaction	Product
Electron transport chains → H+	Electron leakage
$O_2 + H^+ \rightarrow O_2^-$	Superoxide
$2O_2^- + 2H^+ \rightarrow O_2 + HOOH$	Hydrogen peroxide
$O^- + Fe^{3+} \rightarrow O_2 + Fe^{2+}$	Reduced iron
$HOOH + Fe^{2+} \rightarrow Fe^{3+} + O_2 + {\cdot}OH$	Hydroxyl radical
${\cdot}OH + RH \rightarrow H_2O + R{\cdot}$	Oxidizing radical
${\cdot}OH + LH \rightarrow H_2O + L{\cdot}$	Fatty acid radical
$R{\cdot} + LH \rightarrow RH + L{\cdot}$	Fatty acid radical
$L{\cdot}$ or $R{\cdot} + O_2 \rightarrow LO_2{\cdot}$ or $RO_2{\cdot}$	Peroxy radical
$LO_2 + LH \rightarrow L{\cdot} + LOOH$	Lipid peroxide

[1]Gutteridge, 1987; Halliwell, 1987; Halliwell and Gutteridge, 1990; Slater, 1987.

cytochrome C oxidase and instead leak from the chain to partially reduce O_2 to O_2^- (Levine and Kidd, 1985).
Factors that can elevate rate of respiratory electron transfer (e.g., increases in metabolic rate due to increased production or activity) can elevate rate of ROM production. For example, electron leakage can increase when cellular oxygenation is either abnormally high or low. Increased oxygen consumption during intense exercise has been associated with a higher rate of lipid peroxidation (Packer, 1984). During

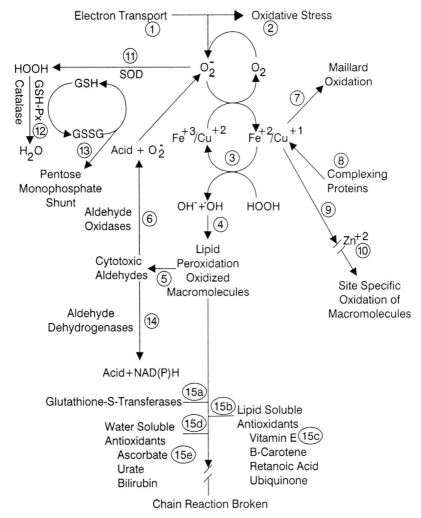

Figure 1. Systems for protection against reactive oxygen metabolites. From Miller et al. (1993). Circled numbers correspond to numbers on Table 2. Numbers 1–9 represent generation of ROM (1–6, generation of ROM in the free pool; 8–9, •OH induced site specific damage to macromolecules); 10, prevention of site specific damage; 11–14, prevention of •OH induced peroxidative chains; and 15a-e, interruption of peroxidative chains

elevated oxygen tension, transfer of electrons may accelerate due to increased availability of oxygen as the electron acceptor. Conversely, during hypoxia, electrons may accumulate at the terminal electron carrier due to insufficient oxygen to accept the electrons. Either condition could elevate electron leakage and increase production of ROM (Levine and Kidd, 1985). When cell respiration is impaired, dehydrogenases may be converted to corresponding oxidases (Suttle, 1991) which transfer electrons to oxygen and thereby increase oxidative load (Figure 1, Table 2).

Table 2. Summary of systems for prevention of and protection against reactive oxygen metabolites[1,2].

1. Superoxide (O_2^-) is generated during normal metabolism.
2. Dietary imbalances, xenobiotics, disease, and solar radiation are likely to increase ROM production above physiological levels.
3. Superoxide reduces Fe^{+3} enabling it to enter into Fenton type reactions which produce hydroxyl radical (•OH).
4. The extremely reactive •OH attacks macromolecules and initiates peroxidative reactions.
5. Cytotoxic aldehydes are end products of lipid peroxidation.
6. Aldehyde oxidases degrade cytotoxic aldehydes but O_2^- is generated which adds to the oxidative load.
7. Catalytic Fe^{+2} can accelerate oxidative Maillard reactions.
8. Catalytic metals may indiscriminately complex on protein SH groups.
9. This can result in •OH production and site-specific damage to the protein. Hydroxyl radical scavengers do not affect initiation of reactions 8 and 9.
10. $Zinc^{+2}$ can mask labile sulfur sites from Fe^{+2} thus protecting the protein.
11. Superoxide dismutases convert O_2^- to peroxides. This conversion retards reduction of Fe^{+3} to Fe^{+2} which catalyzes formation of •OH.
12. Glutathione peroxidase and catalase convert peroxides to forms which do not participate in Fenton reactions. Reduction of peroxides is accompanied by oxidation of glutathione (GSH) to glutathione disulfide (GSSG).
13. GSSG can be reduced to GSH by reducing equivalents from NADPH which is generated by the pentose monophosphate shunt.
14. Aldehyde dehydrogenases generate reducing equivalents in addition to converting aldehydes to less toxic products.
15. Peroxidative chains initiated by ROM that escape enzymatic degradation are interrupted by chain breaking antioxidants
    a. Glutathione-S-transferases dispose of peroxy radicals by conjugating them with GSH. This pathway may be more active when Se or vitamin E is deficient. The resulting destruction of GSH increases consumption of reducing equivalents in competition with other metabolic pathways that depend on NADPH.
    b. Lipid soluble antioxidants are active in membranes and at interfaces with water soluble components.
    c. Vitamin E serves as a chain-breaking antioxidant by reacting directly with ROM. Although vitamin E is oxidized when ROM are quenched, it can be reduced by ascorbic acid. This conserves reducing equivalents in comparison with GSH-S-transferases serving as chain breakers.
    d. Water-soluble antioxidants are active in intra and intercellular fluids.
    e. Ascorbate, in addition to regenerating vitamin E and possibly also GSH, can act in its own right as a water soluble antioxidant.

[1]From Miller et al. (1993).
[2]Numbers correspond to circled numbers in Figure 1.

The cytochrome P-450 enzymes are another significant source of O_2^-. The various P-450 isozymes can be divided into two broad categories according to whether physiological (endogenous) or foreign (xenobiotic) substrates are metabolized (Waterman et al., 1986). Xenobiotics, which elevate activity of cytochrome P-450, are likely to increase ROM production above amounts produced by physiological substrates. Examples of such xenobiotics commonly found in livestock feed include aflatoxins (Swick, 1984), chlorinated water (Levine and Kidd, 1985), endophyte infected fescue (Zanzalari et al., 1989a), gossypol (Willard et al., 1993), nonenzymatically browned feed (Zanzalari et al., 1989b), and phenolic compounds (Levine and Kidd, 1985).

Spontaneous oxidation of biological molecules involving nonenzymatic electrons can contribute to production of O_2^- also (Levine and Kidd, 1985). Examples are autoxidation of glucose in diabetic subjects (Wolff et al., 1989) or during stress-induced increases in circulating glucose. A variety of organic molecules can be activated by sunlight or other radiant energy to generate O_2^-.

Superoxide dismutates spontaneously into hydrogen peroxide (HOOH) (Table 1). Both O_2^- and HOOH are unavoidable products of normal metabolic processes and are not always harmful if properly metabolized. They are involved physiologically in chemistry of several enzymes and prostaglandins and are used by phagocytic cells to kill bacteria (Halliwell, 1987; Levine and Kidd, 1985). Imbalances between production of ROM and their safe disposal, however, can initiate oxidative chain reactions and lipid peroxidation (Table 1).

METALS AS INITIATORS OF MORE REACTIVE OXYGEN SPECIES

There are several situations where nutritionally important transition elements can create more reactive species from less reactive materials. For example, Fe is vital but its presence in the wrong place at a particular time can have extremely damaging effects. Cells normally are protected against harmful effects of transition elements by complexing them in larger molecules that are compartmented away from sites susceptible to damage. Dietary imbalances, inflammation, infection, and environmental stresses all may contribute to "indiscriminate coordination" of transition elements, particularly Fe (Madsen et al., 1990), which increases likelihood of decompartmentalization. Decompartmentalized Fe may be released from hemoglobin by peroxides (Gutteridge, 1986) or from ferritin by O_2^- or lipid peroxides, or escape from the low molecular weight pool involved in transfer of Fe from transferrin to ferritin (Halliwell, 1987). We believe decompartmentalized catalytic metals may create highly reactive oxygen species by at least four different routes: hydroxyl radical (•OH) produced in the free pool by Fenton chemistry, decomposition of lipid peroxides to alkoxyl or peroxyl radicals, site specific damage of macromolecules by •OH, and oxidative changes during the Maillard reaction.

A full chemical description of the transition elements Fe, Cu, and Mn is beyond the scope of this paper. However, an example of Cu will provide the insight necessary to understand the importance of oxidation state and coordination number on the physical chemistry of transition metals and their complexes (Cotton and Wilkinson, 1988). Copper can be found in two and probably three oxidative states (Cu^{+1}, Cu^{+2}, and Cu^{+3}) in warm-blooded animals. $Copper^{+1}$ is a soft Lewis acid, generally displays a coordination number of 4 and can be found mostly in tetrahedral geometric complexes. $Copper^{+2}$ is a soft Lewis acid of intermediate strength. It can display itself in coordination numbers of 4 or 6 with geometries of square (dsp^2) for coordination numbers of 4 and octahedral (d^2sp^3) for coordination numbers of 6. The octahedral geometry has Jahn–Teller effects. The point of the discussion is that it important to know the oxidation states of the transition elements due to the many physical features each element can demonstrate.

Altered oxidation state of these nutritionally important transition elements theoretically could have a bearing on biologically relevant oxidative reactions (Table 3). Coordination character of these elements can profoundly influence ability to transfer electrons. Iron, Cu, and possibly Mn are believed to catalyze reaction of O_2^- and HOOH to generate •OH which is at the focal point of initiation of peroxidative damage (Table 4). Catalytic Fe also initiates peroxidative chains by decomposing lipid peroxides to alkoxyl or peroxyl radicals (Table 5).

Table 3. **Relevant redox cycles of nutritionally important transition elements which may affect oxidative stress.**

Mn^{+2a}	⇔	Mn^{+3}
Cu^{+2}	⇔	$Cu^{+1b,c}$
Fe^{+3}	⇔	Fe^{+2b}

[a]No net crystal field stabilization energy. Mn^{+2} may rapidly oxidize under local basic conditions.
[b]Metals which may participate in Fenton chemistry
[c]Cu^+ is a soft Lewis acid as are most toxic metals
  $Cu^{+3} - Cu^{+2}$ should also be considered as an alternative to
  $Cu^{+2} - Cu^{+1}$ for some biological redox reactions of Cu (Sigel, 1981).

Table 4. **Fenton-type chemistry of copper^{+1} and iron^{+2} which result in site-specific damage.**

Cu^{+1} + target[a]	→	target-Cu^{+1} complex
Target-Cu^{+1} complex + HOOH	→	target–Cu^{+2}-•OH + OH^{-b}
Target-Cu^{+2}-•OH	→	damaged target + Cu^{+2}
Fe^{+2} + target[c]	→	target-Fe^{+2} complex
Target-Fe^{+2} complex	→	target-Fe^{+3}–•OH +OH^{-b}
Target-Fe^{+3}–•OH	→	damaged target +Fe^{+3}

[a]Macromolecules with appropriate Lewis base ligands for Cu^{+1} and Fe^{+2}.
[b]Hydroxyl scavengers do not protect the target.
[c]$Zinc^{+2}$ can play a role in masking labile sulfur site thiols (mercaptide formation) and reduce damage from Fe^{+2} site-specific damage. $Zinc^{+2}$ status of animal is important in the overall protection from oxidative damage (Wilson, 1987).
[d]There is a small pool of "free" Fe which can react directly with HOOH to produce hydroxyl radicals. Hydroxyl radical scavengers can protect from damage generated in this manner.

Table 5. Conversion of less reactive to more reactive species by catalytic metals[1].

Target	Metal species	Reactive products
O_2^- + HOOH	Fe^{+2}/Cu^{+1}	•OH
Lipid peroxides (ROOH)	Fe^{+2}/Cu^{+1}	RO• (alkoxy), RO2 (peroxy), cytotoxic aldehydes
Thiols (RSH)	Fe^{+2}/Cu^{+1}	O_2^-, HOOH, thiyl (RS•), •OH
NAD(P)H	Fe^{+2}/Cu^{+1}	NAD(P)•, •OH, O_2^-
Catecholamines	$Fe^{+2}/Cu^{+1}/Mn$	O_2^-, HOOH, •OH, semiquinones

[1]From Halliwell (1987).

The potential for reactions between O_2^-, •OH, organic molecules, and their oxidized products is evident in Table 1. Once the process has been initiated, ROM tend to propagate in expanding chain reactions (Levine and Kidd, 1985). Change in oxidation state of decompartmentalized Fe ($Fe^{+3} \rightarrow Fe^{+2}$) can accelerate oxidative Maillard reactions (Dills, 1993) resulting in additional damage. An example is the carboxylation of lysine from amadori products.

In addition to being at the focal point of initiation of peroxidative damage (Table 1), •OH can cause site specific damage to macromolecules (Table 4). An Fe atom in the wrong oxidation state associated with protein may catalyze production of •OH, damaging the protein at the site where Fe is located.

Other transition elements such as heavy metals, for example nickel and lead, may also accelerate oxidative stress. These metals may perpetuate ROM production in a fashion similar to Fe^{+2} and Cu^{+1}. Therefore it may be important to consider heavy metal contamination of supplements for livestock.

Defense against ROM

PREVENTIVE SYSTEMS

Normally the body is protected against ROM and their toxic products by a wide range of known defense mechanisms (Figure 1, Table 2). Included among preventive systems are metal-binding macromolecules and antioxidant enzymes.

Examples of macromolecules which keep catalytic transition metals harmlessly complexed in proper oxidation states and away from Fenton chemistry in extracellular fluids are transferrin, ceruloplasmin, and albumin (Halliwell, 1987). Ceruloplasmin maintains Fe in the +3 oxidation state and inhibits both xanthine oxidase and ferritin-dependent peroxidation of phospholipids and O_2^- mediated mobilization of Fe from ferritin (Samokyszyn et al., 1989). Free coordination sites on Fe are thought to permit it to produce OH and hindrance of these coordination sites prevents Fe initiated •OH formation (Graf et al., 1984), although there are complexes which transfer electrons by outer sphere processes and therefore do not require binding sites. Certain endogenous materials such as creatinine (Glazer, 1988) bind metals and prevent indiscriminate coordination of proteins by Cu^{+1} and possibly Fe^{+2} which can lead to oxidation of the protein. Creatinine does not quench peroxy radicals but rather it protects against metal-mediated damage by a complexing mechanism (Glazer, 1988). Iron does not seem to promote peroxide decomposition when it is bound to the specific Fe binding sites of transferrin or lactoferrin (Halliwell and Gutteridge, 1990).

Metals which are not involved in electron transfer but yet have the correct outer electron makeup and are strong Lewis acids would be ideal to retard site specific damage by catalytic Fe. $Zinc^{+2}$ has been shown to play a role in masking labile sulfur sites from Fe^{+2} (Wilson, 1987) thereby affording protection from decompartmentalized Fe. Site specific damage from catalytic transition elements is different from damage initiated by •OH in that it cannot be prevented by •OH scavengers. It seems important, therefore, that Zn nutrition be examined carefully when excess Fe is being fed (this happens often in practical cattle diets) or when Fe is likely to be decompartmentalized such as in Cu deficiency, inflammation, and infection. Cellular injury makes transition elements more available. Situations such as calving or other tissue insults may indeed increase oxidative stress.

Within cells, superoxide dismutases (SOD), glutathione peroxidase (GSH-Px), and catalase remove O_2^- and HOOH before they approach available promoters of Fenton chemistry (Halliwell, 1987). Reduction of peroxides is accompanied by oxidation of reduced glutathione (GSH) to glutathione disulfide (GSSG), but GSSG can be reduced back to GSH by reducing equivalents from NADPH (Golden and Ramdath, 1987; Wilson, 1987). These reactions can alter redox potential in the cell and affect flux rates in metabolic pathways.

CHAIN-BREAKING SYSTEMS

Some O_2^- and HOOH may escape despite the preventive systems to be catalyzed by decompartmentalized Fe into ROM capable of initiating peroxidative chains (Table 1). These are dealt with by chain-breaking antioxidants including lipid soluble vitamin E, ubiquinone, β-carotene, retanoic acid and water soluble ascorbate, GSH, phenolics, and urate (Kinsella et al., 1993; Machlin and Bendich, 1987; Sies, 1985). Vitamin E reacts directly with a variety of peroxy radicals by giving up an electron, thus stabilizing the reactive molecule (Machlin and Bendich, 1987; Wilson, 1987). Vitamin E itself becomes a radical but in this form does not react with organic molecules so the chain reaction is terminated.

Although vitamin E is oxidized when ROM are quenched (Golden and Ramdath, 1987), it can be regenerated by ascorbate (Wilson, 1987). If vitamin E is inadequate, however, glutathione-S-transferases conjugate peroxy radicals with GSH (Golden and Ramdath, 1987). Since GSH is consumed by this process, reducing equivalents are conserved when vitamin E, rather than GSH, serves as a chain breaker. Therefore, physiological or tissue oxidative stress can affect metabolic flux differently depending on which chain breaker is used to terminate the process.

In addition to regenerating vitamin E, ascorbate can act directly as a water soluble antioxidant (Wilson, 1987).

Toxic aldehydes produced when peroxidized lipids decompose can crosslink proteins, phospholipids, and nucleic acids much in the same way that formaldehyde fixes tissue specimens for histological examination (Levine and Kidd, 1985). An additional defense against ROM damage involves aldehyde dehydrogenases which oxidize cytotoxic aldehydes by transferring electrons to NAD^+ (Commission in Biochemical Nomenclature, 1972). An example is xanthine dehydrogenase, which catalyzes, but is not specific to, production of urate and NADH from xanthine, NAD^+, and H_2O. As an additional benefit, xanthine dehydrogenase helps keep Fe in the less reactive oxidized form (Emery, 1991), thereby conserving reducing equivalents, vitamin E, and other chainbreaking nutrients.

Several essential nutrients are involved in manufacture or structure of known components of antioxidant defense (Sies, 1985; Slater et al., 1987). Metal chelators, ubiquinone, urate, GSH, and ascorbate may be of endogenous or dietary origin but the diet also should contain adequate protein (N and S for ruminants), energy, vitamin E, β-carotene, vitamin A, Cu, Fe, Mg, Mn, Mo, selenium (Se) and Zn.

Plasma fast-acting antioxidants as an index of an animal's ability to protect itself against ROM damage

A relatively simple, reproducible, noninvasive procedure for evaluating oxidant-antioxidant status in animals would be useful in relating oxidative stress to disease conditions. Two potentially useful assays have been proposed. Wayner et al. (1985) measured total radical-trapping antioxidant potential (TRAP) in human plasma subjected in vitro to controlled lipid peroxidation. Subsequently they (Wayner et al., 1987) reported the sum of urate, plasma proteins, ascorbate, and – tocopherol accounted for all of TRAP in most samples. Plasma TRAP was increased (P<0.01) over 29 d in human volunteers by 1 g/d of ascorbic acid, α-tocopherol, or a combination, but not by a placebo (Mulholland and Strain, 1992). The combined ability of ascorbate, vitamin E, urate, and protein thiols in plasma to trap peroxy radicals in vitro has been used to assess antioxidant capacity in Fe overload (Berger et al., 1990), acute myocardial infarction (Mulholland and Strain, 1991), rheumatoid arthritis (Thrunham et al., 1987), and malaria (Thurnham et al., 1990).

Glazer (1988) quantified antioxidants in biological fluids by their capacity to prolong fluorescence of phycoerythrin exposed to peroxy radicals generated in vitro. Two distinct fractions in blood plasma have been identified (DeLange and Glazer, 1989). A slow-acting fraction consists primarily of plasma proteins. A fast-acting fraction remaining after precipitation of proteins contains antioxidants including vitamin E, ascorbate, billirubin, and urate. Plasma fast-acting antioxidants (FAA) have not, to our knowledge, been used previously in clinical comparisons.

Wayner et al. (1987) fractionated TRAP of human plasma and found proteins accounted for up to half of the total peroxyl radical quenching capacity compared with only 5–10% for α-tocopherol. Plasma protein also appears to protect phycoerythrin as a "slow-acting antioxidant" since it shares the attack by peroxy radicals (DeLang and Glazer, 1989). Fast-acting antioxidants in plasma protect phycoerythrin almost completely until they are consumed, after which plasma proteins offer only partial protection. This is because on an equimolar basis, FAA react with peroxy radicals at least 100 times faster than phycoerythrin which is turn reacts with peroxy radicals about 60 times faster than plasma proteins (DeLange and Glazer, 1989).

Included among properties of an ideal free radical scavenger as listed by Rose and Bode (1993) are speed of oxidation and suitability for regeneration. Antioxidants included in FAA appear to satisfy these requirements. These antioxidants react much more rapidly with ROM than the molecules they are protecting, do not perpetuate chain reactions, and can be regenerated in vivo.

In contrast to FAA which terminate chain reactions already underway, plasma protein should prevent ROM reactions by complexing catalytic transition elements. Control of catalytic transition elements by protein and other endogenous sequestering agents before •OH can be generated reduces pressure on the cell to maintain reducing equivalents because ROM reactions are stopped at the initial focal point. Although plasma protein appears to function as a slow-acting antioxidant in vitro, sacrificial quenching of ROM should not be its primary function in vivo. Proteins are oxidized when they react with peroxy radicals (Glazer, 1988; 1993) and can now alter metabolism as well as enter in peroxidative chain reactions (Gebicki and Gebicki, 1993). In addition, protein's important function of removing catalytic transition elements may be compromised. Oxidative stress has become severe indeed when FAA have been exhausted and the animal must resort to slow-acting antioxidants for protection against ROM. For these reasons, we have chosen FAA rather than total antioxidants in plasma as an index of oxidative status.

Association between low plasma fast-acting antioxidants and periparturient disorders in dairy cows

For five years we have investigated relationships between oxidative stress and periparturient disorders in dairy cows. During two years FAA (Glazer, 1988; 1990) were measured in plasma samples from 183 cows. Results were related to incidence of retained fetal membranes (RFM) and udder edema.

RETAINED FETAL MEMBRANES

Cows that failed to shed membranes within 12 h after parturition were considered to have RFM. Incidence of RFM was higher (P<0.025) in 33 cows with below average FAA than in 37 cows with above average FAA (Table 6). The odds ratio (Fletcher et al., 1988) for below average plasma FAA as a risk factor for RFM was 3.4. This means that cows with

Table 6. Numbers of cows with or without retained fetal membranes (RFM) or udder edema which were above or below average in fast-acting antioxidants (FAA) at parturition.

FAA and disease status	Disorder	
	RFM	Edema
FAA average or above		
apparently normal	29	24
affected	8	8
FAA below average		
apprently normal	17	7
affected	16	11
Odds ratio	3.4	4.7
95% confidence limits	1.08–11.14	1.16–19.58
Chi square	5.59	6.38
Probability	0.025	0.012

below average plasma FAA near parturition are 3.4 times more likely to retain placentas than cows with average or above plasma FAA.

UDDER EDEMA

Percentage decrease in udder floor area due to removal of milk (udder shrink) on the 2nd and 3rd days after calving was used as an index of edema in 50 heifers. The more edematous and rigid the udder, the less udder floor area decreased after milking. Udder floor area was measured before and after each milking by touching a piece of paper against wet teat ends and calculating area of the quadrilateral formed connecting the four spots.

Eighteen heifers with udder shrink less than the average of 50 heifers were classified as edematous; 32 heifers with greater than average udder shrink were considered nonedematous (Table 6). Odds ratio (Fletcher et al., 1988) was calculated to test the hypothesis that low plasma FAA is a risk factor for udder edema. The odds ratio indicated that heifers with below average plasma FAA in late gestation are 4.7 times more likely to have udder edema than heifers with average or above FAA.

Response of indices of oxidative stress to supplementation with vitamin E

Oxidative stress is the condition resulting when ROM are produced faster than they are effectively removed (Sies, 1985). Decreases in antioxidant defense systems, increases in end products of ROM production, or a combination could thus be considered as indices of oxidative stress. Six such measurements in dairy cows as influenced by supplementation with the antioxidant vitamin E (1000 IU/cow daily) during the last 40 d of gestation are in Table 7. Sixty-three near term dairy cows were classified as either above or below the average of all cows for each index

Table 7. Numbers of cows supplemented or unsupplemented prepartum with vitamin E with indices of oxidative stress above or below the average of all cows in a comparison[1].

Supplementation and oxidative stress index status	Plasma FAA	Serum α-tocopherol	RBC GSH-Px Se	RBC GSH-Px Non-Se	RBC GSH	RBC TBARS
Supplemented						
Average or above	30	21	20	18	22	4
Below average	2	11	12	10	14	28
Unsupplemented						
Average or above	2	11	13	14	10	27
Below average	29	20	18	21	17	4
Odds ratio	217.5	3.47	2.31	2.7	2.7	0.02
95% Confidence limits	23.4–2599.5	1.1–11.2	0.8–7.2	0.9–8.6	0.8–8.5	0–0.1
Chi square	48.01	5.72	2.67	3.67	3.58	35.06
Probability	0.001	0.02	0.10	0.06	0.07	0.001

[1] Lower than average plasma fast-acting antioxidants (FAA) serum α-tocopherol, erythrocyte Se and non-Se dependent glutathione peroxidase (GSH-Px) and erythrocyte reduced glutathione (GSH) or higher than average erythrocyte thiobarbituric acid reactive substances (TBARS) were considered as indicative of impaired ability to combat oxidative stress.

of oxidative stress. It was considered an index of oxidative stress if a cow was below average in plasma FAA, serum α-tocopherol, erythrocyte (RBC) Se and non-Se dependent GSH-Px, and RBC reduced glutathione (GSH), or above average in RBC thiobarbituric acid reactive substances (TBARS).

Odds ratios were calculated to measure strength of the association between supplementation with vitamin E and improvement of each index of oxidative stress. All indices of oxidative stress were improved ($P<0.10$ to $P<0.001$) by vitamin E supplementation. However, plasma FAA (antioxidant protection) and RBC TBARS (end products of lipid peroxidation) were most strongly influenced by vitamin E. By odds ratio, cows supplemented with vitamin E were 218 times ($P<0.005$) more likely to have above average plasma FAA and only 2% ($P<0.005$) as likely to have above average RBC TBARS as unsupplemented cows.

Response of periparturient disorders in dairy cows to supplementation prepartum with an antioxidant (vitamin E) or iron (a potential proxidant)

During investigations with a total of 479 dairy cows spanning five years (Brzezinska-Slebodzinska and Miller, 1992; Mueller et al., 1988; 1989a,b; Thomas et al., 1990a,b; P. K. Hillyer and J. K. Miller, unpublished data), half of the cows were each supplemented daily for six weeks before calving with 1000 IU of vitamin E each year, and half were unsupplemented. In three of the years, half of the cows were also given 1200 ppm Fe in a 2 x 2 factorial arrangement with vitamin E.

RETAINED FETAL MEMBRANES

Incidence of RFM was reduced in the comparison with 479 cows over five years when vitamin E was supplemented prepartum (Table 8). Odds ratio for RFM in relation to supplemental vitamin E calculated from five year totals was 0.32 (P<0.001). This means that cows given supplemental vitamin E during the nonlactating period were only about one-third as likely to have RFM as unsupplemented cows. Excess dietary Fe appeared to increase incidence of RFM (Table 8). When odds ratio was calculated using total numbers of cows for three years, cows fed excess Fe were twice as likely (P<0.05) as unsupplemented controls to retain placentas.

Table 8. Numbers of periparturient dairy cows testing the hypothesis that incidence of retained fetal membranes (RFM) is likely to be decreased by supplementation with an antioxidant (vitamin E) but increased by feeding a prooxidant (excess dietary iron).

Supplementation and RFM status	Treatment	
	Vitamin E	Excess iron
Supplemented		
Retained	24	32
Non-retained	161	70
Unsupplemented		
Retained	62	19
Non-retained	132	82
Odds ratio	0.32	2.0
95% confidence limits	0.18–0.55	1.0–4.0
Chi square	19.47	4.26
Probability	0.001	0.04

UDDER EDEMA

Effects of daily supplementation with 1000 IU/d of vitamin E during the last six weeks of gestation on severity of udder edema were investigated with a total of 125 primigravid heifers over a three year period. Udder

Table 9. Incidence of udder edema in cows supplemented or unsupplemented prepartum with vitamin E.

Disorder and year	Number of cows in comparison	Supplemented	Unsupplemented	P value%,
		% shrink[1]		
1989[2]	40	23.7	16.9	.05
1990[3]	61	23.2	17.3	.05
1992	49	17.9	18.5	NS

[1]Percentage decrease in udder floor area after removal of milk was used as an index of edema: The more edematous and rigid the udder, the less udder floor area decreased after milking.
[2]Mueller et al. (1989).
[3]Thomas et al. (1990).

shrink was greater in heifers receiving supplemental vitamin E in two of the three years (Table 9), indicating supplementation prepartum with vitamin E reduced severity of udder edema.

OXIDATIVE STRESS AND STERIODOGENESIS

Steroid hormones are synthesized from cholesterol through a series of steps catalyzed by individual enzymes (Bhagavan, 1978). Enzymes specific to cytochrome P-450 appear to differ in their vulnerability to ROM attack, with 17 α–hydroxylase being one of the most susceptible (Takayanagi et al., 1987). Since steroidogenesis proceeds by different pathways, inadequacy of a key enzyme for one pathway may misdirect the reaction. Deficiency of 17 α–hydroxylase impairs synthesis of androgens and estrogens but increases plasma concentrations of corticosterone in a recognized human congenital defect (Yanase et al., 1991).

We measured corticosterone and 17 β-estradiol (E_2) in plasma from 112 periparturient cows to test our hypothesis that oxidative stress could alter steroidogenesis through unequal vulnerability of different steroidogenic enzymes (Miller et al., 1993). Corticosterone and E_2 were chosen as representative of 17 α–hydroxylase independent and dependent pathways, respectively. Thirty-one primigravid heifers and 28 multiparous cows with RFM, udder edema, below average FAA, or a combination were classified as under "oxidative stress". Nineteen primigravid heifers and 34 multiparous cows without udder edema or RFM and with above average plasma FAA were considered apparently normal.

The expected average increase in corticosterone or decrease in E_2 in "oxidatively stressed" relative to nonstressed groups was not seen (Table 10). Blood was collected for FAA and steroid hormone assay the same day each week so differences in sampling day in relation to parturition could have contributed to variations in corticosterone and E_2 concentrations. In future comparisons it would be desirable to bleed cows daily near expected parturition so comparable samples in relation

Table 10. Representative 17 α–hydroxylase dependent (17β-estradiol) and independent (corticosterone) hormones in plasma of oxidatively stressed or apparently normal Periparturient Dairy Cows.

Hormone	Apparent oxidative status[1]			
	Stressed	Unstressed	Pooled SEM	P value
Cows, number	59	53	–	–
Corticosterone, pg/ml	872	864	23	NS
17β-estradiol, pg/ml	227	241	17	NS
Corticosterone: 17β-estradiol ratio	12.2	5.5	2.4	.06

[1]Thirty-one primigravid heifers and 28 multiparous cows with udder edema, RFM, below average fast-acting antioxidants, or a combination were classified as under "oxidative stress".

to calving could be chosen. However, within cow corticosterone to E_2 ratios averaged over twice as high (P<0.06) in oxidatively stressed than in apparently normal cows (Table 10). This is consistent with our hypothesis that oxidative stress could alter steroidogenesis through selective damage to steroidogenic enzymes.

Discussion

Levine and Kidd (1985) have hypothesized four stages through which adaptation to oxidative stress results in chronic disease. In the first stage, the individual is healthy and able to cope with oxidative stress. Chronic deficiency of antioxidant nutrients or exposure to prooxidants introduces the second stage in which the individual adapts to oxidative stress by developing a tolerance. Adaptative changes depend upon availability of sufficient antioxidant factors. As antioxidants are depleted by continued oxidative stress, the individual passes into the third stage in which cell components are subjected to peroxidative damage. Protection of target organs against ROM at the cost of antioxidant defenses produces a state in which absorption is impaired and availability of antioxidant nutrients becomes even more limiting. This introduces the fourth stage of the progression through antioxidant adaptation to chronic disease by establishing a vicious cycle in which rate of deterioration of antioxidant defense continually exceeds the rate of recovery. At this stage, oxidative stress has progressed to established disease.

Incidence of periparturient disorders such as RFM and udder edema appears to be increased by potential prooxidants such as excess dietary Fe but decreased by antioxidants such as vitamin E (Tables 8, 9). Below average plasma FAA content, a possible risk factor in these disorders of dairy cows (Table 6), responds favorably to dietary supplementation with vitamin E (Table 7). Below average plasma FAA may be an indication of oxidative stress resulting from excessive exposure to prooxidants, inadequate dietary antioxidants, or a combination. If oxidative stress is a risk factor for disease, and if plasma FAA reflect an animal's ability to protect itself against oxidative stress, then measurement of FAA may have diagnostic value.

Implications

Modern livestock consume diets which may expose them to excessive intakes of prooxidants; and antioxidant requirements may be greater than generally recognized. Antioxidant intakes needed to control ROM balance effectively may exceed amounts contained in average feedstuffs. Vitamin E contents of hay and silage decrease progressively with storage, and long-term feeding of stored feed without access to green forage can reduce vitamin E status of dairy cows (Schingoethe et al., 1978). This is evidenced by lower serum α–tocopherol and higher incidence of retained placenta in the spring prior to the growing season than in the fall when new crop feeds are available (Schingoethe et al., 1982). The importance of vitamin E supplementation for dairy cows has thus increased with widespread dependence on stored feeds.

Supplementation with all known nutrients required for antioxidant defense in adequate and balanced amounts would be beneficial. Based on present knowledge for dairy cows, and under average conditions, we suggest 0.3 ppm Se, 20 ppm Cu, and 60 ppm each of Zn and Mn in dietary DM plus 1000 IU/d of vitamin E for dry cows or 500 IU/d of vitamin E for lactating cows. Cellular redox balance can vary greatly depending on environment, diet, disease, and other factors. It may be necessary, therefore, to adjust amounts of these nutrients accordingly. Under conditions of stress and especially if Fe or molybdenum (Mo) are excessive, it may be beneficial to supply one-third of total supplemental Cu, and Zn in organically chelated form. Iron is more likely to be excessive than deficient in feedstuffs (Miller and Madsen, 1992), so supplementation with Fe, if done at all, should be considered with care to avoid disrupting cellular redox balance.

Molybdenum content of the diet should be monitored and appropriate action should be taken if the ratio of Mo to Cu is excessive (i.e., 1:4 Mo:Cu). Reduced absorption and utilization of Cu can contribute to decompartmentalization and accumulation of Fe in the liver with resulting oxidative stress. Reduced body content of Cu can also adversely affect enzymes in antioxidant defense and reducing molecules such as GSH. Other nutrients (Mg, Zn) known to be involved in stabilizing membranes are also critical in optimizing cellular oxidation–reduction potential.

Acknowledgements

Parts of the research reported herein were supported by the BASF Corporation, Parsippany, NJ. Appreciation is also expressed to Dr M. B. Coelho and Dr G. L. Lynch for the opportunity to participate in BASF Technical Symposia, for which many of the ideas herein were developed; to Dr A. N. Glazer, University of California, Berkeley, for his encouragement and guidance in measuring fast-acting antioxidants; to Dr Ewa Brzezinska-Slebodzinska and Patsy Hillyer for their contributions to this research; to C. R. Holmes and M. H. Campbell, University of Tennessee dairy farm, for their cooperation; and to Nikki Bell for preparation of the manuscript.

References

Berger, H.M., J.H.N. Lindeman, D. Van Zoren-Grabben, E. Houdkamp, J. Schrijver, and H.H. Kanhi. 1990. Iron overload, free radical damage, and rhesus haemolytic disease. Lancet 335:933.

Bhagavan, N.V. 1978. Biochemistry. 2nd ed. J.B. Lippincott Co., Philadelphia, PA.

Brzezinska-Slebodzinska, E. and J.K. Miller. 1992. Antioxidant status of dairy cows supplemented prepartum with vitamin E and selenium. FASEB J. 6:A1953 (abstr).
Commission on Biochemical Nomenclature. 1972. Enzyme Nomenclature. Elsevier Sci. Publ. Co., Amsterdam, Neth.
Cotton, F.A., and G. Wilkinson. 1988. Advanced Inorganic Chemistry, 5th ed. 1411 pp. John Wiley and Sons, New York, NY.
Delange, R.J., and A.N. Glazer. 1989. Phycoerythrin fluorescence-based assay for peroxy radicals: A screen for biologically relevant protective agents. Anal. Biochem. 177:300.
Dills, W.L., Jr. 1993. Protein fructosylation: fructose and the Maillard reaction. Am. J. Clin. Nutr. 58(Suppl):7795.
Emery, T.F. 1991. Iron and Your Health: Facts and Fallacies. 116 pp. CRC Press, Inc., Boston, MA.
Fletcher, R.H., S.W. Fletcher, and E.H. Wagner. 1988. Clinical Epidemiology, 2nd ed. Williams and Wilkins, Baltimore, MD.
Gebicki, S. and J.M. Gibicki. 1993. Formation of peroxides in amino acids and proteins exposed to oxygen free radicals. Biochem. J. 289:743.
Glazer, A.N. 1988. Fluorescence-based assay for reactive oxygen species: A protective role for creatinine. Fed. Am. Soc. Exp. Biol. J. 2:2487.
Glazer, A.N. 1990. Phycoerythrin fluorescence-based assay for reactive oxygen species. Page 161 In: Methods in Enzymology. Vol. 186. Oxygen Radicals in Biological Systems. Academic Press, San Diego, CA.
Golden, M.H.N., and D. Ramdath. 1987. Free radicals in the pathogenesis of kwashiorkor. Proc. Nutr. Soc. 46:53.
Graf, E., J.R. Mahoney, R.G. Bryant, and J.W. Eaton. 1984. Iron-catalyzed hydroxyl radical formation. Stringent requirement for free iron coordination site. J. Biol. Chem. 259:3620.
Gutteridge, J.M.C. 1986. Iron promoters of the Fenton reaction and lipid peroxidation can be released from haemoglobin by peroxides. Fed. Eur. Biol. Soc. Lett. 201:291.
Gutteridge, J.M.C. 1987. Lipid peroxidation: Some problems and concepts. Page 9 In: Proc. Upjohn Symp. Oxygen Radicals and Tissue Injury. Fed. Am. Soc. Exp. Biol. Bethesda, MD.
Halliwell, B. 1987. Oxidants and human disease: some new concepts. Fed. Am. Soc. Exp. Biol. J. 1:358.
Halliwell, B., and J.M.C. Gutteridge. 1990. Role of free radicals and catalytic metal ions in human disease. An overview. Page 1 In: Methods of Enzymology. Vol. 186. Oxygen Radicals in Biological Systems. L. Packer and A.N. Glazer, eds. Academic Press, San Diego, CA.
Kinsella, J.E., E. Frankel, B. German, and J. Kanner. 1993. Possible mechanisms for the protective role of antioxidants in wine and plant foods. Food Technology 47(4):85.
Levine, S.A., and P.M. Kidd. 1985. Antioxidant Adaptation. Its Role in Free Radical Pathology. 367 pp. Biocurrents Div., Allergy Research Group, San Leandro, CA.

Machlin, L.J., and A. Bendich. 1987. Free radical tissue damage: protective role of antioxidant nutrients. FASEB J. 1:441.
Madsen, F.C., R.E. Rompala, and J.K. Miller. 1990. Effect of disease on the metabolism of essential trace elements: A role for dietary coordination complexes. Feed Management 41:20.
Miller, J.K., E. Brzezinska-Slebodzinska, and F.C. Madsen. 1993. Oxidative stress, antioxidants and animal function. J. Dairy Sci. 76:2812.
Miller, J.K., and F.C. Madsen. 1992. Trace minerals. Large Dairy Herd Management. Van Horn, H. H., and C. J. Wilcox, eds. p. 287. American Dairy Science Association, Champaign, IL.
Mueller, F.J., J.K. Miller, N. Ramsey, R.C. DeLost, and F.C. Madsen. 1989. Reduced udder edema in heifers fed vitamin E prepartum. J. Dairy Sci. 72:2211 (abstr).
Mulholland, C.W., and J.J. Strain. 1991. Serum total free radical trapping ability in acute myocardial infarction. Clin. Biochem. 24:437.
Mulholland, C.W., and J.J. Strain. 1992. Total radical-trapping antioxidant potential (TRAP) of plasma: Effects of supplementation of young healthy volunteers with large doses of α–tocopherol and ascorbic acid. Internat. J. Vit. Nutr. Res. 63:27.
Packer, L. 1984. Vitamin E, physical exercise and tissue damage in animals. Med. Biol. 62:105.
Powell, D.W. 1991. Immunophysiology of intestinal electrolyte transport. Page 591 In: Handbook of Physiology 6. The Gastrointestinal System, IV. Intestinal Absorption and Secretion. Am. Phys. Soc. Bethesda, MD.
Rose, R.C., and A.M. Bode. 1993. Biology of free radical scavengers: An evaluation of ascorbate. FASEB J. 7:1135.
Samokyszyn, V.M., D.M. Miller, D.M. Feif, and S.D. Aust. 1989. Inhibition of superoxide and ferritin dependent lipid peroxidation by ceruloplasmin. J. Biol. Chem. 264:21.
Schingoethe, D.J., C.A. Kirkbride, I.S. Palmer, M.J. Ownes, and W.L. Tucker. 1982. Response of cows consuming adequate selenium to vitamin E and selenium supplementation postpartum. J. Dairy Sci. 65:2338.
Schingoethe, D.J., J.G. Parson, F.C. Ludens, W.L. Tucker, and H.J. Shave. 1978. Vitamin E status of dairy cows fed stored feeds continuously or pastured during summer. J. Dairy Sci. 61:1582.
Sies, H., ed. 1985. Oxidative Stress. 507 pp. Academic Press, Orlando, FL.
Sigel, H., ed. 1981. Metal Ions in Biological Systems. 12. Properties of Copper. 341 pp. Marcel Dekker, New York.
Slater, T.F., K.H. Cheeseman, M.J. Davies, K. Proudfoot, and W. Xin. 1987. Free radical mechanisms in relation to tissue injury. Proc. Nutr. Soc. 46:1.
Suttle, N.F. 1991. The interactions between copper, molybdenum, and sulphur in ruminant nutrition. Ann. Rev. Nutr. 11:121.
Swick, R.A. 1984. Hepatic metabolism and bioactivation of mycotoxins and plant toxins. J. Anim. Sci. 58:1017.

Takayanagi, R., K.I. Kato, and H. Ibayashi. 1986. Relative inactivation of steroidogenic enzyme activities of in vitro vitamin E-depleted human adrenal microsomes by lipid peroxidation. Endocrinology. 119:464.
Thomas, D.G., J.K. Miller, F.J. Mueller, and F.C. Madsen. 1990b. Udder edema reduced by prepartum vitamin E supplementation. J. Dairy Sci. 73(Suppl.1):271 (abstr).
Thurnham, D.I., R. Singkamani, R. Kalivichit, and K. Wangworapat. 1990. Influence of malaria infection on peroxyl radical trapping capacity in plasma from rural and urban Thai adults. Brit. J. Nutr. 64:257.
Thurnham, D.I., R.D. Situnayake, B. Kootathep, B. McConkey, and M. Davis. Antioxidant status measured by the TRAP assay in rheumatoid arthritis. Rice-Evans, C., ed. Free Radicals, Oxidant Stress and Drug Action. p. 169. Richelieu Press, London.
Waterman, W.R., M.E. John, and E.R. Simpson. 1986. Regulation of synthesis and activity of cytocrome P-450 enzymes in physiological pathways. Page 345 In: Cytochrome P-450. Structure, Mechanism, and Biochemistry. P.R. Ortiz de Montellano, ed. Plenum Press, New York, NY.
Wayner, D.D.M., G.W. Burton, K.U. Ingold, L.R.C. Barclay, and S.J. Lake. 1987. The relative contributions of vitamin E, urate, ascorbate and proteins to the total peroxyl radical trapping antioxidant activity of human blood plasma. Biochem. Biophys. Acta. 924:408.
Wayner, D.D.M., G.W. Burton, K.U. Ingold, and S. Locke. 1985. Quantitative measurement of the total peroxyl radical trapping antioxidant capability of human blood plasma by controlled lipid peroxidation. FEBS Lett. 187:33.
Willard, S.T., R.L. Stuart, and R.D. Randel. 1993. Free gossypol in the diet may affect plasma vitamin A, E and β-carotene in prepartum and postpartum Brahman cows and suckling calves. J. Anim. Sci. 71(Suppl.1):23 (abstr).
Wilson, R.L. 1987. Vitamin, selenium, zinc and copper interactions in free radical protection against ill-placed iron. Proc. Nutr. Soc. 46:27.
Wolff, S.P., Z.A. Bascal, and J.V. Hunt. 1989. "Autoxidative glycosylation": free radicals and glycation. Page 259 In: The Maillard Reaction in Aging, Diabetes, and Nutrition. J.W. Baynes and V.M. Monnier, eds. Alan R. Liss, Inc., New York, NY.
Yanase, T., E.R. Simpson, and M.R. Waterman. 1991. 17 β-hydroxylase/17, 20–lyase deficiency: from clinical investigation to molecular definition. Endocrine Rev. 12:91.
Zanzalari, K.P., R.N. Heitmann, J.B. McLaren, and H.A. Fribourg. 1989a. Effects of endophyte-infected fescue and cimetidine on respiration rates, rectal temperatures and hepatic mixed function oxidase activity as measured by hepatic antipyrine metabolism in sheep. J. Anim. Sci. 67:3370.
Zanzalari, K.P., J.T. Smith, J.K. Miller, N. Ramsey, and F.C. Madsen. 1989b. Induction of the mixed-function oxidase system of rats and mice by feeding Maillard reaction products. J. Am. Coll. Nutr. 8:445 (abstr).

CHOOSE A NEW PATH: BORON BIOLOGICALS

BERNARD F. SPIELVOGEL
Boron Biologicals, Inc., Raleigh, North Carolina, USA

Introduction

The element boron, next to carbon in the periodic table, is actually a rare element in the context of its relative abundance in the earth's crust (Muertterties, 1967). Its concentration, 3 ppm in the earth's crust, is low; it is even rarer in the universe, about a thousand times less than its abundance in the earth's crust. Nevertheless, on earth, geochemical processes have resulted in formation of a number of large, relatively accessible deposits of boron ores, resulting in its ready availability (in economic terms).

All boron ores are boron-oxygen compounds, e.g. borax, $Na_2B_4O_5(OH)_4 \cdot 8H_2O$. Major boron minerals, refined borates, and boric acid have found many industrial and commercial uses as a result of ready availability. Uses include detergents and bleaches, fiberglass, glazes, fire retardants, and many others. Because of this widespread use, concern has been expressed as to whether boron presents a health hazard to those working with or using boron compounds. In September, 1992, the First International Symposium on the Health Effects of Boron and its Compounds was held at the University of California, Irvine to address these possible concerns. The major conclusion of the Symposium was that there appears to be a wide margin of human safety at typical exposure levels. The proceedings of this Symposium are to be published in the Spring Supplement (1994) in Environmental Health Perspectives.

While it has been known for over 50 years that boron is essential for plant life, until recently, the requirement for boron in animal life was unknown or thought to be extremely small. Since the early 1980's, however, evidence has been mounting that boron is a rather dynamic trace element that has major effects on the metabolism of minerals in higher animals including humans (Nielsen, 1988).

Much of the interest in boron as an essential trace element has focused on "inorganic" boron (borates and derivatives). We have been exploring the biological activity of "organic" boron, and more specifically, boron analogues of important biologically active molecules based upon carbon. Life on this planet is based upon the element carbon, and the chemistry of carbon is commonly referred to as organic chemistry. Carbon atoms, chemically combined with nitrogen, hydrogen, oxygen, etc., make up the fundamental molecules of life such as amino acids, proteins, and nucleic acids (DNA). Boron Biologicals, Inc. (BBI) has invented ways to prepare organic-like bio-molecules that are based on the element boron rather than carbon. These entirely new boron biochemicals enable us to create powerful commercial products for human therapeutic and diagnostic, agricultural, and animal health care applications. BBI is the only company in the world focused on the creation of a platform technology based on biologically active boron compounds.

Less than 30 years ago, Muetterties(1967), in the preface of the book on the chemistry of boron, wrote, "If it were not for the fact that boron compounds, unlike those of carbon, nitrogen, oxygen, are of little significance biologically and physiologically, boron might be one of the most important elements in the periodic table." The purpose of this article is to inform and convince the reader that the investigation of the biological activity of boron and its compounds is quite worthwhile and presents a vast and virtually unexplored area of considerable technological and commercial value.

PROPERTIES OF BORON

Our attention to the element boron and its compounds, particularly in its biological activity, arises from its fascinating intrinsic chemical, structural, and nuclear properties.

Chemical and structural properties of boron
Boron is next to carbon on the periodic table of elements, and boron and carbon compounds are very similar structurally. However, there are some important differences, including the fact that boron, atomic number 5, has three electrons in its valence shell while carbon, atomic number 6, has four valence electrons. Both boron and carbon form simple compounds with hydrogen. In terms of elementary chemical theory, these compounds obey the Lewis octet rule and four pairs of electrons surround the central boron and carbon atoms (Figure 1a). The species are termed isoelectronic and isostructural (both species have their bonds to hydrogen distributed in a tetrahedral geometry, Figure 1b).

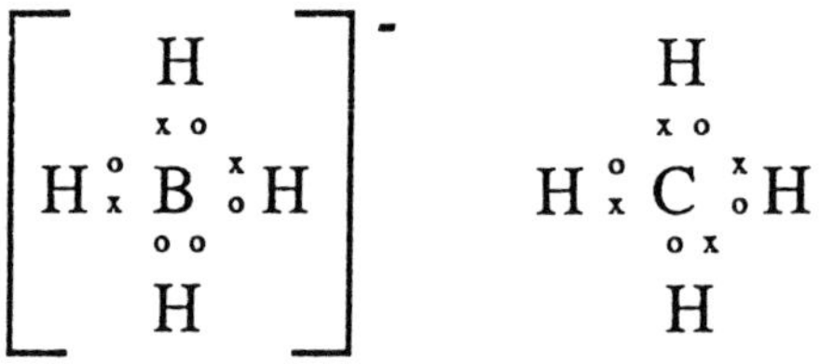

$$\begin{bmatrix} \mathrm{H} \\ \mathrm{H} : \mathrm{B} : \mathrm{H} \\ \mathrm{H} \end{bmatrix}^{-} \qquad \mathrm{H} : \mathrm{C} : \mathrm{H}$$
$$\qquad\qquad\qquad\qquad\ \mathrm{H}$$

Figure 1a. Electron configurations of boron and carbon.

Bernard F. Spielvogel

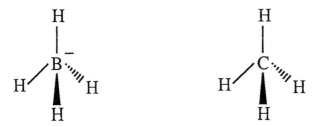

Figure 1b. Isoelectroinic and isostructural borohydride (BH_4^-) and methane.

However, we can see that the boron species has an overall negative charge while the carbon species is electrically neutral (Figure 1b).
Chemists have long made analogies between boron compounds and carbon compounds. Frequently, though, these analogies have involved typical organic compounds and inorganic compounds based on three-coordinate carbon and boron. While of chemical and theoretical interest, the three-coordinate boron compounds are usually quite sensitive to water, hydrolyzing rapidly to give boric acid (Figure 2). Therefore, biologically, the three-coordinate boron compounds are not useful except possibly as precursors to the biologically relevant boric acid.

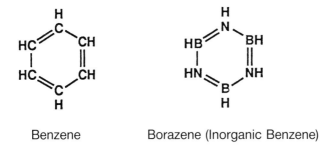

Benzene   Borazene (Inorganic Benzene)

Figure 2. Three coordinate compounds of carbon and boron: benzene and inorganic borazene.

At BBI, we have advanced and exploited the concept that "molecules of life" built around four-coordinate boron would have considerable biological activity because of their close structural relationship to four-coordinate carbon compounds. Four-coordinate boron compounds, like their carbon counterparts, are hydrolytically and oxidatively stable which is essential for their survival and utilization in biological systems.

Our first efforts were directed toward the synthesis of the boron analogue of glycine. Glycine is the simplest amino acid and was first isolated and studied many decades ago. We synthesized the first isoelectronic and isostructural boron analogue of glycine (Figure 3) and indeed found it to be biologically active (Spielvogel et al., 1980). We found some important similarities as well as some important differences

305

Choose a new path: boron biologicals

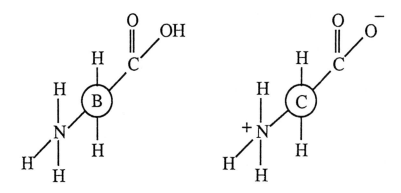

Figure 3. Boroglycine™ and Glycine (Schematic Representations).

between glycine and its boronated counterpart. The results were similar for most amino acids we have boronated. Both glycine and boronated glycine (Boroglycine™) are white crystalline solids at room temperature, soluble in H_2O, and nearly identical in molecular weight (75.07 and 74.88, respectively). However, Boroglycine™ has several unique properties which merit exploitation. Boroglycine™ can more easily cross membrane barriers because it is uncharged. It has a six-fold higher pK than glycine, and it is remarkably stable in air and water.

From that beginning, we have now prepared hundreds of boron analogues of biologically important molecules. More than fifty patents and patents pending have resulted, most of which cover composition of matter. The important classes include: analogues of α–amino acids and their amides and esters; peptides; cholines and thiocholines; phosphonates; and nucleosides and nucleotides including DNA and will be discussed later.

Nuclear properties of boron

The nuclear properties of boron represent a second area of keen interest to BBI. Naturally occurring boron consists of two *non-radioactive* isotopes of boron, boron-10 and boron-11. The boron-10 isotope makes up about 20% of this mixture. Both nuclei possess magnetic moments and structure and bonding in boron compounds is facilitated by NMR.

Boron-10 has a profound tendency to "capture" a neutron. This propensity for neutron capture is what makes boron so valuable in the nuclear industry both in nuclear reactors and in nuclear weapons. However, this same neutron-capturing ability of boron-10 can also be applied to the treatment of cancer. This form of therapy, termed Boron Neutron Capture Therapy (BNCT) consists of localization of boron-10 in tumor cells, followed by irradiation by non-ionizing thermal neutrons. Upon capture of neutrons by B-10, the nuclear reaction $^{10}B(n,\alpha.)^7Li$ is unleashed in the tumor cells, killing them but sparing nearby normal cells. This aspect will not be discussed in this article but the interested reader is referred to a recent book on the subject (Soloway et al., 1993).

INORGANIC BORON IN BIOLOGY

In nature, boron almost exclusively exists as three-coordinate or four-coordinate boron oxygen compounds and complexes. Such species have a propensity to form complexes with hydroxy species in biological compounds. Boric acid itself spontaneously ionizes to form the hydroxyborate ion (Edwards and Ross, 1967):

$$B(OH)_3 + 2H_2O \rightarrow B(OH)_4^- + H_3O^+$$

The equilibria between boric acid, monoborate ions, and polyborates in aqueous solution is rapidly reversible. Because of the rapid equilibrium, boric acid forms many complexes, some quite strong, especially with polyhydroxy biological compounds such as certain sugars, polysaccharides, (ribo)nucleic acids, etc. Boric acid can readily form strong monocyclic and bicyclic ring complexes with certain molecules containing cis-hydroxy groups as shown in Scheme 1.

Plants
Numerous publications and reviews exist on the role of boron in plants. Absence of boron has been shown to affect nucleic acid (especially RNA processes) and carbohydrate metabolism; and there is also a relationship between boron, auxin, and phenolic compounds (Pilbeam and Kirby, 1983). Boron is essential to maintain the structural integrity of plant membranes and boron deficiency in plants gives rise to many symptoms which are actually the result of membrane permeability changes.

Only two biologically synthesized compounds containing boron are known, aplasmomycin and boromycin, both antibiotics.

Animal and human experiments
Although boron had been generally accepted as essential for plants, it was not until 1981 that conclusive evidence was obtained that boron deprivation depressed growth and elevated plasma alkaline phosphatase activity in chicks with inadequate Vitamin D_3 (Hunt and Nielson, 1981). From studies carried out on both chicks and rats, Hunt (1993) has suggested that boron has an effect on at least three separate metabolic sites. These conclusions were reached because dietary boron supplementation compensated for perturbation in energy substrate utilization induced by vitamin D_3 deficiency, enhanced the mineral content in bone in concert with vitamin D_3, and enhanced some indices of growth cartilage maturation independently of vitamin D_3.

One of the few toxic effects studied in detail is the reproductive toxicity of boric acid in rodents (Chapin and Ku, 1992). High dose boric acid exposure produces testicular lesions in adult rats characterized by inhibited spermiation that may progress to atrophy.

Neilsen et al. (1987) carried out the first nutritional study involving boron with humans. In a metabolic unit, 12 post-menopausal women were fed a diet that provided 0.25 mg B/2000 kcal for 119 days, then were placed on the same diet with boron supplementation at 3 mg/day for 48 days. With boron supplementation, the total plasma concentration

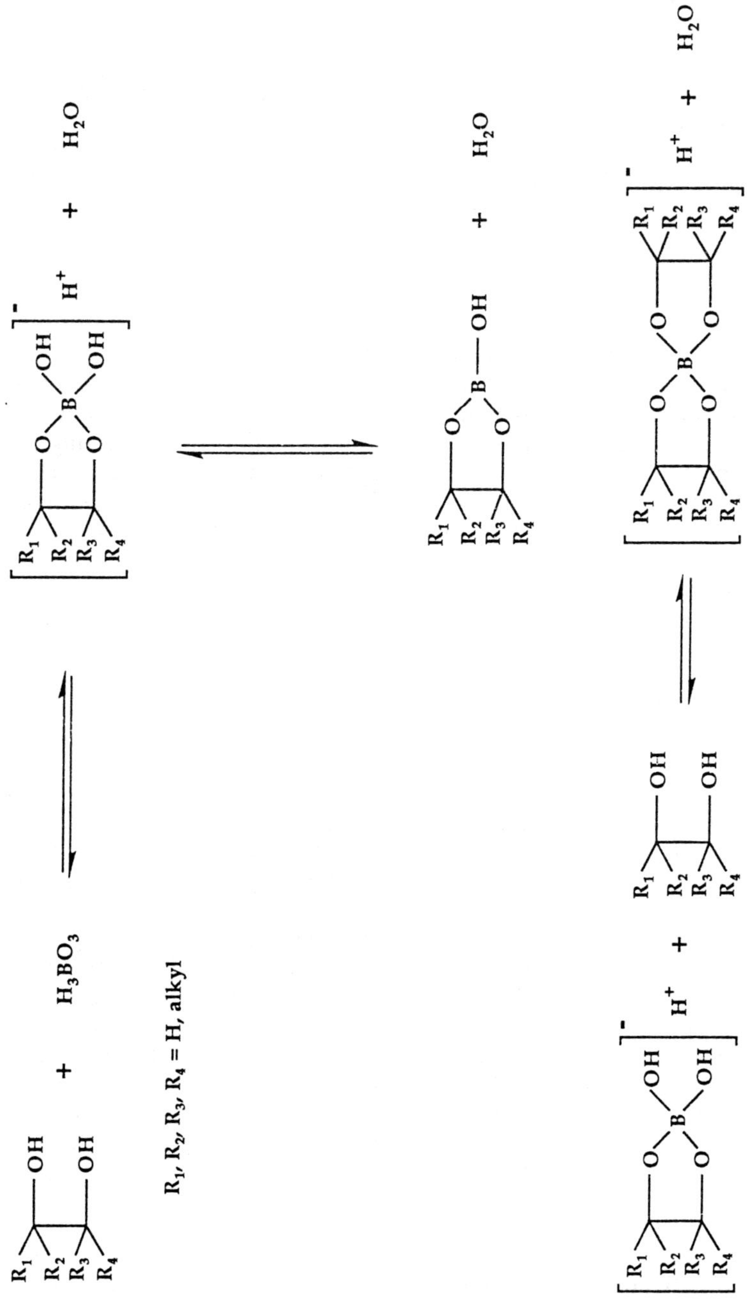

Scheme 1.

of Ca was reduced as was the urinary excretion of Ca and Mg. Serum concentration of β–estradiol and testosterone were elevated.

Two additional studies were carried out involving men, post-menopausal women, some receiving estrogen and some not (Nielson, 1993). The results of these studies indicated that boron can both enhance and mimic some effects of estrogen ingestion. Boron deprivation was shown to increase platelet counts, red blood cell counts, and hematocrit. The mean corpuscular hemoglobin content and blood hemoglobin concentration were reduced. Thus, boron appears to have an essential function that affects macromineral and cellular metabolism.

BIOLOGICALLY ACTIVE ORGANIC BORON

Background: organic vs. inorganic trace minerals
In the Proceedings of Alltech's Ninth Annual Symposium, the advantages of using organic or complex mineral species rather than inorganic were discussed. An entire section was devoted to this topic indicating how incorporation of a mineral into an organic form or complex affects its biological availability, biological activity, toxicity, biodistribution, etc.

The biological activity of inorganic boron in animals, although of considerable potential importance in nutrition, will most likely be limited compared to that of organic boron. Additionally, it will be more difficult to determine the mechanism of action of inorganic boron at the molecular level. As discussed above, boric acid and borate complexes undergo rapid equilibrium reactions with water and hydroxy and polyhydroxy compounds. These equilibria vary with pH. Thus, it becomes very difficult to determine and unravel all the biochemical functions of the boron species involved. For organic boron, especially of 4–coordinate analogues of biomolecules, the behavior of the molecules can be expected to be much more like their organic counterparts. Thus, the study of organic boronated biomolecules should be greatly facilitated by application of the well developed methodology used to elucidate the behavior of organic biomolecules.

Another analogy contrasting the biological activity of inorganic vs organic boron would be to compare the limited biological activity of C in species such as CO_2, H_2CO_3, and carbonates with the vast scope and biological activity of C in its chemical or organic compounds. Indeed, although studies of organic boron analogues of biomolecules are yet in their infancy, the breadth of the biological activity already found is quite large and will be briefly reviewed below after a survey of the synthetic approaches to boron analogues of biomolecules is presented. The description of syntheses also illustrates the wide variety of species that can be made.

SYNTHESIS OF BORON ANALOGUES OF BIOMOLECULES

The following is a brief review of synthetic routes to boron analogues and some of their chemical properties. Detailed synthetic procedures can be found in the references cited. Many additional syntheses are also described in the references cited under biological and pharmacological behavior.

Amino acid analogues
The first amino acid analogue we prepared was Borobetaine™ (trimethylamine-carboxyborane), Me_3NBH_2COOH (Spielvogel et al., 1976). It was prepared from Me_3NBH_2CN by alkylation of the cyano nitrogen with $Et_3O^+BF_4^-$ followed by hydrolysis. Borobetaine™ is air stable and hydrolytically very stable. For example, no changes in its ir spectra were noted after exposure to the atmosphere for two months. No decomposition of an aqueous solution was noted after a period of one month. In 1N HCl, it decomposed about 50% after three weeks. Boroglycine™, H_3NBH_2COOH, can be prepared from Borobetaine™ by amine exchange. (Spielvogel et al., 1976)

The other N-methylglycine analogues have also been prepared by amine exchange (Spielvogel, 1980). Although all the N-methylamines are less hydrolytically stable, they possess sufficient stability to carry out biological experiments.

Replacement of the α–carbon in an amino acid by boron results in a six fold increase in pKa (Scheller et al., 1982). In Table 1, the pKa's for glycine and its N-methyl glycine analogues are shown. The pKa for the carboxylic acid group is six fold higher than in the corresponding amino acids (pKa~2). The pKa for deprotonation of NH is greater than 11. Thus, the boron amino acids are very weak acids and exist primarily in the molecular form. Such molecular species should cross cell membranes more readily than the dipolar normal amino acids.

Table 1. Acid strength of amine-carboxyboranes.

Amino acid	pK_1	pK_2	Boron analog	pK_1	pK_2
$H_3N^+CH_2CO_2^-$	2.36	9.6	$H_3NBH_2CO_2H$	8.33	>11
$MeN^+H_2CH_2CO_2^-$	2.14		$MeNH_2BH_2CO_2H$	8.23	
$Me_2N^+HCH_2CO_2^-$	1.94		$Me_2NHBH_2CO_2H$	8.14	
$Me_3N^+CH_2CO_2^-$	1.83		$Me_3NBH_2CO_2H$	8.38	

Derivatization of amine-carboxyboranes to their esters is readily achieved (Spielvogel et al., 1984a; Spielvogel et al., 1986a). Amide derivatives (Spielvogel et al., 1984b) as well as boron analogues of common amino acids, e.g. $H_3NBH(CH_3)C(O)NHEt$ (Sood and Spielvogel, 1989b) have been prepared.

Boron containing peptides
Typical condensation methods can be used to prepare peptides at the carboxy terminus of amine-carboxyboranes. Although condensing agents such as dicyclohexyl-carbodiimide can be used, the yield is quite low and a condensation method using Ph_3P and CCl_4, similar to that used for coupling normal amino acids, produces good yields of di- and tri-

peptides (Sood et al., 1990). Because of the very high pKa of the NH on boroamino acids, typical peptide bond formation procedures on the amine terminus are unsatisfactory and suitable preparative methods are under development. Examples of the variety of dipeptides that can be prepared and method of synthesis are shown below in Scheme 2.

$$Me_3NBH_2COOH + HCl \cdot NH_2CH(R)COOR \xrightarrow[NEt_3/CH_3CN/RT]{PPh_3CCl_4} Me_3NBH_2C(O)NHCH(R)COOR' \quad (1)$$

1	R	R'
a	H	CH_3
b	CH_3	t-Bu
c	$CH(CH_3)_2$	CH_3
d	$CH_2CH(CH_3)_2$	CH_3

1	R	R'
e	$CH(CH_3)CH_2CH_3$	CH_3
f	CH_2OH	CH_3
g	$p\text{-}CH_2C_6H_4OH$	CH_3
h	$CH_2CH_2SCH_3$	CH_3

$$(1) + NH_3 \xrightarrow[\text{or } 58\,°C/15\,hrs]{RT/3\,weeks} NH_3BH_2C(O)NHCH(R)COX \quad (2)$$

2	R	X
a	H	NH_2
b	$CH(CH_3)_2$	OCH_3
c	$CH_2CH(CH_3)_2$	NH_2

2	R	X
d	$CH(CH_3)CH_2CH_3$	OCH_3
e	CH_2OH	NH_2
f	$CH_2CH_2SCH_3$	NH_2

Scheme 2. Method of synthesis for a variety of boronated dipeptides.

Boron analogues of amino acids and peptides are generally stable in water and have amphiphilic character, i.e., they are soluble in water as well as most organic solvents and lipids. This amphiphilic character may be useful for the transport of these compounds across cell membranes. Studies with $Me_3NBH_2C(O)NHCH(Ph)C(O)OMe$, a dipeptide of borobetaine and phenylalanine, indicate that boronated dipeptides can cross cell membranes without hydrolysis of the amide bond as is found with the majority of normal peptides (Elkins et al., 1993). Such stability may be useful in overcoming problems associated with peptide or protein drug delivery.

Boron analogues of neurotransmitters
The acetylcholine (ACH) cation, $H_3CN^+(Me)_2CH_2CH_2OC(O)Me$, is one of the most important neurotransmitters. We have made an isoelectronic and isostructural boron analogue of ACh (Spielvogel et al., 1986b). This analogue, $H_3BNMe_2CH_2CH_2OC(O)Me$, is a molecular compound since the negative charge on boron is cancelled by the formal positive charge on nitrogen. This analogue can be easily made in high yield in a one pot synthesis.

Boron analogues of phosphonates

Since phosphate and phosphonate groups are present in a vast variety of biologically important molecules, e.g., DNA, phospholipids, etc., and synthetic phosphonates have been found to possess significant antiviral activity, we set out to prepare a variety of molecules with at lease one P-B bond (Sood et al., 1991a). Such derivatives may be considered as boron analogues of phosphonates. Scheme 3 presents the overall preparative scheme used to obtain $(EtO)_3PBH_2COOH$, **3** which is a boron analogue of phosphonoacetic acid. The synthesis begins essentially with $(EtO)_3PBHCN$, **1** which is readily prepared as shown. Attempts to prepare **3** directly from the N-ethylated derivative were not successful but it can be readily prepared from the amide **2**. A variety of derivatives of **3** of both the carboxylic acid and the POR (e.g. conversion to POH) group have been prepared and characterized.

Heterocyclic amine derivatives

To carry out structure activity relationships, we have prepared and characterized a large variety of heterocyclic amine adducts of BH_2CN, $BH_2C(O)NEtH$, BH_2COOH, and BH_2COOR (Sood et al., 1991c; Hall et al., 1991a). Two methods were used (Scheme 4). The heterocyclic amine (saturated and unsaturated) derivatives included imidazoles, piperazines, pyridines, etc.

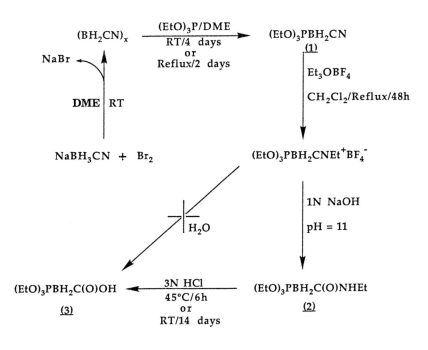

Scheme 3. Preparative scheme for a boron analogue of a phosphonate.

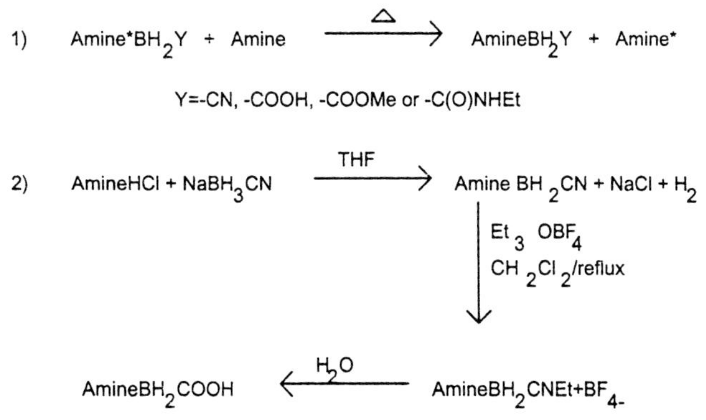

1) Amine·BH$_2$Y + Amine $\xrightarrow{\Delta}$ AmineBH$_2$Y + Amine·

    Y= -CN, -COOH, -COOMe or -C(O)NHEt

2) AmineHCl + NaBH$_3$CN $\xrightarrow{\text{THF}}$ Amine BH$_2$CN + NaCl + H$_2$

$\Big\downarrow$ Et$_3$OBF$_4$ / CH$_2$Cl$_2$/reflux

AmineBH$_2$COOH $\xleftarrow{\text{H}_2\text{O}}$ AmineBH$_2$CNEt+BF$_4^-$

Scheme 4. Methods used for preparing heterocyclic adducts of BH$_2$CN, BH$_2$C(O)NEtH, BH$_2$COOH, AND BH$_2$COOR.

Nucleoside and nucleic acid derivatives

With experience gained in the preparation of boron analogues of phosphonates and heterocyclic amines, attention was focused on related very biologically relevant species, nucleosides, and nucleic acids. The boronated (BH$_2$CN) nucleosides were prepared by an exchange reaction of silylated nucleosides with triphenylphosphine-cyanoborane followed by deprotection of the silylated products (Sood et al., 1989b). The compounds prepared and characterized are shown in Figure 4.

All of the nucleoside cyanoboranes except Ino* are quite stable to hydrolysis (>94% over a one week period). Ino* decomposes by 50% after 1 week. Variable temperature ^{1}H nmr studies on a Gua*.Cyt base pair showed the H-bonding to be approximately as strong as in a normal Gua.Cyt base pair, which indicates that when the coordination site for

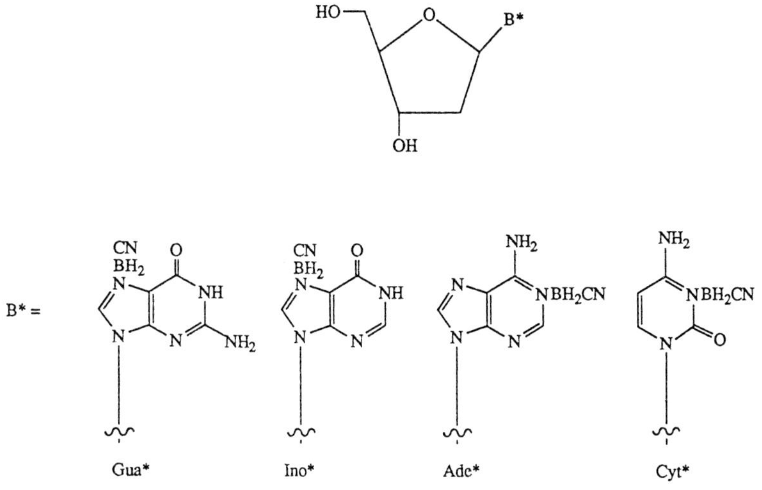

Figure 4. Nucleoside cyanoboranes.

BH$_2$CN (i.e. Gua* and Ino*) is away from the sites required for Watson-Crick base pairing, that normal base pairing is feasible.

The "boronophosphate" oligonucleotides (Scheme 5) are prepared by formation of an intermediate phosphite which is then converted into the phosphite-borane by reaction with Me$_2$SBH$_3$ (Sood et al., 1990b). Treatment with base gives the boranophosphate.

The hydrolytic and nuclease stability is of considerable importance for use of these species in antisense therapy and for BNCT. We have found

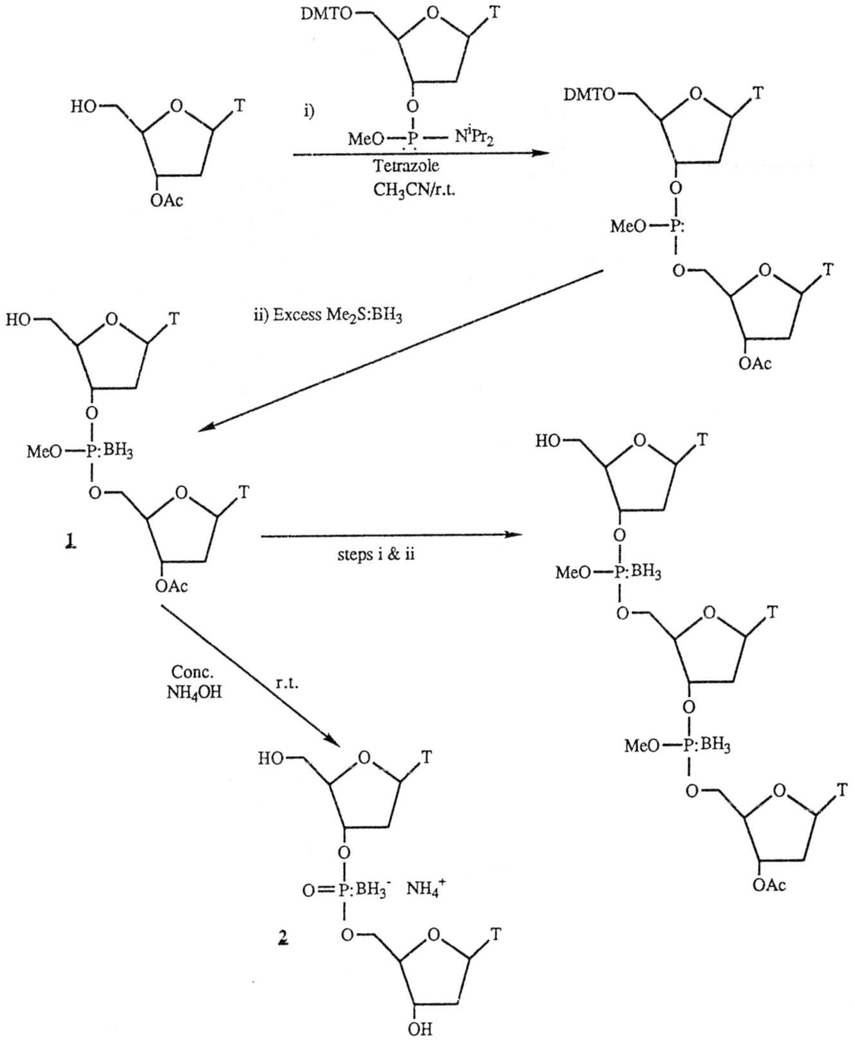

Scheme 5. Preparation of boronophosphate oligonucleotides.

Bernard F. Spielvogel

that the internucleotide boranophosphate group is remarkably stable to basic and acidic hydrolysis and is also quite stable to nucleases (Sood et al., 1990b; Shaw et al., 1993).

Thus, heating 1 at 55°C overnight in concentrated NH_4OH (conditions used for deprotection of bases in normal oligonucleotide synthesis) does not result in any change other than the deprotection described above. The boranophosphate group is also remarkably stable under acidic conditions. When 1 is shaken at room temperature overnight in a mixture of 1N HCl and MeOH (1:1 v/v), 10% of the boranophosphate group is converted into phosphate (by ^{11}B and ^{31}P NMR). The 3'-acetate group is hydrolyzed although the POMe group appears to remain intact (by ^1H NMR).

Finally, the boranophosphate internucleotide linkage in dimer 2 is quite stable toward cleavage by both calf spleen phosphodiesterase and snake venom phosphodiesterase. Thus, under the conditions where normal dithymidyl phosphate is 97% cleaved, dimer 2 is 92% stable.

Related to the species in Figure 4 and Scheme 5 are nucleoside triphosphates that have been recently prepared (Spielvogel et al, 1993; Tomasz et al, 1992) and are shown in Figure 5.

Both 1 and 2 are substrates for DNA polymerases and can be enzymatically incorporated into DNA. Indeed, 5'-αP-borane triphosphate of all four nucleoside bases T, A, C, G, have been prepared and can be incorporated, base specifically, into DNA during the polymerase chain reaction (PCR) and may simplify PCR sequencing (Porter et al., 1993).

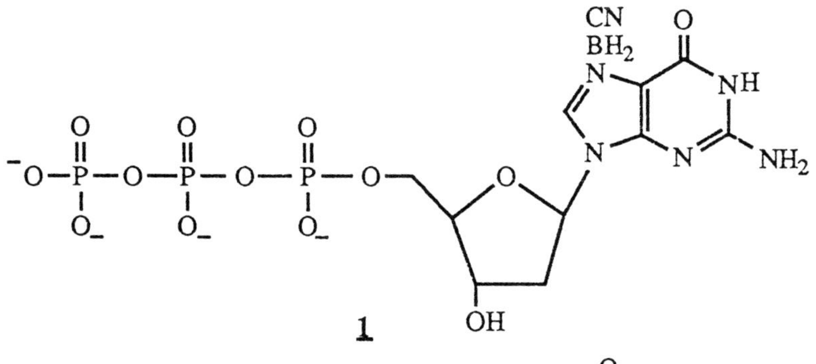

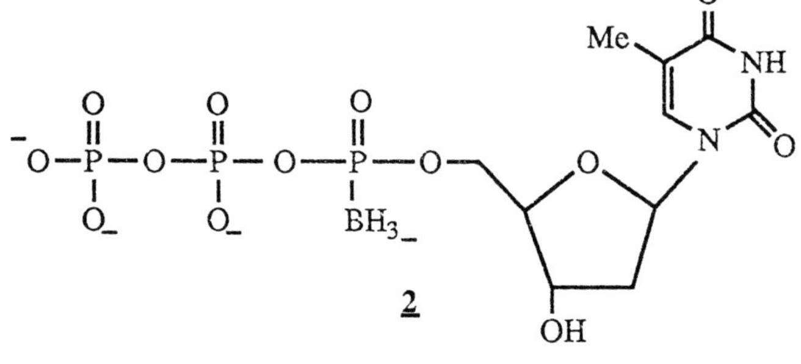

Figure 5. N^7-Cyanoborane-2'-deoxyguanosine 5'-triphosphate (1), –5'-P-borane thymidine triphosphate (2).

BIOLOGICAL AND PHARMACOLOGICAL BEHAVIOR

Hypolipidemic activity
Boron analogues of selected biomolecules and their derivatives cause significant reduction of serum cholesterol and triglyceride levels in rodents (Hall et al., 1981; 1987; 1989; 1991a; 1991b; 1992b; 1992e; Sood et al., 1991) These compounds reduce LDL and VLDL cholesterol levels while elevating HDL cholesterol. This type of cholesterol modulation has been shown to protect humans against myocardial infarctions. Based on animal studies, these compounds are very promising hypolipidemic agents, comparable or superior to currently available drugs. For example, in rats, trimethylamine-boranecarboxylic acid methyl ester reduces serum cholesterol by 37% (as compared to control) and serum triglycerides by 34%. In contrast, lovastatin, a drug used clinically, reduces cholesterol by only 18% and triglycerides by 14%. Additionally, the effect on high density lipoproteins (HDL, "good cholesterol") is also markedly different. Trimethylamine-boranecarboxylic acid methyl ester increases HDL to 195% of control while lovastatin increases HDL to only 127%. These dramatic effects on reduction of lipids appear to be via inhibition of Acyl CoA cholesterol acyl transferase, sn-glycerol-3–phosphate acyl transferase, and stimulation of neutral cholesterol ester hydrolase. Thus, less cholesterol ester is stored in the aorta wall.

Anti-inflammatory activity
Boron analogues have expressed potent anti-inflammatory, anti-arthritic, antipleurisy, and analgesic activity in animal model studies in rodents (Hall et al., 1980; 1993; Sood et al., 1990a; 1992a). With activities similar to the commercial drug indomethacin, select boron analogues are potent inhibitors of induced edema in mice and rats. These compounds, however, have a much larger therapeutic index than indomethacin. For example, methylamine-boranecarboxylic acid (borosarcosine™) at 8 mg/kg (×2) results in 46% inhibition of the carrageenan-induced edema in mice and has an LD_{50} of >1000 mg/kg. Indomethacin, on the other hand, although somewhat better in inhibiting induced edema (78% inhibition) at 10 mg/kg (×2), has an LD_{50} of only 28 mg/kg in mice. They also inhibit prostaglandin synthesis and lysosomal enzyme activities in PMNs, macrophages, and fibroblasts. Modification of cytokine behavior is another area of activity found for boron analogues.

Antineoplastic activity
Boron analogues of amino acids and peptides have shown significant cytotoxic activity against a number of murine and human tumor cell lines, and additionally, in vivo antitumor activity has been demonstrated in a variety of tumors (Hall et al., 1992a; 1979; 1984; 1985; 1991b; 1992c; 1992d; Sood et al., 1990a; 1991& 1991c; 1992; Spielvogel et al., 1991). Activities observed include inhibition of nuclear DNA polymerase, 5–phosphoriboxyl-l-pyrophosphate (PRPP) amidotransferase, dihydrofolate reductase, and Topo II isomerase and other activities as described in the cited references.

Osteoporosis

Osteoporosis is associated with reduced bone volume leading to increased frequency of bone fractures. This process is due to a metabolic imbalance between rates of new bone formation and bone resorption. The process of bone resorption is divided into two concurrent processes. Phase I involves inorganic mineral metabolism conducted by osteoclasts, macrophages, monocytes, PMNs, and fibroblasts. Phase II involves organic metabolism where proteolytic destruction of the bone matrix collagen releases hydroxyproline to the extracellular compartment.

Evidence that boron, as inorganic borate, at 3 mg/kg/day in humans may prevent osteoporosis has been presented (Nielson, 1993). We have found evidence that select boron analogues of biomolecules increases the rate of bone formation and decreases bone resorption (Hall et al., 1993) Incorporation of B into biomolecules improves solubility, bioavailability, and transport into the cell or across natural barriers in the body.

Toxicology

Recently, the first International Symposium on Health Effects of Boron and Its Compounds was held and the proceedings published (1994 Spring Supplement, Environmental Health Perspectives). Most data available is on inorganic borate, and it appears that there is a wide safety margin for inorganic boron for humans. Indeed, it is now clear that very small amounts of boron have primarily beneficial effects.

Acute toxicology studies on boron analogues we have prepared (Hall et al., 1991c; 1992c) have shown no toxicity at 1, 3, 5 times the therapeutic dose. Thus, the boron agents tested are safe in their therapeutic range based on organ weights, histological tissue sections, clinical chemistry, and hematopoietic parameters. Additional studies at much higher doses are underway to uncover possible toxic effects.

In general, simply on the basis of LD_{50} values in mice, toxicity varies with the substituents attached to boron and for BH containing compounds, with the reducing power. The LD_{50} (mice) for the following series illustrates this (Table 2).

Table 2. LD_{50} dosages of boron and BH-containing compounds.

Compound	H_3BCN^-	BH_4^-	Me_3NBH_2CN	$Me_3NBH_2C(O)NEtH$	Me_3NBH_2COOH
LD_{50} (mg/kg)	<1	~50	70	320	>1800

Conclusions

Boron, in either the inorganic or organic form has now been shown to possess biological activity in animals. Boron, in the inorganic form, appears to be an essential trace element for animals, including humans. Organic boron in the form of boron analogues of biomolecules has been shown to have a very broad range of biological activity, including potentially valuable pharmacological value on the basis of animal model studies. Boron analogues have been shown to have potent cellular activity with effects on enzymes and cell receptors. Some boronated molecules inhibit enzymes while others are actually efficient enzyme substrates. Many other kinds of biological and pharmacological activity (antiviral) are being discovered and will be reported in the future.

The opportunity to optimize specific activity through structure activity relationship studies, as in organic chemistry, is clearly present in these organic boron compounds as opposed to inorganic boron. Boron analogues of biomolecules represent an entirely new approach to the study and expression of biological activity and promises to impact many areas of technology.

ACKNOWLEDGEMENTS

The author is greately indebted to his principal collaborators: Dr. Anup Sood, Boron Biologicals, Inc., Dr. Iris H. Hall, The University of North Carolina; Dr. Barbara Shaw, Duke University. Support by the NIH, DOE, and N.C. Biotechnology Center is gratefully acknowledged.

References

Chapin, R.E. and W.W. Ku. 1992. The reproductive toxicity of boric acid. Abstracts, Intl. Symp. on Health Effects of Boron and Its Compounds. Published 1994 Spring Supplement, Environmental Health Perspectives.

Edward, J.O. and V.F. Ross. 1967. The structural chemistry of the borates, Chapt. 3 in Muertties, E. (Ed) 1967. The Chemistry of Boron and Its Compounds. John Wiley and Sons, New York.

Elkins, A.L., M. Cho, R.P. Shrewsbury, A. Sood, B.F. Spielvogel, I.H. Hall and M. C. Miller, III, Eighth Intl. Meeting on Boron Chemistry, Knoxville, TN, Abstracts, p. 103, July 11–15, 1993.

Hall, I. H., E. S. Hall, L. K. Chi, B. R. Shaw, A. Sood and B. F. Spielvogel. 1992a. Antineoplastic activity of boron-containing thymidine nucleosides in Tmolt3 leukemic cells. Anticancer Res. 12:1091.

Hall, I.H., A. Sood and B.F. Spielvogel. 1991a. The hypolipidemic activity of heterocyclic amine-boranes in rodents. Biomed & Pharmacother 45:333.

Hall, I.H., M.K. Das, F. Harchelroad, Jr., P. Wisian-Neilson, A.T. McPhail and B.F. Spielvogel. 1981. The antihyperlipidemic activity of amine cyano- and carboxyboranes and related compounds. J. Pharm. Sci. 70:339.

Hall, I.H., W. Williams, C.J. Gilbert, A.T. McPhail, B.F. Spielvogel. 1984. The hypolipidemic activity of tetrakis-μ–(trimethylamine-borane-carboxylato)-bis trimethylamine-carboxyborane)dicopper(II) in rodents and its effect on lipid metabolism. J. Pharm. Sci. 73:973.

Hall, I.H., B.F. Spielvogel, A. Sood, F. Ahmed and S. Jafri. 1987. Hypolipidemic activity of trimethylamine-carbomethoxyborane and related derivatives in rodents. J. Pharm. Sci. 76:359.
Hall, I.H., B.F. Spielvogel, T.S. Griffin, E.L. Docks and R.J. Brotherton. 1989. The effects of boron hypolipidemic agents on LDL and HDL receptor binding and related enzyme activities of rat hepatocytes, aorta cells and human fibroblasts. Research Communications in Chemical Pathology and Pharmacology 65:297.
Hall, I.H., O.T. Wong, A. Sood, C.K. Sood, B.F. Spielvogel, R.P. Shrewsbury and K.W. Morse. 1992b. Hypolipidaemic activity in rodents of boron analogs of phosphonoacetates and cyanoborane adducts of dialkyl aminomethylphosphonates. Pharmacological Research 3:259.
Hall, I.H., C.O. Starnes, A.T. McPhail, P. Wisian-Neilson, M.K. Das, F. Harchelroad, Jr. and B.F. Spielvogel. 1980. The anti-inflammatory activity of amine-cyanoboranes, amine-carboxyboranes and related compounds. J. Pharm. Sci. 69:1025.
Hall, I.H., C.O. Starnes, B.F. Spielvogel, P. Wisian-Neilson, M.K. Das and L. Wojnowich. 1979. Boron betaine analogs: antitumor activity and effects on Ehrlich ascites tumor cell metabolism. J. Pharm. Sci. 68:685.
Hall, I.H., B.F. Spielvogel and A.T. McPhail. 1984. Antineoplastic activity of tetrakis-µ–(trimethylamine-boranecarboxylato)-bis(trimethylamine-carboxyborane) dicopper(II) in Ehrlich ascites carcinoma. J. Pharm. Sci. 73:222.
Hall, I.H., C.J. Gilbert, A.T. McPhail, K.W. Morse and B.F. Spielvogel. 1985. The antineoplastic activity of a series of boron analogues of amino acids. J. Pharm. Sci. 24:765.
Hall, I.H., B.F. Spielvogel and A. Sood. 1991b. The antineoplastic activity of trimethylamine carboxyboranes and related esters and amides in murine and human tumor cell lines. Anti-Cancer Drugs 1:133.
Hall, I.H., E.S. Hall, L.K. Chi, M.C. Miller, III, K.F. Bastow, A. Sood and B.F. Spielvogel. 1992c. The anti-neoplastic activity of ethylamine-carboxyborane and triphenylphosphine-carboxyborane in L_{1210} lymphoid leukemia cells. Applied Organometallic Chemistry 6:229.
Hall, I.H., E.S. Hall, L.K. Chi, B.R. Shaw, A. Sood and B.F. Spielvogel. 1992d. Antineoplastic activity of boron-containing thymidine nucleosides in $Tmolt_3$ leukemic cells. Anticancer Research 12:1091.
Hall, I.H., S.Y. Chen, K.G. Rajendran, A. Sood, B.F. Spielvogel and J. Shih. 1992e. Hypolipidemic, anti-obesity, anti-inflammatory, antiosteoporotic, and anti-neoplastic properties of amine-carboxyboranes, Abstracts, Intl. Symp. on Health Effects of Boron and Its Compounds (in press).
Hall, I.H., D.J. Reynolds, J. Chang, B.F. Spielvogel, T.S. Griffin and E.L. Docks. 1991c. Acute toxicity of amine-boranes and related derivatives in mice. Arch. Pharm. (Weinheim) 324:573.
Hunt, C.D. 1993. The biochemical effects of physiological amounts of

boron in animals in Abstracts, Intl. Symp. on Health Effects of Boron and Its Compounds (in press).
Muertterties, E. (Ed.) 1967. The Chemistry of Boron and its Compounds. John Wiley and Sons, New York.
Nielsen, F.H. 1988. Boron – an overlooked element of potential nutrition importance. Nutrition Today, Jan/Feb p. 4.
Nielsen, F.H. 1993. Abstracts, Intl. Symp. on Health Effects of Boron and Its Compounds. To be published in Environmental Health Perspectives 1994 Spring Supplement.
Pilbeam, D.D. and E.A. Kirkby. 1983. The physiological role of boron in plants. J. of Plant Nutrition 6:563.
Porter, K., D. Briley, F. Huang, A. Sood, B.F. Spielvogel and B.R. Shaw. One-step PCR sequencing, Genome Sequencing and Analysis Conference V, Abstracts, Hilton Head Island, S.C., Oct. 23–27, 1993.
Scheller, K.H., R.B. Martin, B.F. Spielvogel and A.T. McPhail. 1982. Basicity and metal ion binding capability of amine-carboxyboranes R_3NBH_2COOH, boron analogs of glycine and N-methylated glycines. Inorg. Chemica Acta 57:227.
Shaw, B. Ramsay, J. Madison, A. Sood and B.F. Spielvogel. 1993. Oligonucleoside boranophosphate In: Methods in Molecular Biology. S. Agrawal (Ed.) Vol. 20, Humana Press Inc., Totowa, N.J., 225.
Soloway, A.H., R.F. Barth and D.E. Carpenter (Eds.) *Advances in Neutron Capture Therapy*, Plenum Press, New York, N.Y., 1993. This is the published proceedings of the Fifth International Symposium on Neutron Capture Therapy, held Sept. 14–17, 1992 in Columbus, Ohio.
Sood, A., B.F. Spielvogel. 1989a. Boron analogus of amino acids VII. Synthesis of ammonia-(N-ethylcarbomoyl)methylborane, $H_3NBH(CH_3)C(O)NHEt$, a boron analogue of the N-ethylamide of alanine. Main Group Metal Chemistry 12:143.
Sood, A., C.K. Sood, B.F. Spielvogel, I.H. Hall. 1990a. Boron analogues of amino acids VI. Synthesis and characteristics of di- and tripeptide analogues as antineoplastic, anti-inflammatory and hypolipidemic agents. European Journal of Medicinal Chemistry 25:301.
Sood, A., C.K. Sood, I.H. Hall and B.F. Spielvogel. 1991a. Characterization and antitumor properties of sodium diethylphosphite-carboxyborane and related compounds. Tetrahedron No. 34 47:6915.
Sood, A., B.R. Shaw and B.F. Spielvogel. 1989b. Boron containing nucleic acids: Synthesis of cyanoborane adducts of 2′-deoxynucleosides. J. Amer. Chem. Soc. 111:9234.
Sood, A., B.R. Shaw and B.F. Spielvogel. 1990b. Boron containing nucleic acids 2: Synthesis of oligodeoxynucleoside boranophosphates. J. Am. Chem. Soc. 112:9000.
Sood, A., C.K. Sood, B.F. Spielvogel, I.H. Hall and O.T. Wong. 1992a. Synthesis, cytotoxicity, hypolipidemic and anti-inflammatory activities of amine-boranes and esters of boron analogues of choline and thiocholine. J. of Pharm. Sci. 81:458.
Sood, A., C.K. Sood, B.F. Spielvogel, I.H. Hall and O.T. Wong. 1991b.

Synthesis and hypolipidemic activity of amine-carboxyboranes, and their amides and esters in rodents. Arch der Pharm 324:423.
Sood, A., B.F. Spielvogel, B.R. Shaw, L.D. Carlton, B.S. Burnham, E.S. Hall and I.H. Hall. 1992b. The xynthesis and antineoplastic activity of 2'-deoxynucleoside- cyanoboranes in murine and human culture cells. Anticancer Research 12:235.
Sood, C.K., A. Sood, B.F. Spielvogel, J.A. Yousef, B. Burnham and I.H. Hall. 1991c. Synthesis and antineoplastic activity of some cyano-, carboxy-, carbomethoy-, and carbamoylborane adducts of heterocyclic amines. J. Pharm. Sci. No. 12 80:1133.
Spielvogel, B.F., L. Wojnowich, M.K. Das, A.T. McPhail and K.D. Hargrave. 1976. Boron analogues of amino acids. Synthesis and biological activity of boron analogues of betaine. J. Am. Chem. Soc. 96:5702.
Spielvogel, B.F., A.T. McPhail, M.K. Das and I.H. Hall. 1980a. Boron analogues of amino acids. Synthesis and biological activity of boron analogues of betaine. J. Amer. Chem. Soc. 102:6343.
Spielvogel, B.F. 1980b. Synthesis and biological activity of boron analogues of the α–amino acids and related compounds, Boron IV, R.W. Parry and G. Kodama, ed. Pergamon Press, 119.
Spielvogel, B.F., F.U. Ahmed, G.L. Silvey, P. Wisian-Neilson and A.T. McPhail. 1984a. Boron analogues of amino acids. 4. Synthesis of glycine and N-methylated glycine ester analogues. Inorg. Chem. 23:4322.
Spielvogel, B.F., F.U. Ahmed and A.T. McPhail. 1986a. Boron analogues of amino acids. 5. A simple, convenient method for the esterification of amine-carboxy-boranes. Synthesis 833.
Spielvogel, B.F., F.U. Ahmed, A.T. McPhail. 1986b. Boron analogues of choline 2. Efficient syntheses of boron analogue of choline, acetylcholine and substituted acetylcholines. Inorg. Chem. 25:4395.
Spielvogel, B.F., F.U. Ahmed, K.W. Morse and A.T. McPhail. 1984a. Boron analogues of amino acids. 3. Synthesis of (ethylcarbomoyl)borane adducts of Me_3N, Me_2N, Me_2NH, $MeNH_2$, and NH_3. Inorg. Chem. 23:1776.
Spielvogel, B.F., A. Sood, W. Powell, J. Tomasz, K. Porter and B.R. Shaw. 1993. Chemical and enzymnatic incorporation of boron into DNA. Soloway, A., R.F. Barth and D.E. Carpenter (Eds.) Advances in Neutron Capture Therapy, Plenum Press, New York, N.Y. p. 389.
Spielvogel, B.F., A. Sood, K.W. Morse, O.T. Wong and I.H. Hall 1991. The cytotoxicity of amine-cyanoboranes, amine-cyanoalkylboranes and aminomethyl-phosphonate cyanoborane adducts against the growth of murine and human tissue culture cells. Pharmazie 46:592.
Tomasz, J., B.R. Shaw, K. Porter, B.F. Spielvogel and A. Sood. 1992. 5'-P-borane-substituted thymidine monophosphate and triphosphate. Angew. Chem. Int. Ed. Engl. 31:1373.

ORGANIC SELENIUM SOURCES FOR SWINE – HOW DO THEY COMPARE WITH INORGANIC SELENIUM SOURCES?

D.C. MAHAN

Animal Science Department, Ohio State University, 2121 Fyffe Road, Columbus, OH

Introduction

Most selenium in the animal is found associated in some manner with proteinaceous tissue. However, the quantity of selenium retained and how it is utilized by the body depends on whether dietary selenium is from an inorganic or organic source.

Inorganic selenium provided as selenite or selenate can be effective in the production of glutathione peroxidase, but the absorbed oxidized element must be initially converted to the selenide (Se^o) form before selenoproteins are synthesized. Cysteine combines with selenide forming selenocysteine which serves as the biologically active precursor for the synthesis of the various selenoproteins including glutathione peroxidase. Additional inorganic selenium can be excreted through the kidney or bound by muscle tissue generally in a nonspecific manner. The magnitude of tissue retention from inorganic selenium appears limited.

Organic selenium found in feed ingredients is comprised of several organic forms, but selenomethionine generally appears to represent around 50% of the selenium in cereal grains (Olson and Palmer et al., 1976) and in a selenium-enriched yeast product produced by Alltech, Inc. (Headon, 1993). Although swine and other nonruminants require the amino acid methionine in the diet, there is no evidence that a biological need exists for selenomethionine, nor does the pig apparently distinguish between methionine or its corresponding seleno-amino acid during absorption or its incorporation into body proteins. Consequently, methionine or selenomethionine can be incorporated directly into body protein tissue in direct proportion to the availability of the two amino acid analogs from the diet. Much of the organically bound selenium in muscle tissue thus exists as a nonfunctional but storage form of the element (Burk and Hill, 1993). Subsequent tissue release of selenium is variable, but the degree of release from body tissue is dependent upon factors which affect tissue catabolism or turnover.

Selenomethionine from either dietary or tissue sources can therefore serve as a selenium source for the synthesis of glutathione peroxidase and other selenoproteins. The selenium from this seleno-amino acid must be

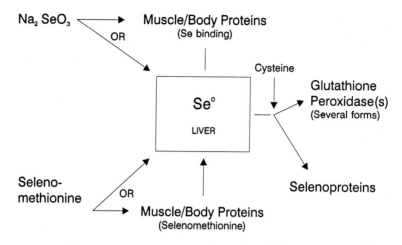

Figure 1. General pathway for selenium utilization and incorporation in tissue selenoproteins when provided from inorganic or organic selenium sources (adapted from Olson and Palmer, 1976; Burk and Hill, 1993).

released from the amino acid and the selenium converted to selenide, as with inorganic selenium, before the synthesis of selenocysteine (Olson and Palmer, 1976). A simplified pathway outlining selenium utilization from inorganic or organic selenium is presented in Figure 1.

Natural organic forms of selenium

There are clearly regional differences in the United States, Canada, and the rest of the world regarding the natural selenium content in grains and forages (Arthur, 1991; Jenkins and Winter, 1973; Young et al., 1977; Kurkela and Kääntee, 1984). Perhaps one of the better data sets depicting regional differences in the United States and the subsequent effects on tissue Se concentrations of pigs, prior to the time when dietary selenium supplementation was approved, was determined by Ku et al. (1972). Their data evaluated the effects of natural dietary selenium levels in feedstuffs grown within various regions and the resulting longissimus muscle Se concentrations when the feedstuffs were fed to pigs. The selenium content of these diets ranged from 0.027 to 0.493 ppm. The resulting selenium concentrations of the pig's longissimus muscle at market weight were higher when the higher dietary selenium levels were fed (Table 1). A correlation of 0.95 resulted between the dietary selenium levels and subsequent tissue selenium concentrations.

Because of regional differences in feed grain selenium content, the incidence of the vitamin E-selenium deficiency was regionalized in confined pigs during the 1960s and 70s. Those areas in the United States where the natural dietary selenium were lowest were found to have the highest incidence of the selenium deficiency, frequently resulting in

Table 1. Effect of dietary selenium on subsequent selenium content in longissimus muscle of market pigs.[a,b]

State	Dietary Se (ppm)	Longissimus Se (ppm)
Arkansas	0.152	0.212
Idaho	0.086	0.111
Illinois	0.036	0.059
Indiana	0.052	0.063
Iowa	0.235	0.278
Michigan	0.040	0.052
Nebraska	0.330	0.313
New York	0.036	0.046
North Dakota	0.412	0.386
South Dakota	0.493	0.521
Virginia	0.027	0.034
Wisconsin	0.178	0.125
Wyoming	0.158	0.311

[a] Diets were formulated using grain (corn, barley, wheat oat)-soybean meal mixture with the basal feeds grown within the region. The dietary Se analyses were generally considered to have reflected the geological soil and crop growing factors within the region.
[b] Source: Ku et al. (1992).

herd mortalities of 10% and morbidities as high as 25%. Areas that had the highest natural selenium level in the grains had the lowest incidence of the deficiency. It was therefore clearly recognized during the early 1970s that there was a dietary need for supplemental selenium in certain regions, whereas there was less of a need in other regions. Consequently, the level for swine was initially approved at 0.1 ppm (FDA, 1974), but then later raised to 0.3 ppm (FDA, 1987a,b). The supplemental form of selenium approved by FDA was inorganic sodium selenite or sodium selenate, and could be supplemented in addition to that which was naturally present in feed grains. The historical development of this process was recently reviewed by Ullrey (1992).

Recently (September 13, 1993), FDA announced that the 1987 approved selenium level has been stayed. The ruling effectively retains the supplemental selenium level in weanling pig diets at 0.3 ppm, but reduces the level to 0.1 ppm selenium for other phases of swine production. This ruling may return us to the preexisting deficiency conditions encountered during the 1970s, particularly in reproducing swine in areas of the United States where natural grain selenium contents are low. Although the FDA restriction relates to inorganic selenium supplementation, it has no effect on the natural selenium contributed from grains. Organic selenium provided from grain normally grown in selenium-adequate areas can be a source of selenium, but it would not be cost effective to transport the necessary quantities of grain to raise the dietary selenium level.

Using the concept of replacing the sulfur with selenium (i.e., similar chemical properties) in the sulfur-containing products synthesized during the growth of a specific yeast strain, a highly enriched selenium source has been produced by Alltech, Inc. (Nicholasville, KY). Headon (1993) characterized the compounds in the selenium-enriched yeast product and reported that over 50% of the selenium in the yeast product was present as selenomethionine with the other seleno-analogs including selenocysteine comprising the remainder of the selenium components. The efficacy of the enriched selenium-yeast product has been reported to result in higher milk selenium contents in ruminants compared with when inorganic selenium was fed (Pehrson, 1993).

Tissue retention of organic and inorganic selenium

The retention of selenium in animal tissue depends on the form of selenium provided in the diet. A rat study where titrated levels of inorganic or organic selenium were fed for a 42–day period resulted in higher whole body selenium and heart muscle selenium concentrations when selenomethionine was fed when compared with selenate (Table 2). Tissue retention of selenium increased sevenfold when graded levels of organic selenium were provided, but increased only threefold when selenate was fed (Salbe and Levander, 1990). This clearly demonstrated the higher retention of the organic form of selenium, particularly in muscle tissue. Although such a study has not been conducted with swine, there are data available which imply the same conclusion.

In a Canadian study when pig starter diets contained a naturally high (0.46 ppm) selenium content, and the grower finisher diets contained 0.78 ppm selenium, the resulting tissue selenium levels were quite high (Table 3). When 0.1 ppm selenium as selenate was added to the diets, there was no further increase in blood or tissue selenium concentrations (Table 3; Jenkins and Winter, 1973). Consequently, the inorganic selenium when added on "top" of a diet which contained a naturally high organic selenium level did not appear to be effectively retained. In agreement with this are the data of Ku et al. (1973) who fed grower-finishing swine grains which were grown either in Michigan (a recognized selenium-deficient area) or in South Dakota (a recognized selenium-adequate area) to market weight. The Michigan scientists increased the selenium content of their diet (0.04 ppm natural selenium) by the addition of inorganic selenium to the isoselenium level of the South Dakota diet. Their results

Table 2. Effect of dietary inorganic selenium or selenomethionine on tissue selenium concentration in rats[a].

Dietary Se level[b]	Selenate	Selenomethionine
	Total body Se (mcg/rat)	
0.1	33	38
0.5	41	71
2.5	93	275
	Heart Se	
0.1	0.21	0.22
0.5	0.29	0.51
2.5	0.41	1.42

[a] Salbe and Levander (1990).
[b] Fed for a 42–day period.

Table 3. Efficacy of supplemental inorganic selenium in diets with naturally high selenium contents on resulting tissue selenium in market pigs[a].

Basal feeds:	Barley, Wheat, Soybean meal, Fishmeal			
Natural Se, ppm[b]:	0.46/0.78	0.46/0.78	0.46/0.78	
Selenate, ppm[c]:	0/0	0.10/00	0.10/0.10	SE
Number of pigs	8	8	8	
Tissue Se, ppm				
Blood	0.35	0.36	0.30	0.03
Longissimus[d]	0.48	0.55	0.50	0.04
Liver[d]	0.80	0.85	0.79	0.05
Kidney[d]	2.61	2.71	2.75	0.16
Heart[d]	0.66	0.72	0.71	0.04

[a]Adapted from the data of Jenkins and Winter (1973).
[b]The starter diet analyzed 0.46 ppm Se; the grower-finisher diet analyzed 0.78 ppm Se.
[c]The addition of selenate (ppm Se) during the starter and grower-finisher periods, respectively.
[d]Values were adjusted to a wet tissue basis by assuming a 75% water content of the tissue.

demonstrated that pig muscle tissue selenium concentration at market weight was higher when the South Dakota feeds were provided than when the Michigan diet was supplemented with selenite (Table 4). They further demonstrated that when additional selenite was added to the South Dakota diet there was no further increase in muscle tissue selenium content compared with the unsupplemented group. The combined data of Jenkins and Winter (1973) and Ku et al. (1973) suggests that dietary organic selenium is incorporated into muscle tissue at a higher concentration than inorganic selenium, but that when added inorganic selenium is "top-dressed," the added selenium is not as effectively retained in pig muscle tissue.

The effects noted above are the responses to relatively long-term feeding trials when diets were fed from weaning to market weight. Although the evidence is clear that inorganic selenium is effective in increasing the synthesis of glutathione peroxidase in younger swine

Table 4. Effect of basal feed origin and the efficacy of supplemental inorganic selenium on tissue selenium content in market pigs[a].

Feed origin:	Michigan[b]	South Dakota[c]	
Item Selenite, ppm:	0.40	0	0.1
No. pigs	4	4	4
Tissue Se, ppm			
Longissimus	0.12	0.48	0.45
Liver	0.61	0.84	0.92
Kidney	2.14	2.17	2.33

[a]Adopted from the data of Ku et al. (1973).
[b]Natural Se content (0.04 ppm Se).
[c]Natural Se content (0.40 ppm Se).

Table 5. Effect of supplemental inorganic selenium to diets differing in organic selenium on resulting tissue selenium concentration in weaned pigs.

Item	Basal feeds: Organic Se ppm: Selenite Se ppm:	Corn/SBM[a] 0.07 0.10	0.07 0.50	Complex[b] 0.18 0.10	0.18 0.50	SE
No. pigs		12	12	12	12	–
Initial weight, kg		7.1	7.0	7.0	7.2	0.15
Serum Se, ppm						
14–day		0.063	0.168	0.061	0.171	0.010[c]
28-day		0.061	0.146	0.062	0.140	0.005[c]
Liver Se, ppm[d]						
14–day		0.219	0.467	0.223	0.532	0.021[c]
28-day		0.204	0.557	0.228	0.525	0.030[c]
Longissimus Se, ppm[d]						
14–day		0.049	0.078	0.058	0.081	0.004[c]
35-day		0.056	0.091	0.065	0.115	0.004[c]
Heart Se, ppm[d]						
14–day		0.120	0.165	0.123	0.187	0.007[c]
35-day		0.106	0.187	0.126	0.210	0.006[c]

[a] C-SBM = corn soybean meal diet (0.07 ppm Se).
[b] Complex = a mixture of corn, soybean meal, brewers grains, distillers grains, fishmeal and alfalfa meal (0.18 ppm Se).
[c] Selenite level response ($P < 0.01$).
[d] Each mean represents six observations.

(Meyer et al., 1981); its retention in muscle tissue may be more dependent upon selenium source and perhaps the age of the pig. The data in Table 5 (Mahan, unpublished data) evaluated two starter diets with weaned pigs. One diet (complex) was formulated using several feed ingredients that were higher in their natural organic selenium content (brewers grains, distillers grains, fishmeal, alfalfa) resulting in a dietary natural selenium content of 0.18 ppm, whereas a second corn–soybean meal diet was supplemented with 0.1 ppm selenium as selenite (total dietary level 0.17 ppm selenium). Both diets were fed to weanling pigs for a 28-day period. The results demonstrated a similar tissue selenium concentration at both 14 and 35 days postfeeding when either the corn–soybean meal plus inorganic selenium diet (0.17 ppm) or when the complex diet was provided (0.18 ppm). When feeds were supplemented with additional inorganic selenium at 0.5 ppm, there was an increase in tissue selenium concentrations when both diets were fed (Table 5). These results suggest that both selenium forms were similarly retained by younger swine when fed at relatively low dietary levels. It should be noted that the dietary level of selenium (organic or inorganic) was below the weanling pig's requirement (NRC, 1983).

EXCRETION OF INORGANIC SELENIUM

The excretion, retention, and apparent digestibility of inorganic selenium during the postweaning period (5 weeks) are presented in Figures 2,

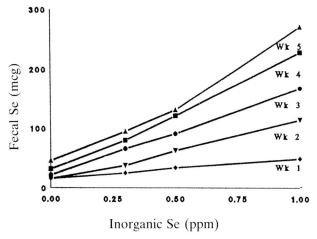

Figure 2. Weekly fecal selenium excretion (expressed on a per day basis) when weanling pigs were fed inorganic selenite (0, 0.3, 0.5, 1.0 ppm Se).

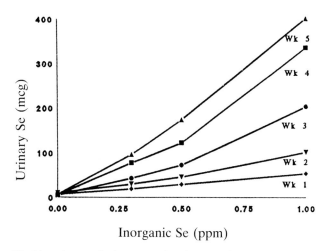

Figure 3. Weekly urinary selenium excretion (expressed on a per day basis) when weanling pigs were fed inorganic selenite (0, 0.3, 0.5, 1.0 ppm Se).

3, and Table 6. The information presented in Figures 2 and 3 clearly shows that as the supplemental level of inorganic selenium increased, there was a corresponding increase in fecal and urinary excretion of selenium during each week postweaning. This increased more when the dietary level exceeded 0.5 ppm selenium toward the end of the starter period. The weekly increase of excreted fecal and urinary selenium was attributed to the higher selenium intake each week postweaning. When the excretion and digestibility data were averaged for the 5–week

Organic selenium sources for swine

Table 6. Effect of dietary inorganic selenium levels on various balance measurements averaged over a 5 week period in weaned pigs[a].

Measurement[b]	Selenite (ppm)			
	0	0.3	0.5	1.0
	% of Se intake			
Urinary Se	14.4	26.2	29.2	35.2
Fecal Se	68.0	29.9	28.9	26.4
Se Retention	15.3	44.1	42.5	40.5
Apparent Se digestibility, %	36.2	69.3	70.1	73.3

[a]Source: Derived from the data of Mahan (1985).
[b]Values were averaged over a 5-week postweaning period.

Table 7. Efficacy of dietary selenium source at two levels on serum and tissue selenium concentrations with weaned pigs[a].

Selenium source	Selenite		Fishmeal		Brewer's grains		SE
Total Se, ppm	0.1	0.4	0.1	0.4	0.1	0.4	
No. pigs	7	6	7	6	6	7	
– Serum Se, ppm							
14–day	0.067	0.130	0.056	0.103	0.080	0.155	0.006[b]
35-day	0.064	0.115	0.053	0.135	0.084	0.151	0.006[b]
Tissue Se, ppm (35–day)							
Liver	0.233	0.464	0.167	0.436	0.276	0.554	0.036[bc]
Longissimus	0.062	0.094	0.058	0.122	0.093	0.116	0.008[bd]
Kidney	1.129	1.218	0.904	1.339	1.038	1.458	0.067[b]
Heart	0.120	0.194	0.117	0.214	0.144	0.239	0.009[bc]

[a]Source: Mahan and Moxon (1978).
[b]Main effect of 0.1 vs 0.4 ppm Se (P < 0.01).
[c]Brewers grains vs selenite, fishmeal (P < 0.01).
[d]Brewers grains, fishmeal vs selenite (P < 0.01).

period, the results demonstrated an increased percentage of the inorganic selenium consumed was excreted through the kidney, while a decreasing percent of selenium was excreted through the feces (Table 6). The apparent digestibility of inorganic selenium appears to be approximately 70%. Although the percentage apparent selenium digestibility had a tendency to increase as dietary selenium level increased, much of this was reflected by the higher absorption of the inorganic selenium which was subsequently excreted through the kidneys and not retained by the tissue.

Bioavailability of organic selenium

Organic selenium sources vary in the bioavailability of the element and its subsequent retention in tissue. The data presented in Table 7 where inorganic selenium (selenite) was added at 0.1 or 0.4 ppm to a corn–soybean meal diet fed to weanling pigs were compared with diets formulated using either fishmeal or brewer's grains, both added to provide dietary selenium levels of 0.1 or 0.4 ppm. The results demonstrated that the serum and tissue selenium concentrations were influenced by the organic selenium source. When the diet contained brewers grains, there were higher serum and tissue selenium concentrations of the element compared with when the same dietary selenium levels were provided either as selenite or fishmeal. The fishmeal diet resulted in similar tissue selenium concentration to that of the selenite treatment group. This response was attributed to the higher mercury content of fishmeal (Mahan and Moxon, 1978). It has been previously established that certain heavy metals can reduce the availability of selenium, suggesting a different bioavailability of organic selenium sources, depending on other elements, particularly the heavy metals.

Organic versus inorganic selenium toxicosis

Because higher tissue selenium levels will be retained when organic selenium sources are fed, the effects of high dietary levels of either selenium form may exacerbate the incidence of selenium toxicosis. During the early era of selenium nutrition this was the predominant recognition of selenium's role in animal nutrition. The plants and(or) grains consumed in these situations were, however, extremely high in selenium, much higher than that used as feed grains normally fed to livestock. Recent research with rats (Salbe and Levander, 1990) demonstrated that methionine deficient rats had more severe incidences of selenium toxicosis when fed inorganic selenium than selenomethionine. Research with pigs (Herigstad et al., 1973) demonstrated a higher tissue selenium concentration when selenomethionine was fed compared with sodium selenite (Table 8), but they also reported that the degree of toxicosis appeared to be similar at each of the dietary levels of selenium, regardless of source.

Table 8. Effect of inorganic Se on swine tissue selenium[a,b,c].

			Tissue Se (ppm)	
Se source	Dietary Se, ppm	Days on test	Liver	Kidney
Selenite	0.0	84[d]	0.20	1.27
Selenomethionine	0.0	91[d]	0.20	0.90
Selenite	0.1	37	0.41	1.23
Selenomethionine	0.1	37	0.23	0.90
Selenite	5.0	39	2.36	2.42
Selenomethionine	5.0	38	4.90	6.61
Selenite	10.0	56	3.62	3.61
Selenomethionine	10.0	56	9.92	10.40

[a]Weaned pigs (28 days of age) were fed a torula yeast-based diet.
[b]Adapted from the data of Herigstad et al. (1973).
[c]Each mean reflects two observations except as noted.
[d]One observation.

Summary

The feeding of organic selenium increases the tissue levels of selenium, particularly muscle tissue. This is attributed to the substitution of selenomethionine for methionine. Selenomethionine, although not the biologically active form of selenium, appears to be stored in proteinaceous tissue, and upon muscle catabolism or turnover can be a source of selenium for the subsequent synthesis of various body selenoproteins. When nonruminant animals are fed organic selenium, higher tissue selenium concentrations are attained than when inorganic selenium is fed, particularly when animals are fed the diets for an extended time period. When grower-finisher swine are fed diets high in organic selenium, supplementing additional inorganic selenium to the diet does not appear to increase tissue selenium concentration by market weight, whereas with younger pigs the effects of feeding organic and inorganic selenium during the postweaning period appear to be additive. As dietary levels of inorganic selenium increase for the young pig, increasing amounts are excreted through the feces and urine, with a higher proportion excreted through the kidney. The data suggest that although organic selenium sources are more efficaciously retained by the growing animal than inorganic selenium, selenium toxicosis occurs at similar dietary selenium levels regardless of organic or inorganic origin.

References

Arthur, D. 1971. Selenium content of some feed ingredients available in Canada. Can. J. Anim. Sci. 51:71.
Burk, R.F. and K.E. Hill. 1993. Regulation of selenoproteins. Annu. Rev. Nutr. 13:65.
FDA. 1974. Food additives: Selenium in animal feed. Federal Register 39:1355 (Tuesday, January 8).
FDA. 1987a. Food additives permitted in feed and drinking water of animals: Selenium. Federal Register 52:10887 (Monday, April 6).
FDA 1987b. Food additives permitted in feed and drinking water of animals: Selenium; Correction. Federal Register 52:21001 (Thursday, June 4).
Headon, D.R. 1993. The production and characterization of selenium-enriched yeast. (Seminar, Tokyo, Japan)
Herigstad, R.R., C.K. Whitehair, and O.E. Olsen. 1973. Inorganic and organic selenium toxicosis in young swine: Comparisons of pathologic changes with those in swine with vitamin E-selenium deficiency. Am. J. Vet. Res. 34:1227.
Jenkins, K.J. and K.A. Winter. 1973. Effects of selenium supplementation of naturally high selenium swine rations on tissue levels of the element. Can. J. Anim. Sci. 53:561.
Ku, P.K., W.T. Ely, A.W. Groce, and D.E. Ullrey. 1972. Natural dietary selenium, α–tocopherol and effect on tissue selenium. J. Anim. Sci. 34:208.

Ku, P.K., E.R. Miller, R.C. Wahlstrom, A.W. Groce, J.P. Hitchcock, and D.E. Ullrey. 1973. Selenium supplementation of naturally high selenium diets for swine. J. Anim. Sci. 37:501.

Kurkela, P. and E. Kääntee. 1984. Effects of barley-bound organic selenium compared with inorganic selenite on selenium concentration and structure of tissues in pigs. J. Agr. Sci. in Finland 56:61.

Mahan, D.C. 1985. Effect of inorganic selenium supplementation on selenium retention in postweaning swine. J. Anim. Sci. 61:173.

Mahan, D.C. and A.L. Moxon. 1978. Effects of adding inorganic or organic selenium sources to the diets of young swine. J. Anim. Sci. 47:456.

Meyer, W.R., D.C. Mahan, and A.L. Moxon. 1981. Value of dietary selenium and vitamin E for weanling swine as measured by performance and tissue selenium and glutathione peroxidase activities. J. Anim. Sci. 52:302.

NRC. 1983. Selenium in Nutrition. National Academy Press, Washington, D.C.

Olson, O.E., and I.S. Palmer. 1976. Selenoamino acids in tissues of rats administered inorganic selenium. Metabolism 25:299.

Pehrson, B.G. 1993. Selenium in nutrition with special reference to the biopotency of organic and inorganic selenium compounds. In: Biotechnology in the Feed Industry, Proc. 9th Symposium. T.P. Lyons (Ed.) Alltech Technical Publications.

Salbe, A.D., and O.A. Levander. 1990. Comparative toxicity and tissue retention of selenium in methionine deficient rats fed sodium selenate or L-selenomethionine. J. Nutr. 120:207.

Ullrey, D.E. 1992. Basis for regulation of selenium supplements in animal diets. J. Anim. Sci. 70:3922.

Young, L.G., K.J. Jenkins, and D.M. Edmeades. 1977. Selenium content of feedstuffs grown in Ontario. Can. J. Anim. Sci. 57:793.

INDEX

Aflatoxin,
 strategies for detoxification, 236–237
 degradability of aflatoxin in presence of Yea-Sacc[1026], 240–242
 effect of Yea-Sacc[1026] in diets containing, 37, 235
  broilers, 237–239
  ducklings, 239–240
Allzyme BG (see enzymes β-glucanase)
 poultry, 39–40
Allzyme PT (see enzymes: pentosanase)
 poultry, 40
Amino acids,
 ideal dietary profiles, 68–70
 feedstuff profiles, 70
 digestibility, 70–71
 feathermeal vs. enzyme-treated feather protein, 72–75
 fishmeal vs. enzyme-treated fish protein, 75–67
Ammonia,
 control with *Yucca schidigera* extract, 4
 ammonia-binding, 4
 ascites, 5
 prawn culture, 7
Antioxidant nutrients (see also copper, zinc, manganese, selenium),
 roles in immunocompetence,
  trace minerals, 275–276
  vitamins, 275
  vitamin E and oxidative stress, 290–291, 293–296, 297–298
Ascites,
 clinical findings, 199–201
 vs. SDS, 199–200
 acid/base balance,
  lactic acid metabolism, 201–203
  higher sulfur and nitrogen in the diet, 204
  weight gain vs. dietary meq, 204–206
  effect of synthetic amino acid use, 208
 De-Odorase, effect of, 5, 206–207

Aquaculture,
 bio-Mos, effects of: see Bio-Mos, 11
 furunculosis, 11
 shrimp and prawns: water quality and De-Odorase

Beef cattle,
 stress factors, 275
 stress physiology, 255–262
 chromium and stress, 27
 bioplex Cu and hepatic Cu content, 34–35
β-Glucans and β-Glucanase (see also plant polysaccharides)
 β-glucan structure, levels, 52, 104
 effects on digestion, 104
 effects of enzyme supplementation,
  pigs, 39, 105
  poultry, 39–40, 104–105
Bio-Mos (see also oligosaccharides)
 effects on performance, health,
  aquaculture, 11, 173
  calves, 16–18, 172–173
  pigs, 17–18, 172
  poultry,
   broilers, 16–17, 171–172
   turkeys, 171–172
   rabbits, 19
Bioplex mineral proteinates,
 Bioplex copper,
  Effect on hepatic Cu in beef cows, 34–35
  comparative bioavailability of $CuSO_4$, Cu Bioplex and Cu-lysine, 35
 Bioplex Manganese,
  effect on fattening pigs, 27–29
 Bioplex zinc,
  vs. zinc oxide: incidence of mastitis, 33–34
  somatic cell count, 33
  quality control, 36–37

335

Index

Boar taint,
  androstenone, indole and skatole, 175
  composition of boar odor, 175
  skatole, 176
  methods for testing, 175
  effects of Yucca extract, 176–180
  effect of end weight, 177–178
  boars vs. barrows, 179–180
Boron,
  boron occurence, industrial uses, 303
  chemical, structural and nuclear properties, 2–3, 304–306
  comparison with carbon, 305
  inorganic boron in biology,
    role in plants, 307
    animals and humans, 307–309
  organic boron, 309
  boron analogues,
    boron analogues of amino acids, 3, 305–306, 310
    boron-containing peptides, 310–311
    neurotransmitters, 311
    phosphonates, 312
    heterocyclic amines, 312
    nucleoside and nucleic acid derivatives, 313
    biological and pharmacological behavior,
      cholesterol modulation, 316
      anti-inflammatory, 316
      antineoplastic, 316
      osteoporosis, 317
      toxicology of analogues, 317
By-products,
  animal by-products and recycling, 67–68
  blood meal, 70–71
  feathermeal, 70–71
  enzyme-treated feather protein, 72–75
  enzyme-treated fish protein, 75–76
  milling by-products,
    rice bran nutritive value, limitations, 58–60

Cancer,
  therapeutic compounds,
    boroglycine, 3
    taxol, 3
  selenium and cancer, 21
Chromium,
  absorption of inorganic vs. organic, 26–29, 276
  glucose tolerance factor, 25–26, 276–277
  deficiency symptoms, 268–269
  effect of stress on requirements, 267
    humans, 269–271
    livestock, 271–277
  effects on health, performance, of stressed beef cattle, 27, 277–278
  effects in dairy cattle, 27, 278, 280
  effects on finishing pigs, 27–29, 271–272, 279

  immune and endocrine function, 279–280
  sources of Cr, 25–26, 276–277
  toxicity of Cr^{+6}, 277
Competitive exclusion,
  undefined cultures,
    effects on Salmonella colonization in poultry, 156–158
  In ovo application, 157–158
  defined cultures: regulatory needs, 158
  selection and culture systems, 159–160
Copper,
  effect of Bioplex Cu in correcting deficiency in beef cows, 34
  bioavailability of $CuSO_4$, Cu Bioplex and Cu lysine to rats, 35
  effect of Cu source on Zn bioavailability in rats, 35
  copper sulfate: pharmacological levels, 8

Dairy cattle,
  stress, 275
  periparturient disorders and oxidative stress, 292–293
  response to Vitamin E, 293–297
  potential role for supplemental Cr, 27
  selenium yeast and milk Se, 23, 25
  selenium yeast and services per conception, 25
  water quality, 183–196
De-Odorase,
  litter quality, effects on, 5
  fertilizer value, effect on, 5
  pigs,
    ammonia reduction, 5
    boar taint, 6
    performance, 5
  poultry,
    ascites, 5
    performance, 5
  shrimp and prawns, effects on water quality, 248–252

Enzymes,
  α-amylase, 103
  α-galactosidase and antinutritional factors in legumes, 42
  arabinoxylanase (see pentosans and pentosanase)
  assessing enzyme activity,
    problems with analytical methods for enzymes in complete feed, 43–44, 62, 121
    new assay development, 121–126
    direct detection of activity, 122–123
    radial diffusion, 124–126
  pelleting survival, 62–64
  β-glucanase (see β-glucanase)
  cellulase, 103 (see also fiber)
  enzyme activity in liquid feeding systems, 231–232

Index

enzyme addition to ruminant diets: (see also fiber, enzymatic degradation of)
  proteolytic attack, 83–84
  effects of cell-free *A. niger* extracts on cellulose digestion, 84–85
  feed industry use of enzymes, 102–104, 117–118
  US market potential, 100–101
  cereal grains used, 100
  matching enzyme and substrate, identifying the limiting NSP in by-products, 58
  identifying the structures cross-linking lignin to the cell wall, 88
  formulation of cocktails, 42, 108
  pentosanase (see pentosans and pentosanase)
  phytate and phytase, 40–42, 103–104, 106–108
  endogenous phytase, 107
  using enzymes to upgrade protein biological value,
  bio-treated feather protein, 72–75
  bio-treated fish protein, 75–76
  designing species-specific protein sources: Protagen, 76–81
  regulatory legislation,
  US and Canada, 118
  Europe, 43, 118–120
  role of hydrolases in animal feeds, 39, 58

Fiber,
  biogenesis of lignin, 92
  enzymatic degradation in the rumen,
  fiber digestion in cellulosomes, 83
  rumen fiber digestion and cell wall porosity, 86–87
  limitation by lignin, 88
  targeting appropriate enzyme supplements,
  structures cross-linking lignin to the cell wall, 88–91
  ferulic acid esterases, xylanases, 90
  monocotyledons vs. dicotyledons, 90–91
  oxidative degradation of lignin, 91
  lignin modification to improve degradability, 92–94
  silage additives to modify rumen fiber degradation, 94

GrainGard, 37–38

Horses,
  lactic acidosis, 201–202

Immune function,
  acute phase response in stressed cattle, 258–262
  tissue trauma, 258
  leukocytes, mast cells and macrophages release cytokines, 259
  cytokines affect protein, glucose and lipid metabolism, 259–262
  effects of aflatoxin + Yea-Sacc[1026] on, 243
  overview of cellular and humoral immunity, 169–170
  role of chromium, 279–280
  trace minerals and vitamins: (see antioxidant nutrients)

Liquid feeding systems for pigs,
  advantages of, 229–232
  phosphorus availability, 231
  acidity and lactobacilli, 230–231
  enzyme activity, 231
  ad libitum feeding of weaners, 232

Mannan-oligosaccharides (see oligosaccharides)
Mannose,
  effects on pathogen colonization, 160–162, 167–168
Mycotoxins: see aflatoxin, zearalenone

Oligosaccharides,
  functional foods, 19
  fructo-oligosaccharides, 20,162
  mannan-oligosaccharides,
  effects on livestock and poultry performance, health: (see Bio-Mos)
  effects on colonization by pathogens, 13, 15–16, 167–169, 171
  effects on plant disease resistance, 13–14
  effects on pathogen growth, 16,163
  immunomodulation, 9–13
  aquaculture, effects on: (see Bio-Mos)
  phagocyte activity, 12
  adjuvant and antigen properties, 170–171
  mannose-binding protein, 171
  modified mannan-oligosaccharides: Graingard, 37–38
  sources of, 9–11, 20
  yeast cell wall oligosaccharide, 9–11
Oxidative stress and reactive oxygen metabolites
  reactive oxygen metabolites (ROM), 262
  conditions that increase ROM, 285–286
  decompartmentalized Fe, 287–290, 297

337

Index

evaluating oxidative status,
  plasma fast-acting antioxidants (FAA), 291–292, 297
  association of FAA with periparturient disorders of cows, 292–293
  effects on steroidogenesis, 296–297
  sources of ROM, 284–287
  systems that protect against ROM, 285–286
    metal-binding macromolecules, 289
    chain-breaking antioxidants, 290
    response to Vitamin E, Fe 293–296

Pentosans and pentosanase,
  structure, 53–54
  pentosan antinutritive activity in wheat, rye, 55–58, 105–106
  pentosanase in wheat-based diets,
    poultry, 40, 106
    pigs, 106
Peptides,
  vs. amino acids:absorption, 44–45
Phosphorus,
  volumes excreted in manure, 40–41
  phytate phosphorus, 40–41
  availability in liquid feeding, 231–232
Pigs,
  boar taint (see boar taint)
  castrates vs. intact males,
    performance and FCR, 6–7, 175
    skatole, 176
    effects of De-Odorase, 179–180
  chromium, 27–29
  De-Odorase, effects of: (see De-Odorase)
  enzyme-treated fish protein, 75–76
  enzyme-treated feather protein, 72–75
  ideal amino acid balance, perfect protein, 68–69
  liquid feeding systems, 229–233
  oligosaccharides, 17–18 (see Bio-Mos)
  protagen, 78–79
  selenium sources: inorganic vs. organic, 323–331
  US feed tonnage by region, 100
  water quality, nutrition 211–225
Plant polysaccharides,
  non-starch polysaccharides (see also pentosans, β-glucans)
    structure, 53–54
    physical properties affecting digestion, 55–56, 101
    antinutritive activity of wheat arabinoxylans, 55–58, 86
    how adding glycanase enzymes improves nutrient value, 58
    variability in crop NSP: low AME wheats in Australia, 60–61

NSP composition by plant species,
  rice bran, 58–60
  corn, 100
  barley, 101–102
  wheat, 101–102
  oats, 102
cell wall structure in forages, physical and chemical properties, 85–87
starch structure:amylose and amylopectin, 101
Poultry,
  oligosaccharides, 16–17
  US feed tonnage by region, 101
  ascites,
    and SDS, 199–200
    and acid/base balance, 201–206
    effect of De-Odorase, 206–207
Protein,
  ideal amino acid balance, perfect protein,
    pigs, 68–69
    dairy cattle, 69–70
    atlantic salmon, 69–70
    humans: essential amino acids requirements, 130–131
    idealized essential amino acid ratios for crops, 142
  animal product proteins,
    enzymes to upgrade biological value of animal by-product proteins,
    bio-treated feather protein vs. feathermeal, 72–75
    bio-treated fish protein, 75–76
    bio-treated feather protein, 72
    designing species-specific protein sources: protagen, 76–81
    protagen vs. fishmeal, plasma protein for pigs, 78–79
    protagen for dogs: preference tests, 78–79
  plant proteins,
    design of plant proteins,
      problems with expression of designed proteins, 130
      nutritional quality goals, 136
      De novo design, stabilization, 138–141
      designed nutritional proteins, 141–148
      insertion and expression of the gene in plants, 146–148
      essential amino acids found most limiting, 131, 137
      regulation of storage protein genes, 134–136
      selenoproteins, 323–324
      structure and classification of storage proteins,
        corn, sorghum and millet, 132–134
        legumes, other and dicotyledonous, 132–134

Index

Salmonella,
  annual costs of food-borne salmonellosis, 155
  salmonella colonization of the intestine, competitive exclusion, 156–160
  effect of fructo-oligosaccharides, 162
Selenium,
  deficiency signs in animals, 20
  excretion of inorganic, 328–330
  human health and nutrition, 20
    cancer, 21
    recommended daily intake, 21
    selenium yeast, 21
  metabolism of, 323–324
  organic forms of Se,
    regional variations, 324–325
    selenomethionine, 22, 323
    selenium yeast, 325–326
      production of, 22–23
      pigs, 23
      dairy cows, 23, 25
      aquaculture, 24
      sodium selenite vs. selenium yeast, 25
  vs. inorganic,
    tissue retention, 323, 326–328
    bioavailability, 330
    toxicity, 331
  regulations, requirements,
    US, 20, 22, 325
    Europe, 20
Shrimp and prawns, 7–8, 248–252
Stress,
  defined, 267
  metabolic changes during, 256
    hormonal effects on metabolism, 256–257
    energy regulation,
      energy sources: FFA, 257
      causes and consequences of acidosis, 257
      protein and amino acid metabolism, 257
        cellular and muscle protein catabolism, 257–258
        fates of free amino acids, 258–259
        acute-phase proteins (see also Immune function: acute phase response),
        effects on trace mineral levels in blood,
        roles of cytokines integrating metabolism and immunity, 259–262
        urea cycle and BUN, 258
  nutrition and stress,
    effects on Cr requirements, 267, (see chromium)
  oxidative stress and reactive oxygen metabolites (See oxidative stress)

Water quality,
  criteria, 183–184
  cattle,
    water quality effects on growth, water intake, lactation
    salinity and total dissolved solids, 184–186
      using saline water, 184–185
    sulfates, 185–186
    nitrates and nitrites, 186
    water hardness and pH, 186–187
    heavy metals, 187–188
  water intake and requirements, 188–1
    prediction equation, 188–189
    factors affecting,
      milk yield, 188–189
      feed moisture, 190–191
      metabolic water, 191
    drinking behavior 191
      drinker types, 191–192
      stray voltage, 192
    temperature of water during hot weather, 194–196
    calves, 193
  pigs,
    water consumption,
      drinking behavior and drinker type, 213–215
    water wastage,
      costs and volume involved, 212
      reducing spillage, 212–215
      effect of diet, 215–218
    quality and quantity: effects on production,
      dissolved solids: mineral content, deposits, palatability, 218–219
      suspended solids: organic matter, 219–220
      microbial contamination: sanitization, toxins, 220
      cleanliness at delivery point, 221
      overcoming off-flavors, 221–222
    quantity,
      metabolic and behavioral factors affecting, 222–224
      delivery rate, 222–223
      relationship among volumetric intake, feed and water intake, 224–225
  shrimp and prawns,
    water quality crisis: effects of De-Odorase, 247–252
Water treatment,
  advantages/problems of water vs. feed medication, 225–229

Yeast, 9–10
  cell wall composition, 10–11
  binding capacity for mycotoxins, 37–38

339

Yeast culture,
  diet specific strains, 29–30
    Yea-Sacc^{1026} vs. Yea-Sacc^{8417}, 30
    Yea-Sacc^{8417},
      in finishing cattle, 30–31
      in dairy cattle, 31
      ruminal effects, 31
    identification of with DNA probe, 32
    effects in poultry diets containing
      aflatoxin, 37, 235–240
    ruminant mode of action,
      ruminal effects, 29–30
Yucca schidigera extract (see De-Odorase, ammonia, ascites, shrimp and prawns)

Zearalenone,
  binding by mannan fractions, 37–38
Zinc,
  pharmacological levels of Zn oxide, 8
    dermatitis and Zn oxide, 33
  physiological roles, 33
  oxide vs. proteinate: see also Zn bioplex
    incidence of mastitis, 33–34
    somatic cell count, 33
    teat keratin, 33–34

DISTRIBUTORS AROUND THE WORLD

ARGENTINA
LABORATORIOS SCOPE S.A.
Hector Petersen & Rodolfo Ruiz
Av. San Juan 2866
1932 Buenos Aires
Tel: 541 941-6774 • Fax: 541-941-5016

AUSTRALIA
RHÔNE-POULENC ANIMAL
  NUTRITION
Dr. Ray Johnson
19-23 Paramount Road • West Footscray,
Victoria 3012
Tel: 61 3 316 9744 • Fax: 61 3 314 9386
Telex AA33906

BELGIUM
ALLTECH BELGIUM
Responsible importer: VEPROLIM
Boshoekstraat 45
3665 As
Tel: 65-82-50 • Fax: 65-94-56

BRAZIL
ALLTECH DO BRAZIL
Aidan Connolly
Caixa Postal 10808
C.P. 1080-CEP 81.170.610
CIC - Curitiba - Paraná
Tel: 55 41 246-6515 • Fax: 55 41
  246-5188

CANADA
ALLTECH CANADA
Dr. A.E. (Ted) Sefton
1 Air Park Place • Guelph, Ontario
  N1H6H8
Tel: 519 763 3331 • Fax: 519 763 5682

CHILE
ALLTECH CHILE
Dr. Mario Román
Atenas 7542
Las Condes
Santiago, Chile
Tel: 56 2 212-7356 • Fax: 56 2 201-2986

CHINA
ALLTECH CHINA
Dr. Lan Luo
Institute of Animal Science
Chinese Academy of Agricultural
  Sciences
CASS
Malianwa Haidian
100094 Beijing
Fax: 86 1 258-2332

COLOMBIA
INVERSIÓNES AMAYA
  (INVERAMAYA)
Luis Londonō J.
Calle 85 No. 20-25 - Oficina 401 A
Bogotá
Tel: 57 1 218-2829 • Fax: 57 1 218-5317

COSTA RICA
NUTEC, S.A.
Carlos Lang
Apartado 392 P.O. Box 392 • Tibas
Tel/Fax: 506 333-110

CZECH REPUBLIC
ALLTCH CZECH REPUBLIC
Vladimír Siske, MSc, Ph.D.
Moldavská 7
625 00 Brno
Tel/Fax: 42 5 324 417

Distributors around the world

DENMARK
NUTRISCAN A/S
Richard B. Hansen
Rørsangervej 8 • Postboks 141
DK-8300 Odder
Tel: 45 86 542 488 • Fax: 45 86 560 359

ECUADOR
PROPEC, S.A.
Gonzalo Arguello
Av. 12 de Noviembre # 24-85 y Av.
El Rey • Ambato
Tel: 593 2 823-180 • 593 2 829-267

EGYPT
EGYTECH
Dr. Mohy Z. Sabry
P.O. Box 442 - Dokki • Cairo 12311
Tel: 202 361 0605/703 436
Fax: 202 361 5909

FINLAND
BERNER LTD.
Jukka Klemola
Eteläranta 4B
00130 Helsinki
Tel: 358 0 134 511 • Fax: 358 0 13451380
Telex 124447

FRANCE
ALLTECH FRANCE
Marc Larousse
2-4, avenue du 6 juin 1944
95190 Goussainville
Tel: 33 1 398 86351
Fax: 33 1 398 80778

GERMANY
ALLTECH DEUTSCHLAND GmbH
Rolf Neelsen
Esmarchstraße 4
23795 Bad Segeberg
Tel: 04551/8870-0 • Fax: 04551/8870-99

GREAT BRITAIN
ALLTECH UNITED KINGDOM
Jem Clay
Unit 16-17, Abenbury Way • Wrexham
 Ind. Estate
Wrexham, Clwyd LL13 9UZ
Tel: 44 978 660 198
Fax: 44 978 661 136
Telex 61628

GREECE
LAPAPHARM, INC.
A. Mantis
73 Menandrou Str • 10437 Athens
Tel: 30 1 522 7208 • Fax: 30 1 299 367
Telex 215158

HONDURAS
S.B.F. INTERNATIONAL
Sigfrido Burgos
Colonia Palermo, No. 1862
Tegucigalpa
Tel/Fax: 504 32-3964

HONG KONG
PING SHAN ENTERPRISE CO., LTD.
Oscar Lam
19 Queen Street
Tel: 852 858 1188 • Fax: 852 858 1452
Telex 86016

HUNGARY
ALLTECH HUNGARY
Levente Gati
Kresz Geza utca 16
H-1132 Budapest
Tel: 36 1 202 1755 • Fax: 36 1 201 5215

INDIA
VETCARE
Bharat Tandon
No. 90, 3rd Cross, 2nd Main
Gangenahalli, Bangalore 560 032
Tel: 91 812 332 174 • Fax: 91 812 334
 041
Telex 953 845 8703

INDONESIA
P. T. ROMINDO PRIMAVETCOM
Dr. Lukas
Dr. Saharjo No. 266
Jakarta 12870
Tel: 62 21 830 0300 • Fax: 62 21 828
 0678

IRELAND
ALLTECH IRELAND
Aidan Brophy
Unit 28, Cookstown Industrial Estate •
 Tallaght, Dublin 24
Tel: 353 1 510276 • Fax: 353 1 510131
Telex 91863

Distributors around the world

ITALY
ALLTECH ITALIA
c/o Ascor Chimici s.r.l.
Via Piana, 265
47032 Capocolle di Bertinoro (Forlì)
Tel: 39 543 448070
Fax: 39 543 448644

JAMAICA
MASTER BLEND FEEDS LTD.
Julio Forbes
P.O. Box 24 • Old Harbour • St. Catherine
Tel: 809 983 2305/2306 • Fax: 809 983 2338

JAPAN
MITSUI & CO., LTD.
Nick Koyama
2-1 Ohtemachi 1-Chome • Chiyoda-Ku
Tokyo
Tel: 81 33 285 5326 • Fax: 81 33 285 9958
Telex J22253

KOREA
YOONEE CHEMICAL CO., LTD.
Dr. H.T. Park
C.P.O. Box 6161 • Seoul
Tel: 82 2 585 1801 • Fax: 82 2 584 2523
Telex K29937

MALAYSIA
FARM CARE SDN. BHD.
Ong Seng Say
No. 48-3, Jalan Radin Tengah
Seri Petaling • 57000 Kuala Lumpur
Tel: 60 3 957 3669 • Fax: 60 3 957 3648

MEXICO, D.F.
APLIGEN SA DE CV
Dra. Gladys Hoyos
Palestina #67-A
Col. Claveria
02800 México, D.F.
Tel: 52 5 396-3840 • Fax: 52 5 396-3565

NEPAL
NEPA PHARMAVET PVT. LTD.
Suvas Bishet
GA-1-481, Wotu Tole
Katmandu-3
Tel: 977 1 217 952 • Fax: 977 1 224 627

NETHERLANDS
ALLTECH NEDERLAND
Timm Neelsen
Luk Van Noon
Holandsch Diep 63
2904 EP CAPELLE AAN DEN IJSSEL
PO Box 103
Tel: 31 10 450-1038 • Fax: 31 10 442-3798

NEW ZEALAND
CUNDY TECHNICAL SERVICES LTD.
M. E. Cundy
P.O. Box 69 - 170 Glendene • Auckland 8
Tel: 64 9 837 3243 • Fax: 64 9 837 3214

PERU
ALLTECH SUCURSAL PERUANA
Dr. Ricardo León
Ramón Zavala #249
Lima 18
Tel/Fax: 51 14 470-690

PHILIPPINES
FERMENTATION INDUSTRIES CORP.
Rodney Yu
P.O. Box 440 Greenhills, Metro Manila
Tel: 63 2 400 888 • Fax: 63-2-530-0596
Telex 40001

POLAND
ALLTECH POLSKA
Dr. Wojciech Zalewski
00-950 Warszawa, Wspolna 30
gmach MINISTERSTWA ROLNISTWA
Tel: 48 2 623 1005
Fax: 48 2 623 1006

PORTUGAL
ALLTECH PORTUGAL
Orlando de Sousa
Rua Álvaro de Brée No. 6
Leceia 2745 Queluz
Tel: 351-1-4395101 • Fax: 351-1-4395100

SPAIN
PROBASA
Juan Rosell Lizana
c/o Argenters, 9 Nave 3
Pol Ind Santiga, Sta. • Perpetua de la Moguda
Barcelona
Tel: 34 3 718 2215 • Fax: 34 3 719 1307

Distributors around the world

SWITZERLAND
INTERFERM AG
Dr. Fritz Näf
Postfach 112, Hardturmstrasse 175 • 8037 Zürich
Tel: 41 1 272 8024 • Fax: 41 1 273 1844
Telex 822379

TAIWAN
JARSEN CO., LTD.
Wen Ho Lin
12 Fl., No. 1377. Chung Cheng Rd.
Tao-Yuan City. Tao-Yuan Hsien
R.O.C.
Tel: 886-3-356 6678 • Fax: 886-3-356 5527

THAILAND
DIETHELM TRADING CO., LTD.
Dr. Chai
2533 Sukhumvit Road • Bangchack, Prakhanong
Bangkok 10250
Tel: 662 332 7140 • Fax: 662 332 7164
Telex 20595

TRINIDAD
ALLTECH TRINIDAD
V. Peter Chin Cheong
44 St. Michaels Terrace • Blue Range Diegos • Martin
Tel: 809 632 4519 • Fax: 809 628 0971
Telex 387 22539

USA
ALLTECH, INC.
3031 Catnip Hill Pike • Nicholasville, KY 40356
Tel: 606 885 9613 • Fax: 606 885 6736
Telex 218425

VENEZUELA
SIDECA - SIDELAC
Dr. Rafael Alonso
P.O. Box 1813 • Maracaibo
Tel: 58 62 47079 • Fax: 58 61 918889

VIET NAM
AGRITECH SAIGON
Nguyen Tan Manh
Nicholas Brand
13 Xom Moi Hamlet, Phuoc Long Village
Thu Duc District
Ho Chi Minh City
Tel: 84-8-960127 • Fax: 84-8-298540